10
Topics in Heterocyclic Chemistry

Series Editor: R. R. Gupta

Topics in Heterocyclic Chemistry
Series Editor: R. R. Gupta
Recently Published and Forthcoming Volumes

Bioactive Heterocycles V
Volume Editor: M. T. H. Khan
Volume 11, 2007

Bioactive Heterocycles IV
Volume Editor: M. T. H. Khan
Volume 10, 2007

Bioactive Heterocycles III
Volume Editor: M. T. H. Khan
Volume 9, 2007

Bioactive Heterocycles II
Volume Editor: S. Eguchi
Volume 8, 2007

Heterocycles from Carbohydrate Precursors
Volume Editor: E. S. H. El Ashry
Volume 7, 2007

Bioactive Heterocycles I
Volume Editor: S. Eguchi
Volume 6, 2006

Marine Natural Products
Volume Editor: H. Kiyota
Volume 5, 2006

QSAR and Molecular Modeling Studies in Heterocyclic Drugs II
Volume Editor: S. P. Gupta
Volume 4, 2006

QSAR and Molecular Modeling Studies in Heterocyclic Drugs I
Volume Editor: S. P. Gupta
Volume 3, 2006

Heterocyclic Antitumor Antibiotics
Volume Editor: M. Lee
Volume 2, 2006

Microwave-Assisted Synthesis of Heterocycles
Volume Editors: E. Van der Eycken, C. O. Kappe
Volume 1, 2006

Bioactive Heterocycles IV

Volume Editor: Mahmud Tareq Hassan Khan

With contributions by

A. Ather · R. G. S. Berlinck · E. Branda · P. Buzzini · H. Cerecetto
P. H. Dixneuf · T. Flemming · M. González · M. Goretti · N. Hamdi
C. Hansch · F. Ieri · O. Kayser · M. T. H. Khan · G. S. Kumar
M. Maiti · N. Mulinacci · R. Muntendam · A. Romani · C. Steup
B. Turchetti · R. P. Verma

Springer

The series *Topics in Heterocyclic Chemistry* presents critical reviews on "Heterocyclic Compounds" within topic-related volumes dealing with all aspects such as synthesis, reaction mechanisms, structure complexity, properties, reactivity, stability, fundamental and theoretical studies, biology, biomedical studies, pharmacological aspects, applications in material sciences, etc. Metabolism will be also included which will provide information useful in designing pharmacologically active agents. Pathways involving destruction of heterocyclic rings will also be dealt with so that synthesis of specifically functionalized non-heterocyclic molecules can be designed.
The overall scope is to cover topics dealing with most of the areas of current trends in heterocyclic chemistry which will suit to a larger heterocyclic community.
As a rule contributions are specially commissioned. The editors and publishers will, however, always be pleased to receive suggestions and supplementary information. Papers are accepted for *Topics in Heterocyclic Chemistry* in English.
In references *Topics in Heterocyclic Chemistry* is abbreviated *Top Heterocycl Chem* and is cited as a journal.

Springer WWW home page: springer.com
Visit the THC content at springerlink.com

ISSN 1861-9282

DOI 10.1007/978-3-540-73404-8

ISBN 978-3-642-09245-9 e-ISBN 978-3-540-73404-8

Springer is a part of Springer Science+Business Media

springer.com

Softcover reprint of the hardcover 1st edition 2007

Cover design: *Design & Production* GmbH, Heidelberg

Printed on acid-free paper 02/3100 YL – 5 4 3 2 1 0

Topics in Heterocyclic Chemistry Also Available Electronically

Dedicated to my parents, Hassan and Mahmuda, who made all of my efforts in science possible under their umbrella of love, good wishes, inspiration and prayers.

Preface

This volume contains nine more contributions from expert researchers of the field, providing readers with in depth and current research results regarding the respective topics.

In the first chapter, Flemming et al. review the chemistry, biosynthesis, metabolism and biological activities of tetrahydrocannabinol and its derivatives.

Hansch and Verma contribute to the quantitative structure-activity relationship (QSAR) analysis of heterocyclic topoisomerase I and II inhibitors. These inhibitors, known to inhibit either enzyme, act as antitumor agents and are currently used in chemotherapy and in clinical trials.

In the third chapter, Khan reviews some aspects of molecular modeling studies on biologically active alkaloids, briefly providing considerations on the modeling approaches.

In next chapter, Khan and Ather review different aspects of the microbial transformations of the important nitrogenous molecules, as they have diverse biological activities. This chapter provides a critical update of the microbial transformations reported in recent years, targeting novel biocatalysts from microbes.

In the fifth chapter, Hamdi and Dixneuf describe the synthesis of triazoles and coumarins molecules and their physiological activities.

Maiti and Kumar, in their contribution, review the physicochemical and nucleic acid binding properties of several isoquinoline alkaloids (berberine, palmatine and coralyne) and their derivatives under various environmental conditions.

In chapter seven, Berlinck describes varieties of polycyclic diamine alkaloids such as halicyclamines, 'upenamide, xestospongins, araguspongines, halicyclamines, haliclonacyclamines, arenosclerins, ingenamines and the madangamines, etc., and their synthesis as well as biological activities.

In chapter eight, Buzzini et al. review naturally occurring *O*-heterocycles with antiviral and antimicrobial properties, with paticular emphasis on the catechins and proanthocyanidins. Their modes of action as well as their synergy with currently used antibiotic molecules are also reviewed

In the following chapter, Cerecetto and González review the classical and most modern methods of the synthesis of benzofuroxan and furoxan derivatives, their chemical and biological reactivity, biological properties and mode of action, structure-activity studies and other relevant chemical and biological properties.

Tromsø, Norway 2007

Mahmud Tareq Hassan Khan

Contents

Contents of Volume 9

Bioactive Heterocycles III

Volume Editor: Khan, M. T. H.
ISBN: 978-3-540-73401-7

Contents of Volume 11

Bioactive Heterocycles V

Volume Editor: Khan, M. T. H.
ISBN: 978-3-540-73405-5

Top Heterocycl Chem (2007) 10: 1–42
DOI 10.1007/7081_2007_084

Published online: 14 August 2007

Chemistry and Biological Activity of Tetrahydrocannabinol and its Derivatives

T. Flemming[1,2] · R. Muntendam[2] · C. Steup[1] · Oliver Kayser[3] (✉)

[1]THC-Pharm Ltd., Offenbacher Landstrasse 368A, 60599 Frankfurt, Germany

[2]Department of Pharmaceutical Biology, GUIDE, University of Groningen, Antonius Deusinglaan 1, 9713 AV Groningen, The Netherlands

[3]Department of Pharmaceutical Biology, Groningen Research Institute for Pharmacy (GRIP), University of Groningen, Antonius Deusinglaan 1, 9713 AV Groningen, The Netherlands
o.kayser@rug.nl

T. Flemming and R. Muntendam both contributed equally

Abstract Cannabinoids and in particular the main psychoactive Δ9-THC are promising substances for the development of new drugs and are of high importance in biomedicine and pharmacy. This review gives an overview of the chemical properties of Δ9-THC, its synthesis on industrial scale, and the synthesis of important metabolites. The biosynthesis of cannabinoids in *Cannabis sativa* is extensively described in addition to strategies for optimization of this plant for cannabinoid employment in medicine. The metabolism of Δ9-THC in humans is shown and, based on this, analytical procedures for cannabinoids and their metabolites in human forensic samples as well as in *C. sativa* will be discussed. Furthermore, some aspects of medicinal indications for Δ9-THC and its ways of administration are described. Finally, some synthetic cannabinoids and their importance in research and medicine are delineated.

Keywords Tetryhydrocannabinol · *Cannabis sativa* · Analytical methods · Medicinal applications

1 Chemistry

1.1 Nomenclature

Natural cannabinoids are terpenophenolic compounds that are only biosynthesized in *Cannabis sativa* L., Cannabaceae. For these compounds five different systems of nomenclature are available, well described by Shulgin [1] and by ElSohly [2]. Two of these systems are mainly employed for the description of tetrahydrocannabinol in publications – the dibenzopyrane numbering system (**1.1** in Fig. 1) and the terpene numbering system (**1.2**), based on *p*-cymene. Because of historical and geographical reasons, the missing standardization is not uniform and is the main reason for ongoing confusion in the literature, leading to discussions regarding the numbering and

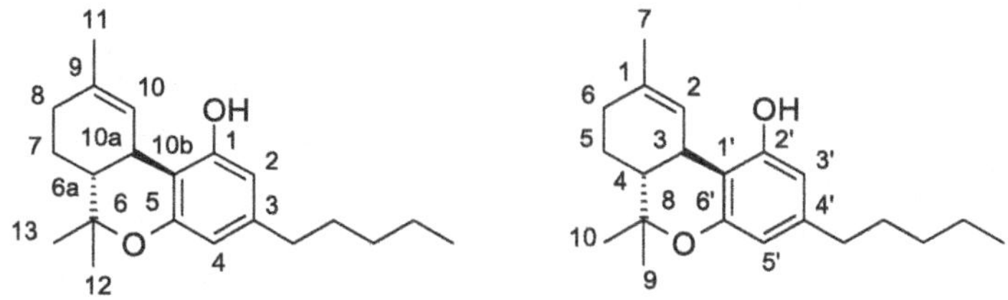

(1.1) Dibenzopyrane numbering system

(1.2) Terpene numbering system

Fig. 1 Commonly used numbering systems for cannabinoids

its order. As an example, the use of the terpene numbering system gives the name Δ1-tetrahydrocannabinol; in contrast, using the dibenzopyrane numbering system leads to the name Δ9-tetrahydrocannabinol for the same compound. The dibenzopyrane numbering system, which stands in agreement with IUPAC rules, is commonly used in North America whereas the terpene numbering system, following the biochemical nature of these compounds, was originally developed in Europe [3]. According to IUPAC rules, the dibenzopyrane system is used despite the fact that this system has a general disadvantage because of a complete change in numbering after loss of the terpenoid ring, as found in many cannabinoids.

The chemical name of Δ9-THC according to the dibenzopyrane numbering system is 3-pentyl-6,6,9-trimethyl-6*a*,7,8,10*a*-tetrahydro-6*H*-dibenzo-[*b*, *d*]pyran-1-ol as depicted in **1.1** (Fig. 1).

Alternatively, Δ9-tetrahydrocannbinol or simply tetrahydrocannabinol is frequently used in the scientific community. When using the short name tetrahydrocannabinol or just THC it always implies the stereochemistry of the Δ9-isomer.

On the market are two drugs under the trade names of Dronabinol, which is the generic name of *trans*-Δ9-THC, and Marinol, which is a medicine containing synthetic dronabinol in sesame oil for oral intake, distributed by Unimed Pharmaceuticals.

1.2
Chemical and Physical Properties of Δ9-THC

Δ9-THC (**2.1** in Fig. 2) is the only major psychoactive constituent of *C. sativa*. It is a pale yellow resinous oil and is sticky at room temperature. Δ9-THC is lipophilic and poorly soluble in water (3 $\mu g\,mL^{-1}$), with a bitter taste but without smell. Furthermore it is sensitive to light and air [4]. Some more physical and chemical data on Δ9-THC are listed in Table 1. Because of its two chiral centers at C-6*a* and C-10*a*, four stereoisomers are known, but only (–)-*trans*-Δ9-THC is found in the *Cannabis* plant [5]. The absolute configuration of the

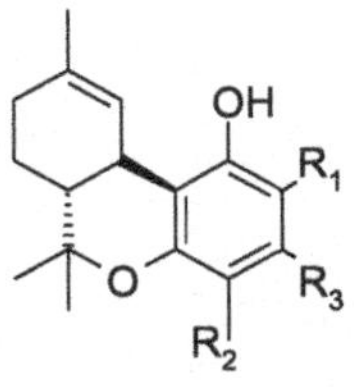

(2.1) R_1=H R_2=H R_3=C_5H_{11}
(2.2) R_1=COOH R_2=H R_3=C_5H_{11}
(2.3) R_1=H R_2=COOH R_3=C_5H_{11}
(2.4) R_1=H R_2=H R_3=C_3H_7
(2.5) R_1=H R_2=H R_3=C_4H_9

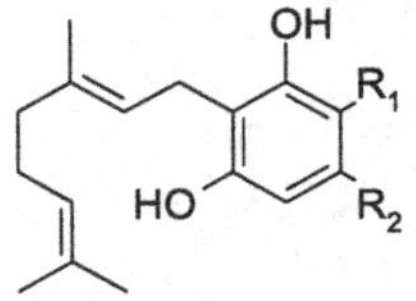

(2.6) R_1=H R_2=C_5H_{11}
(2.7) R_1=COOH R_2=C_5H_{11}
(2.8) R_1=H R_2=C_3H_7

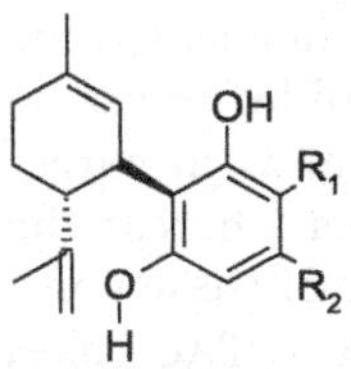

(2.9) R_1=H R_2=C_5H_{11}
(2.10) R_1=COOH R_2=C_5H_{11}
(2.11) R_1=H R_2=C_3H_7
(2.12) R_1=H R_2=C_4H_9

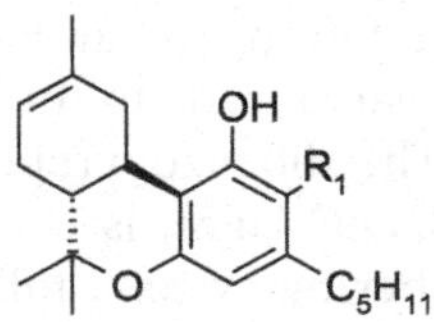

(2.13) R_1=COOH
(2.14) R_1=H

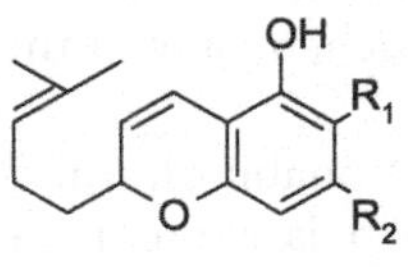

(2.15) R_1=H R_2=C_5H_{11}
(2.16) R_1=COOH R_2=C_5H_{11}
(2.17) R_1=H R_2=C_3H_7

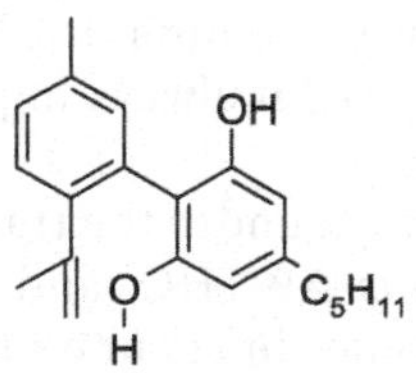

(2.18)

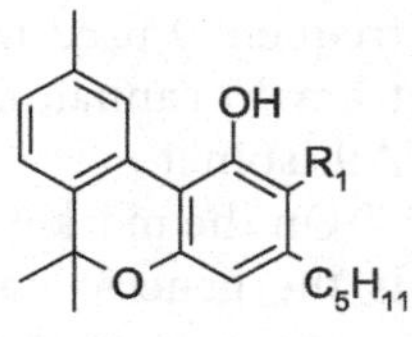

(2.19)

Fig. 2 Chemical structures of some natural cannabinoids

natural product was determined as (6*aR*,10*aR*) [6]. Depending on the position of the double bond in the terpenoid ring six isomers are possible, whereof the Δ9-isomer and the Δ8-isomer are most important. Conformational studies of Δ9-THC using NMR techniques were done by Kriwacki and Makryiannis [7]. The authors found that the arrangement of the terpenoid ring and pyrane ring of this compound is similar to the half-opened wings of a butterfly. An excellent

Table 1 Chemical and physical properties of (–)-*trans*-Δ9-THC [4]

Molecular weight	314.47
Molecular formula	$C_{21}H_{30}O_2$
Boiling point	200 °C (at 0.02 mm Hg)
Rotation of polarized light	$[a]_D^{20}$ = – 150.5° (c = 0.53 in $CHCl_3$)
UV maxima	275 nm and 282 nm (in ethanol)
Mass fragments (m/z) [a]	314 (M+); 299; 271; 258; 243; 231
pK_a	10.6
Stability	Not stable in acidic solution ($t_{1/2}$ = 1 h at pH 1.0 and 55 °C)
Partition coefficient (octanol/water) [b]	12 091
Solubility	Highly insoluble in water (~ 2.8 mg L^{-1} at 23 °C)

[a] These mass fragments were found by our own measurements
[b] In the literature, values between 6000 and 9 440 000 can be found [102]

review by Mechoulam et al. has been published providing more information on this topic and discussing extensively the stereochemistry of cannabinoids and Δ9-THC, with special focus on the structure–activity relationship [8].

It must be noted that Δ9-THC is not present in *C. sativa*, but that the tetrahydrocannbinolic acid (THCA) is almost exclusively found. Two kinds of THCA are known. The first has its carboxylic function at position C-2 and is named 2-carboxy-Δ9-THC or THCA-A (**2.2**); the second has a carboxylic function at position C-4 and is named 4-carboxy-Δ9-THC or THCA-B (**2.3**).

THCA shows no psychotropic effects, but heating (e.g., by smoking of *Cannabis*) leads to decarboxylation, which provides the active substance Δ9-THC. Δ9-THC is naturally accompanied by its homologous compounds containing a propyl side chain (e.g., tetrahydrocannabivarin, THCV, THC-C_3, **2.4**) or a butyl side chain (THC-C_4, **2.5**).

1.3 Further Natural Cannabinoids

Seventy cannabinoids from *C. sativa* have been described up to 2005 [2]. Mostly they appear in low quantities, but some of them shall be mentioned in the following overview – especially because of their functions in the biosynthesis of Δ9-THC and their use in medicinal applications.

1.3.1 Cannabigerol (CBG)

Cannabigerol (CBG, **2.6**) was historically the first identified cannabinoid [9]. It can be comprehended as a molecule of olivetol that is enhanced with

2,5-dimethylhepta-2,5-diene. In plants, its acidic form cannabigerolic acid (CBGA, **2.7**) and also the acid forms of the other cannabinoids prevail. CBGA is the first cannabinoidic precursor in the biosynthesis of Δ9-THC, as discussed in Sect. 2. Although the *n*-pentyl side chain is predominant in natural cannabinoids, cannabigerols with propyl side chains (cannabigerovarin, CBGV, **2.8**) are also present.

1.3.2 Cannabidiol (CBD)

The IUPAC name of cannabidiol is 2-[(1*S*, 6*R*)-3-methyl-6-prop-1-en-2-yl-1-cyclohex-2-enyl]-5-pentyl-benzene-1,3-diol. Cannabidiol (CBD, **2.9**) in its acidic form cannabidiolic acid (CBDA, **2.10**) is the second major cannabinoid in *C. sativa* besides Δ9-THC. As already mentioned for Δ9-THC, variations in the length of the side chain are also possible for CBD. Important in this context are the propyl side chain-substituted CBD, named cannabidivarin (CBDV, **2.11**), and CBD-C_4 (**2.12**), the homologous compound with a butyl side chain. Related to the synthesis starting from CBD to Δ9-THC as described in Sect. 3.1, it was accepted that CBDA serves as a precursor for THCA in the biosynthesis. Recent publications indicate that CBDA and THCA are formed from the same precursor, cannabigerolic acid (CBGA), and that it is unlikely that the biosynthesis of THCA from CBDA takes place in *C. sativa*.

1.3.3 Δ8-Tetrahydrocannabinol (Δ8-THC)

This compound and its related acidic form, Δ8-tetrahydrocannabinolic acid (Δ8-THCA, **2.13**) are structural isomers of Δ9-THC. Although it is the thermodynamically stable form of THC, Δ8-THC (**2.14**) contributes approximately only 1% to the total content of THC in *C. sativa*. In the synthetic production process, Δ8-THC is formed in significantly higher quantities than in plants.

1.3.4 Cannabichromene (CBC)

Among THCA and CBDA, cannabichromene (CBC, **2.15**) and the acidic form cannabichromenic acid (CBCA, **2.16**) are formed from their common precursor CBGA. Besides CBC, its homologous compound cannabiverol (CBCV, **2.17**) with a propyl side chain is also present in plants.

1.3.5 Cannabinodiol (CBND) and Cannabinol (CBN)

Cannabinidiol (CBND, **2.18**) and cannabinol (**2.19**) are oxidation products of CBD and Δ9-THC formed by aromatization of the terpenoid ring. For the dehydrogenation of THC a radical mechanism including polyhydroxylated intermediates is suggested [10, 11]. CBN is not the sole oxidation product of Δ9-THC. Our own studies at THC-Pharm on the stability of Δ9-THC have shown that only about 15% of lost Δ9-THC is recovered as CBN.

2 Biosynthesis of Cannabinoids

The biosynthesis of cannabinoids can only be found in *C. sativa*. These cannabinoids are praised for their medical and psychoactive properties. In addition, the plant material is used for fiber, oil, and food production [12]. For these applications it is important to gain knowledge of the cannabinoid biosynthetic pathway. As an example, fiber production is not allowed if the plant contains more than 0.2% (dry weight) THC. Higher THC content is illegal in most Western countries and cultivation is strictly regulated by authorities. Interestingly, the content of other cannabinoids is of less importance because no psychoactive activity is claimed for them. Furthermore, for forensic purposes the information may be used to discriminate the plants by genotype, which is correlated to the chemotype (see Sect. 2.2), in the early phase of their development. This may help both the cultivator and legal forces. Here the cultivation of illegal plants may be found and controlled by both of them. For the cultivator, to exclude illegally planted plants and for the police to control illegal activities by the cultivators or criminals. Moreover, the information can be used by pharmaceutical companies and scientists. Here it can be used for the studies on controlled production of specific cannabinoids that are of interest in medicine. For instance, THC has been investigated for its tempering effect on the symptoms of multiple sclerosis [13], but CBG and CBD may also have a role in medicine. Both CBD and CBG are related to analgesic and anti-inflammatory effects [14, 15].

In this section, the latest developments and recent publications on the biosynthesis of Δ9-THC and related cannabinoids as precursors are discussed. Special points of interests are the genetic aspects, enzyme regulation, and the environmental factors that have an influence on the cannabinoid content in the plant. Because of new and innovative developments in biotechnology we will give a short overview of new strategies for cannabinoid production in plant cell cultures and in heterologous organisms.

2.1 Biochemistry and Biosynthesis

The biosynthesis of major cannabinoids in *C. sativa* is located in the glandular trachoma, which are located on leaves and flowers. Three known resin-producing glandular trachoma are known, the bulbous glands, the capitate sessile, and the capitate stalked trichoma. It has been reported that the latter contain most cannabinoids [16]. The capitate stalked trichoma become abundant on the bracts when the plant ages and moves into the flowering period. The capitate sessile trichoma show highest densities during vegetative growth [17, 18].

As depicted in Fig. 3, in glandular trichoma the cannabinoids are produced in the cells but accumulate in the secretory sac of the glandular trichomes, dissolved in the essential oil [17–21]. Here, Δ9-THC was found to accumulate in the cell wall, the fibrillar matrix and the surface feature of vesicles in the secretory cavity, the subcutilar wall, and the cuticula of glandular trichomes [19].

As mentioned before, the cannabinoids represent a unique group of secondary metabolites called terpenophenolics, which means that they are composed of a terpenoid and a phenolic moiety. The pathway of ter-

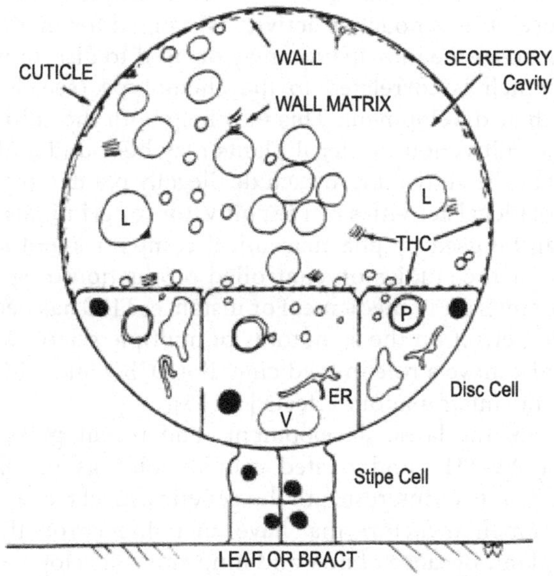

Fig. 3 Representation of mature secretory gland originated from *C. sativa*. The separate compartments of the glandular trichome are clearly shown, and the places where THC accumulates. *Black areas* nuclei, *V* vacuole, *L* vesicle, *P* plastid, *ER* endoplasmic reticulum. Picture obtained from: http://www.hempreport.com/issues/17/malbody17.html

penoid production is already reviewed exhaustively [22–25]. The phenolic unit of cannabinoids is thought to be produced via the polyketide pathway [26–28]. Both the polyketide and terpenoid pathways merge to the cannabinoid pathway and this combination leads to the final biosynthesis of the typical cannabinoid skeleton. Here we will discuss the different aspects of the cannabinoid pathway for most already-found cannabinoids, like cannabigerolic acid (CBGA), tetrahydrocannabinolic acid (THCA), cannabidiolic acid (CBDA), and cannabichromenic acid (CBCA). For convenience the abbreviations of the acidic form will be used through this section because

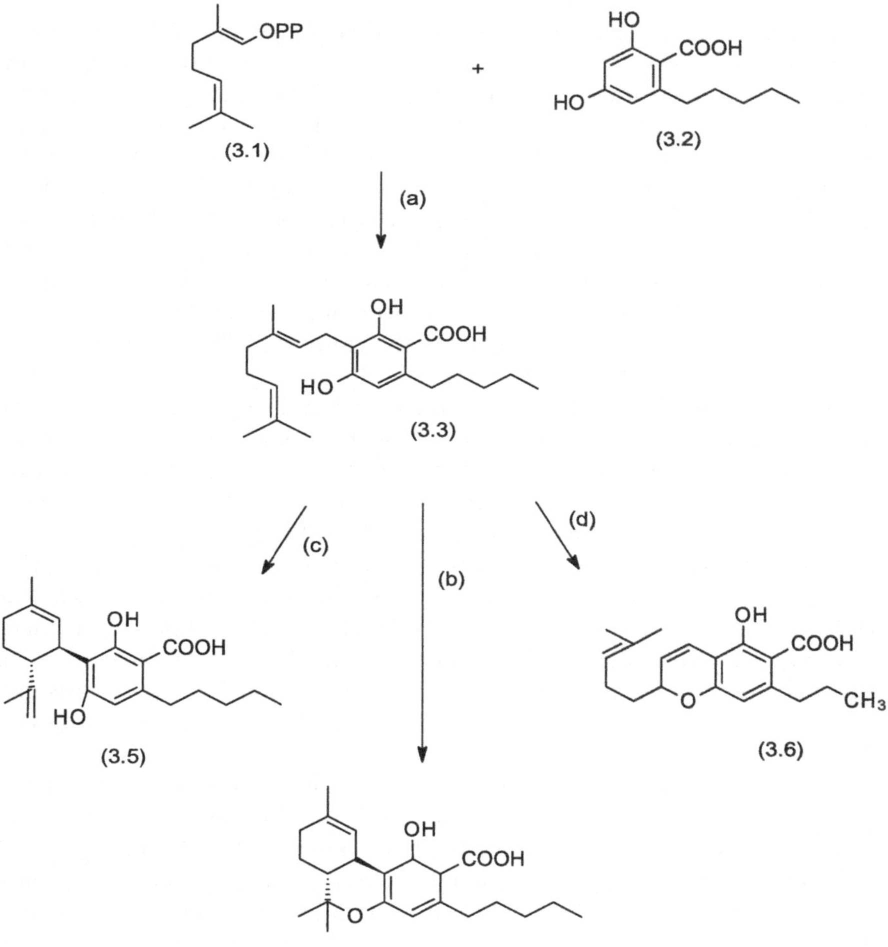

Fig. 4 Biosynthesis of THC and related cannabinoids: *a* GOT, *b* THCs, *c* CBDs, *d* CBCs

they occur as genuine compounds in the biosynthesis. Under plant physiological conditions the decarboxylated products will be absent or be present only in small amounts.

The late cannabinoid pathway starts with the alkylation of olivetolic acid (**3.2** in Fig. 4) as polyketide by geranyl diphosphate (**3.1**) as the terpenoid unit. Terpenoids can be found in all organisms, and in plants two terpenoid pathways are known, the so called mevalonate (MEV) and non-mevalonate (DXP) pathway as described by Eisenrich, Lichtenthaler and Rohdich [23, 24, 29, 30]. The mevalonate pathway is located in the cytoplasm of the plant cells [30], whereas the DXP pathway as major pathway is located in the plastids of the plant cells [29] and delivers geranyl diphosphate as one important precursor in the biosynthesis.

The polyketide pathway for olivetolic acid is not yet fully elucidated. It is assumed that a polyketide III synthase will either couple three malonyl-CoA units with one hexanoyl-CoA unit [26], or catalyze binding of one acetyl-CoA with four malonyl-CoA units [28] to biosynthesize olivetolic acid [26–28, 31, 32]. Olivetolic acid as precursor for Δ9-THC contains a pentyl chain in position C-3 of its phenolic system, but shorter chain lengths have also been observed in cannabinoids [33]. These differences in chain length support the hypothesis of production by a polyketide, as it is a known feature of these enzymes [34]. It was recently described that crude plant cell extracts from *C. sativa* are able to convert polyketide precursors into olivetol [26]; however here no olivetolic acid was detected. On the contrary, Fellermeier et al. [32] showed that only olivetolic acid and not olivetol could serve in the enzymatic prenylation with GPP or NPP. An older article described that both olivetol as olivetolic acid can be incorporated. Here the incorporation of radioactive labeled olivetol has been detected in very low amounts and olivetolic acid in high amounts. These reactions were performed in planta, whereas the previous reactions were performed in vitro [35]. It still remains unclear which structure, olivetol or olivetolic acid, is really preferred. Horper [36] and later Raharjo [26] suggested that the aggregation of the enzymes could prevent the decarboxylation of olivetolic acid. This explanation suggests that the enzymes are either combined or closely located to each other so that the olivetolic acid is placed directly into the site responsible for prenylation. This hypothesis has still to be proven, but supports the fact that olivetolic acid cannot be found in *Cannabis* extracts [35].

Until recently no enzymes able to produce olivetol-like compounds have been isolated. In an article by Funa et al., polyketide III enzymes were responsible for the formation of phenolic lipid compound [34], a natural product group that olivetol belongs to. Although the biosynthesized compounds contained a longer chain, which increased over time, the study supported the hypothesis of olivetolic acid production by a polyketide III synthase. Further studies on the genetic and protein level are essential to elucidate the mode of mechanism by which olivetolic acid is formed in *C. sativa*.

The precursor of the major cannabinoids is proven to be cannabigerolic acid (CBGA, **3.3**) [32, 35]. The formation of this compound is catalyzed by an enzyme from the group of geranyltransferases [28, 32]. This enzyme was studied in crude extracts made from young expanding leafs, were it exhibited activity only with olivetolic acid as the substrate. Despite the fact that no sequence has been published yet, the enzyme was designated geranylpyrophosphate: olivetolate geranyltransferase (GOT). Recently [37] the structure and characterization of a geranyltransferase, named orf-2 and originating from *Streptomyces* CL109, was reported. The authors claimed that the enzyme is able to geranylate both olivetol and olivetolic acid and thus it may be highly similar to the CBGA synthase. Although the authors made this firm statement, they based it on the results obtained by thin layer chromatography. For confirmation of this activity more precise analytical techniques, like LC-MS or NMR, must be performed for structure elucidation of the product produced. Although we have more information about GOT than about polyketide synthase (see Table 2), the mechanism of activity remains uncertain. This means more studies must be performed to obtain the gene sequence.

The last enzymatic step of the cannabinoid pathway is the production of THCA (**3.5**), CBDA (**3.4**) or CBCA (**3.6**). The compounds are produced by three different enzymes. The first enzyme produces the major psychoactive compound of cannabis, THCA [21, 38]; the second and third are responsible for the production of CBDA [39] and CBCA [40], respectively. All of these enzymes belong to the enzyme group oxidoreductases [38–41], which means that they are able to use an electron donor for the transfer of an electron to an acceptor. From these enzymes only the THCA and the CBDA synthase gene sequence have been elucidated. Their product also represents the highest constituent in most *C. sativa* strains.

The enzyme responsible for THCA formation is fully characterized and cloned into several heterologous organisms. When cloned in a host organism, the highest activity was mostly seen in the media. Here the only exception was the introduction of the gene into hairy root cultures made from tobacco [42]. Studies performed on the enzyme sequence indicated that it contained a signal sequence upstream of the actual enzyme. This was found to be 28 amino acids (84 bp) long, suggesting that the enzyme, under native conditions, is localized to another place than where it is produced. Later studies proved that the enzyme is localized in the storage cavity of the glandular trichomes [21]. In the first publication it was determined that no cofactor is used by the enzyme [41], but this research was performed with purified protein from the *C. sativa* extract. Later studies indicated that a flavin adenine dinucleotide (FAD) cofactor was covalently bound to the enzyme. This was later confirmed by nucleotide sequence analysis in silico, revealing the binding motive for the FAD cofactor.

CBDA synthase is though to be an allozyme of THCA synthase and shows 87.9% identity on a nucleotide sequence level. Although the sequence of this

Table 2 Properties of enzymes found in cannabinoid biosynthesis

Enzyme	pH optimal /p*I*	Reaction rate k_{cat} (in vitro) [s^{-1}]	Localization	Metal ion	Cofactors	Mw[a] [kDa]	Comments	Substrates	Refs.
GOT	7/N.R.	N.R.	?	Mg^{2+}				Olivetolic acid with GPP or NPP	[32]
CBDA	5/6.1	0.19[b]	?[d]	None	None[e]	~75		CBGA and CBNRA	[39]
THCA	7.1/6.4	0.2[b]–0.3[c]	Storage cavity of glandular trichome	None	FAD[e]	~74		CBGA and CBNRA	[38, 41]
CBCA	6.5/7.1	0.04[b]	?	None	None[e]	~71	Probably homo-dimeric	CBGA and CBNRA	[40]

[a]Obtained from protein isolation, not heterologously expressed
[b]Determined by purified Cannabis extract
[c]Determined by recombinant proteins isolates
[d]CBDA synthase shown to carry a highly similar N-terminal signal sequence to THCA synthase. It is thus suggested that this enzyme is localized at the same position as THCA synthase. Furthermore, the precursor CBGA has been shown to be toxic for plant cells and is probably localized in the secretory cavity of the glandular trichome. This suggests that CBDA, THCA, CBCA are all localized in the storage cavity
[e]Activity test with crude extract did not show the need for cofactors; however, from analysis performed on THCA synthase, it became clear that FAD is covalently bound to the enzyme. Furthermore, analysis of the enzymes THCA and CBDA showed the motif that is conserved for FAD binding
N.R. not reported

gene is known [43], there are no reports of studies where they produced and characterized it. All information gained about the enzyme was obtained using purified protein from *C. sativa* extracts [39]. Although not tested yet, the deposited sequence shows the same conserved FAD binding motive as found and proven for THCA synthase. Because the CBDA synthase carries the same signal sequence as the THCA synthase it suggests that the CBDA is localized in the same place as the THCA synthase.

For CBCA synthase hardly any information has been published. The enzyme was characterized after it was purified from *C. sativa* extracts and until this moment no sequence has been deposited. After purification of the protein it was found to be a homodimeric enzyme, meaning that enzyme is formed by two identical domains. This was observed after the purification, when the enzyme had a molecular weight of 136 kDa, and after denatured electrophoresis, when it had a molecular weight of ~ 71 kDa. Furthermore, the CBCA synthase has shown to bear higher affinity for CBGA (1717 $M^{-1}S^{-1}$) than THCA synthase and CBDA synthase (respectively 1382 $M^{-1}s^{-1}$ and 1492 $M^{-1}s^{-1}$), which is probably due to its homodimeric nature [40].

From the biosynthetic route a lot of knowledge has been gathered through the years. Up to now only one enzyme has been reasonably characterized, but much information has been gained through crude extract activity studies. This information has already proven to be a solid basis for genetic testing and will be useful for further investigations of the biosynthetic route. Although it must be stated that high polymorphism is detected in the genes [44] and high genetic diversity found within, *C. sativa* can still give unexpected results in other investigations. The information gained from the research reported above is already used frequently in the breeding and detection of certain chemotypes and for the development of new ones, as we will see in the next section.

2.2 Genetics of *Cannabis Sativa*

The majority of *C. sativa* strains exist as a dioeciously (separate sexes) plant species and are wind-pollinated. Under normal conditions it is an annual herb, although longer-living *C. sativa* have been observed [45, 46]. Some *Cannabis* strains appear as monoecious (containing both male and female parts) cultivars, such as the Ukrainian cultivar USO31 [47], or as hermaphrodites. Most of these cultivars are not seen in nature. It is estimated that only 6% of the flowering plants are dioecious and generally they are seen as the most evolved species within the plant kingdom [48, 49]. The *C. sativa* genome is normally a diploid one and contains ten chromosome pairs ($2n = 20$). Here, eighteen are autosomal and two are sex-linked chromosomes. The genome was measured in both female (XX) as well as in male plants (XY). In contrast to animals, the male genome was found to be bigger by 47 Mbp [50, 51]. It must be stated that dioecious plants are able to change sex during their development. This ability is mostly used as a strategy for survival, however it can be chemically induced. Within the *C. sativa* species lots of phenotypes are known. Generally the *C. sativa* plant are believed to be a monotypic species [47] called *Cannabis sativa* L. with further divisions in subspecies. However, Hillig [46] showed, by allozyme analysis in combination with morphological traits, that a separation may be made between *C. sativa* L. and the *C. indica* Lam. He also suggested a putative third one named *C. ruderalis* Janisch. The polytypic species within *C. sativa* was already suggested several years ago when the plants were determined only by their phenotypic traits or drug potential properties [46]. There is still discussion about whether or not the *C. sativa* species are monotypic or polytypic, but in most literature they are referred to as *C. sativa* with further division into the subspecies *indica* or *ruderalis*.

C. sativa is mostly divided into three major chemotypes. The chemotypes boundaries are set by the ratio CBD : THC and are calculated as percentage of dry weight. These three chemotypes consist of the "fiber"-type (CBD > THC), the "intermediate"-type (CBD≈THC) and the "drug"-type (CBD < THC). The chemotypes have been recently shown to be dependent either on one locus on the chromosome, or two closely linked loci [47], but the former theory is the most likely one [52–54]. The locus is called the B locus and until now it is proven to consist of at least two alleles, namely B_t and B_d. There is also an indication for a third allele. This one was named B_0 and seems to be responsible for a CBGA-dominant chemotype [54]. The alleles, B_t and B_d, show co-dominance and the B_0 allele is recessive or an inactive B_d allele. The B_0 is believed to be an inactive B_d allele because it can be indicated by molecular markers specific for the CBDA gene (B_d allele). The evidence for these alleles was gained by breeding with chemotypes and molecular analysis [47]. In crossings made with fiber-type and drug-type, the intermediate chemotype was obtained as offspring. Intercrosses of these F1 plants gave a representative Mendelian ratio (1 : 2 : 1) of chemotypes. This Mendelian ratio suggests that one locus is responsible for the chemotypes. Furthermore, Pacifico et al. [47] proved, with the help of multiplex PCR, that a 100% identification of specific chemotype (from the three accepted chemotypes) could be made. This multiplex PCR was performed with three primers, one of which was designed to anneal with both the THCA synthase and the CBDA synthase gene, while the other two were specific for one or both. The results showed that the intermediate chemotype was heterozygote and thus contained both the CBDA synthase and the THCA synthase genes. The drug- and the fiber-type were shown to be homogeneous for the THCA synthase and the CBDA synthase genes, respectively. Although the genes are not themselves detected, their products are. For instance, the fiber-type group that is shown to be homogenous for the B_d allele, still produces low amounts of THCA. It is thus still possible that the homogeneous type carries the THCA synthase gene; however, it is not detected due to the polymorphisms within the gene, as shown by Kojoma et al. [44].

Recently it has been suggested that there are two more chemotypes. The first (chemotype 4) has a high content of CBGA (B_0 allele) and the second is a strain totally lacking cannabinoids [47]. These strains are of interest because they can serve as good and safe strains for the production of fiber.

2.3 Environmental Factors

Cannabis seems to react to several environmental influences. The most known are hydration, soil nutrients, wounding, competition and UV-B radiation. Proper use of these environmental influences can increase the glandular density and the cannabinoid content. Environmental factors have also been shown to induce sex change in *C. sativa*. Moreover, when some chemotypes are grown in a different environment their cannabinoid content seem to be changed. With genetic analysis it must be possible to determine if a strain is indeed a fiber strain or if it is an intermediate strain that has been suppressed for its cannabinoid content due to the environment of cultivation. Some of the major environmental factors influencing the cannabinoid content are described below. It must be stated that environmental stress also affects the growth of the plant.

2.3.1 Dehydration

In times of less accessibility of water, the plants seem to increase the cannabinoid content. It is suggested that the plant will cover itself with the oily cannabinoids to prevent water evaporation. For instance Sharma (1975) found increased glandular trichome densities in the leaves of Cannabis grown under dry circumstances [55].

2.3.2 Nutrients in Soil

It is clear that the nutrients in soil are important for plant development and that a good nutrient supply within the soil gives healthy plants. However, no profound research results have yet been published on the most optimal soil conditions.

2.3.3 Light

Light has a major influence on plants, and for *Cannabis* plants it is mostly important for growth and flowering. Long daylight induces strong vegetative growth and shorter daylight leads to flowering of the plants. Furthermore, it has been shown by Lydon et al. that the level of THC increase is linear with the increase in UV-B dose [56–58].

2.4
Growing of *Cannabis Sativa* and Optimization of THC Yield

2.4.1
Cultivation of *Cannabis*

C. sativa is cultivated for several purposes. Actually, the main legal purpose is the production of hemp fibers and pulp. From these materials paper, clothes and ropes are made [12] and several Western countries have already legalized the cultivation of *C. sativa* for these purposes. In research, the drug-type of *C. sativa* is also cultivated, however, only for the investigation and determination of forensic studies for chemotype separation. The growth for medicinal purposes is hardly performed. In the Netherlands *C. sativa* is cultivated for medicinal purposes under strictly controlled regulations by the company Bedrocan. In this chapter we discuss basic aspects of the cultivation of *C. sativa* and the optimization of THC content in the plant.

2.4.2
Optimization of THC Yield

The optimization of THC yield is mostly performed through breeding programs. Because of the illegality of the plant in most countries, it is performed on small scale or by illegal drug cultivators. In the previous section we have already discussed the fact that cannabinoid production is mostly genetically determined. This knowledge could thus be used to increase the production of certain compounds and decrease the others. Furthermore, the total THC yield is dependent on the amount of accessible precursors and the level acceptable for the plant.

Within the *C. sativa* strain the total content of cannabinoids varies. In THCA-dominant plants variations have also been noted. Some have low detectable amounts of CBDA whereas others have none. Furthermore, some plants have been shown to contain detectable amounts of CBGA while others had none. There is still a question as to what extent THCA production can be increased in the plant by breeding programs and genetic modifications. From the genetic point of view it should be noted that the yield of THC is not only dependent on the B_t allele, but also depends on the amount of biomass, the density of trichomes, and the production of precursors indicating a complex spectrum of different possibilities for professional plant breeders. The yield of THCA in THCA-dominant plants can be increased by environmental influences. In cultivations of the drug-type, mostly done by illegal cultivators, the male plant is excluded from the field. The background is that this will induce more biomass of the female plant because it cannot be inseminated. But, the exclusion of male plants will not give a more constant increase of THC yield.

Genetic modification seems to be an option for increased yield since the THCA production is mainly dependent on genetic factors (see Sect. 2.1). Here we could think of increasing glandular trichoma densities, increasing precursor production, increasing enzyme activity and knocking out enzymes that use the general precursor of THCA (CBGA). Applying some of these techniques have already shown an increase in the amount of secondary metabolites in microbial organisms.

As stated above, breeding programs could increase the total yield of THCA. Within *C. sativa* there are many phenotypes, e.g., while one has the ability to grow over a few meters high, another stays small. Furthermore, variations in glandular trichoma densities have also been observed in THCA content and ratios. By combining the phenotypes of various plants with each other, a plant could be grown that is large in growth, high in glandular trichoma density, or high in THCA content. Through breeding techniques Meijer et al. [54] have already created a high CBGA-producing plant and in the drug culture the same has been reached for THCA [59]. In the latter, preparations were found containing more than 20% THC, while in the literature and exported Cannabis the normal values lie at 6–10%.

2.4.3 *Cannabis* Standardization

Just like all herbal medicinal preparations, *C. sativa* should be standardized if extracts or whole plant material are to be used for medicinal purposes. Basic requirements are that all detectable constituents should be known, but also a sustainable quality control system must be established to achieve the same quality over all batches. For industrial use of cannabis, standardization could also be necessary to equalize the quality of the product. However, it must be stated that cultivation for this purposes is mostly performed outdoors. Outdoor growth makes standardization of the product difficult due to the environmental changes. For this reason the Dutch medicinal *C. sativa* is grown under strictly controllable conditions, and therefore indoors, by the company Bedrocan. At this company clones are used for breeding to maintain high standards for quantity and quality. After a strictly selective breeding procedure a plant line has been established fulfilling all criteria as a herb for medicinal use.

2.5 Alternative Production Systems for Cannabinoids

It is clear that production of cannabinoids should be controllable to obtain a constant quality of certain cannabinoids. With the knowledge of the biosynthetic route towards cannabinoid production it is now possible to develop biological production systems as an alternative to chemical synthesis. The major advantage of biological systems is not only having the right natural product by structure, but also the only isomer in high yield. Here, three alternative production strategies are introduced. Although two of them are still hypothetical, it should be possible to be realize them in the near future.

2.5.1 Cell Cultures

In the literature several reports can be found on the growth of callus and cell suspension cultures [60–62]. Most of them document that no cannabinoids can be found within these cultures. Although one article by Heitrich and Binder [60] mentioned that variations in media can induce cannabinoid secretion, no second report could confirm these results. Callus and cell suspension are induced by standard techniques for plant cell manipulation. The induction of callus seems to vary per *C. sativa* variant [61]. To obtain cell suspensions from the callus, the same media is used as for callus growth, with the exception of agar as solidifier. In the literature, the cell suspensions made from *C. sativa* callus are mostly used for bioconversion studies. There is one report that described the use of cannabinoid precursors to determine if cannabinoid production can be induced by feeding with specific biosynthetic precursors [60]. When production of cannabinoids can be achieved in cell cultures from *C. sativa* material, it must still be considered that cannabinoids are toxic to the plant cell itself. These compounds induce the apoptotic response [21]. Thus, at high levels of cannabinoid content, techniques have to be developed to extract them from the growth media for continuous production.

2.5.2 Transgenic Plants

Although the use of transgenic plants is not generally accepted for medicinal herbal preparations, transgenic plants could be used to express certain preferable traits. The THCA yield could be increased by manipulation of metabolic pathways or by making knock-outs of biosynthetic genes. With the use of these techniques, the plant could be made resistant to certain parasites and diseases. General plant manipulating strategies can be used to obtain transgenic plants. There is no literature available for the production or use of transgenic *C. sativa* plants.

At the moment, strategies for the production of transgenic plants are already used for maize, tobacco, potato, and rice. The main purpose is to increase their resistance toward diseases [63]. Some plants also get newly introduced products, such as vitamins [64]. Another purpose of transgenic plants is their use for production of vaccines; for instance hepatitis B vaccine in potato plants [65]. The examples shown here are a selection of many to show the possible transgenic plants uses.

2.5.3 Heterologous Expression of Cannabinoid Biosynthetic Genes

Until now there have only been two reports on heterologous expression of *C. sativa* origin genes into host organisms. In these reports the yeast *Pichia pastoris*, hairy root cultures of tobacco, BY-2 tobacco cell cultures, and insect cells were used to produce the THCA synthase enzyme [38, 42]. In the literature, the use of heterologous expression of plant metabolic enzymes have been shown to be useful in the production of several compounds [25]. The same strategy is probably useful for the production of cannabinoids. The production of cannabinoids will probably ask for specific cultivation parameters because some of its constituents may be toxic to the host in certain concentrations. One method could be a constant refreshment of the growth medium. To date, no publications discuss the efforts of using the heterologous production of cannabinoids. This strategy could be, however, of high interest to pharmaceutical companies when some cannabinoids are approved for medicinal use.

3 Chemical Synthesis

3.1 Synthesis Routes for Δ9-THC

After identification of Δ9-THC as the major active compound in *Cannabis* and its structural elucidation by Mechoulam and Gaoni in 1964 [66], a lot of work was invested in chemical synthesis of this substance. Analogous to the biosynthesis of cannabinoids, the central step in most of the Δ9-THC syntheses routes is the reaction of a terpene with a resorcin derivate (e.g., olivetol). Many different compounds were employed as terpenoid compounds, for example citral [67], verbenol [68], or chrysanthenol [69]. The employment of optically pure precursors is inevitable to get the desired (–)-*trans*-Δ9-THC.

A general problem during the syntheses of Δ9-THC is the formation of the thermodynamically more stable Δ8-THC, which reduces the yield of Δ9-THC. It is formed from Δ9-THC by isomerization under acidic conditions. While the usage of strong acids such as *p*-TSA or TFA leads mainly to Δ8-THC, the yield of Δ9-THC can be increased by employment of weak acids, e.g., oxalic acid [70].

Recently the most employed method for the production of Δ9-THC on industrial scale is the condensation of (+)-*p*-mentha-2,8-dien-1-ol (**5.1** in

Fig. 5) with olivetol (**5.2**) in the presence of boron trifluoride etherate, $BF_3{\cdot}OC(C_2H_5)_2$ with CBD as a key intermediate. This one-step synthesis of Δ9-THC is also used for the production of synthetic dronabinol, which is used in the medicinal application named Marinol. The mechanism of this synthesis is particular described by Razdan et al. [71] and is shown in Fig. 5

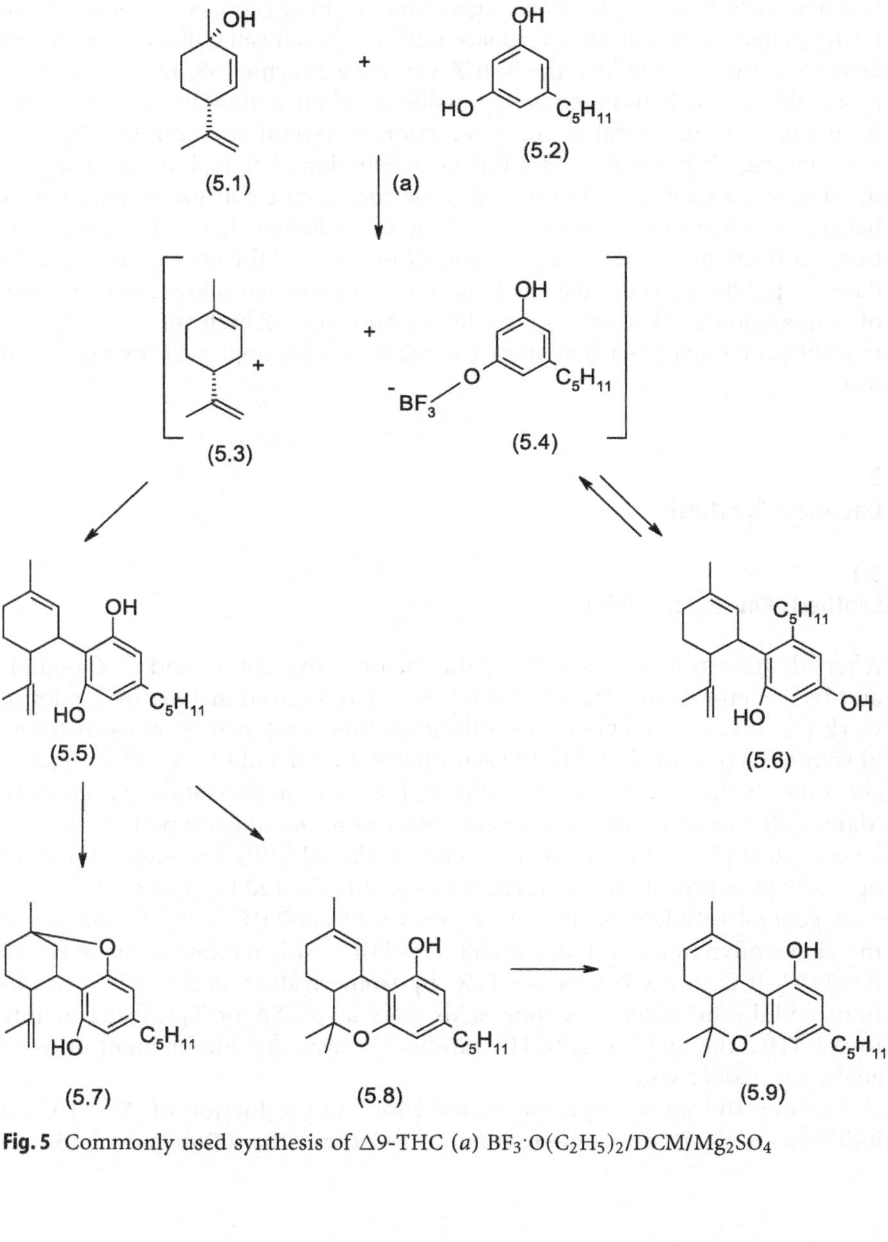

Fig. 5 Commonly used synthesis of Δ9-THC (*a*) $BF_3{\cdot}O(C_2H_5)_2$/DCM/Mg_2SO_4

with the most important side products. There are two possibilities for the condensation of the active terpenoid moiety (**5.3**) with activated olivetol (**5.4**). The fusion of these compounds leads to two intermediates, normal CBD (**5.5**), which has the same structure as natural CBD, and "abnormal" CBD (**5.6**) with transposed positions of the pentyl side chain and a hydroxy group. Fortunately, the latter compound is less stable than the normal CBD and decompensates more easily. The normal CBD directly undergoes a further cyclization to Δ9-THC (**5.7**). If the double bond in the terpenoid ring is used for the cyclization, a isomeric compound named *iso*-tetrahydrocannabinol (*iso*-THC, **5.8**) will be formed. The reaction has to be stopped here otherwise the stable isomer Δ8-THC (**5.9**) arises by decreasing the yield of Δ9-THC. Purification of the reaction mixture is implemented as a liquid chromatographic process using a silica-based stationary phase and a weak polar eluent (e.g., heptane with 2% *tert*-butyl methyl ether). Further cleaning up is possible with vacuum distillation procedures.

3.2 Synthesis of Δ9-Tetrahydrocannabinol from Natural Cannabidiol (Semisynthetic Δ9-THC)

As discussed, the cultivation of *C. sativa* with high content of Δ9-THC (drug-type) is not allowed in many countries. Because of this, there is no opportunity to harvest a high amount of the medicinally important substance Δ9-THC directly from plant material. In the synthesis route for semisynthetic Δ9-THC, natural CBD from fiber hemp plants is employed. It can be extracted with non-polar solvents such as petroleum ether and purified by recrystallization in *n*-pentane. This procedure avoids the formation of "abnormal" CBD and gives the opportunity to produce Δ9-THC from fiber hemp. Semisynthetic Δ9-THC is distinguishable from the synthetic compound because it contains, besides the major product, small amounts of Δ9-THC-C3 and Δ9-THC-C4, which are not available in the synthetic product.

3.2.1 Derivates of Δ9-THC

Most relevant for the affinity for Δ9-THC and analogs to CB-receptors are the phenolic hydroxyl group at C-1, the kind of substitution at C-9, and the properties of the side chain at C-3. Relating to the structure–activity relationships (SAR) between cannabinoids and the CB-receptors, many different modified structures of this substance group were developed and tested. The most important variations include variations of the side chain at the olivetolic moiety of the molecules and different substitutions at positions C-11 and C-9. One of the most popular analogous compounds of Δ9-THC is HU-210 or (–)-*trans*-11-OH-Δ8-THC-DMH, a cannabinoid with a 1′,1-dimethylheptyl side chain (**8.1**). It was constructed in consideration of SAR and has a potency that is about 100 times higher than that of Δ9-THC itself, while its enantiomer HU-211 (Dexanabinol, **8.2**) does not show this property [8]. In the synthesis of HU-210, 5-(1,1-dimethylheptyl)-resorcin is merged with modified [1*S*,5*R*]-myrtenol [72].

Nabilone (**8.3**) is a 11-nor-9-ketohexahydrocannabinoid with a 1′,1′-dimethylheptyl side chain. It is a synthetic analogous compound of THC and is distributed as Cesamet. Usage of diethyl-α-acetoglutarate as "terpenoid" module in the synthesis of Δ9-THC gives nabilone as an intermediate [73]. In spite of the fact that this synthesis was developed for the forming of Δ9-THC it also could be used for the synthesis of nabilone. A newer synthesis route is described by Archer et al. [74]. The pentyl side chain homologous compound of nabilone, 11-nor-9-ketohexahydrocannabinol, is a useful precursor in the chemical synthesis of the major metabolites of Δ9-THC, e.g., 11-nor-9-carboxy-Δ9-THC (THC-COOH) [75].

Direct oxidation of Δ9-THC at position C-11 involves mainly an isomerization to Δ8-THC; another opportunity in the synthesis of Δ9-THC-metabolites is the pretreatment of terpenoid synthons by introduction of protective groups, e.g., 1,3-dithiane (**6.1** in Fig. 6) followed by the condensation with olivetol (**6.2**) [76]. The formed product is a protected derivate

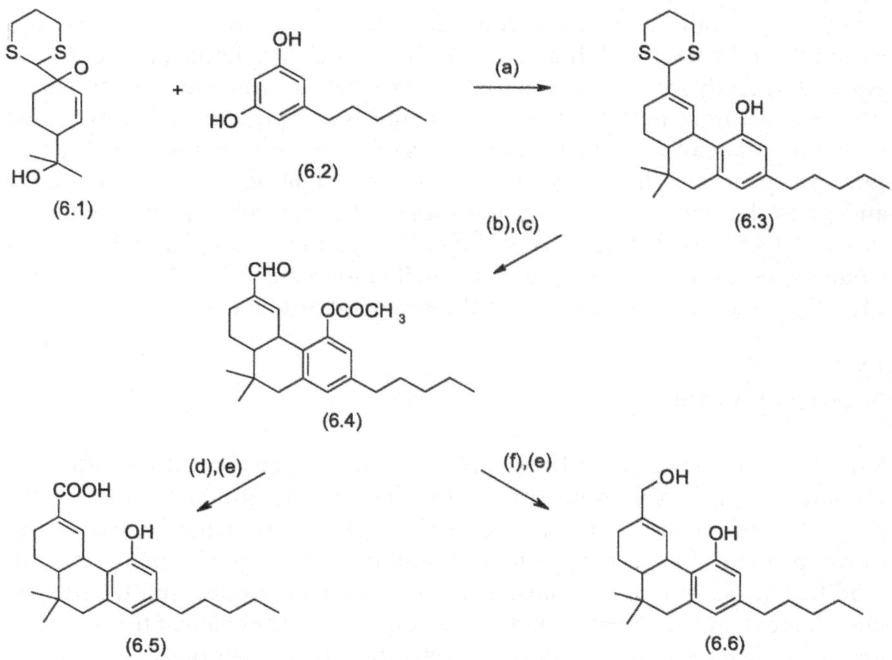

Fig. 6 Synthesis of main metabolites of Δ9-THC: *a* CH_3SO_3H, *b* $(C_2H_5O)_2O$/pyridine, *c* $HgO/BF_3{\cdot}O(C_2H_5)_2$, *d* $NaCN/CH_3COOH/MnO_2$, *e* NaOH/THF, *f* $NaBH_4$/EtOH

of Δ9-THC (**6.3**), which will be modified further. Protection of the phenolic group by esterification, for example, is necessary before the removing of the 1,3-dithiane masking group with mercury oxide. The corresponding aldehyde (**6.4**) can be further oxidized. Deprotection of the phenolic group by alkalic hydrolysis gives the 11-nor-9-carboxy-Δ9-THC (THC-COOH, **6.5**). Under reductive conditions ($NaBH_4$ or $LiAlH_4$) the corresponding alcohol is formed from the aldehyde. This leads to 11-OH-THC (**6.6**), which is the first major metabolite from Δ9-THC formed in humans [77].

When [^{2}H]-labeled precursors are employed the resulting compounds can be used as internal standards for analysis, especially by utilization of mass spectrometric methods. Appropriate deuterated standards are shown in Fig. 7. The introduction of deuterium into the Δ9-THC precursors can be done with Grignard reagents such as $C[^2H_3]MgI$ or reducing substances such as $LiAl[^2H_4]$. The general procedures for the synthesis with these [^{2}H]-labeled precursors are the same as described above for the unlabeled compounds [76, 78].

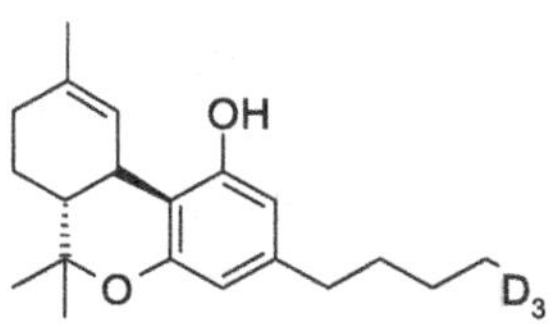

(7.1)

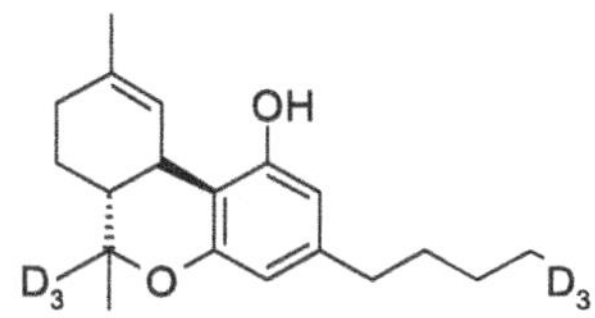

(7.2)

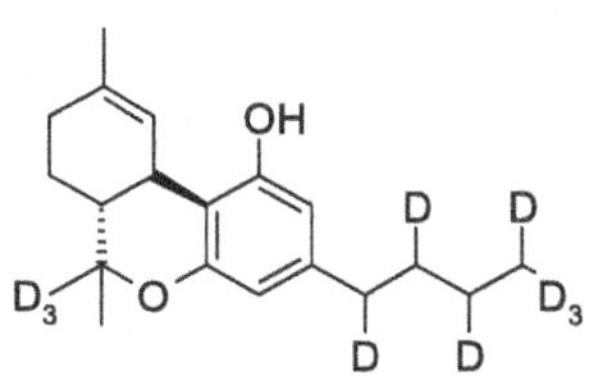

(7.3)

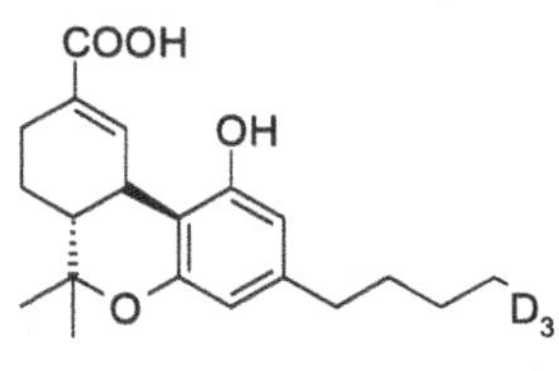

(7.4)

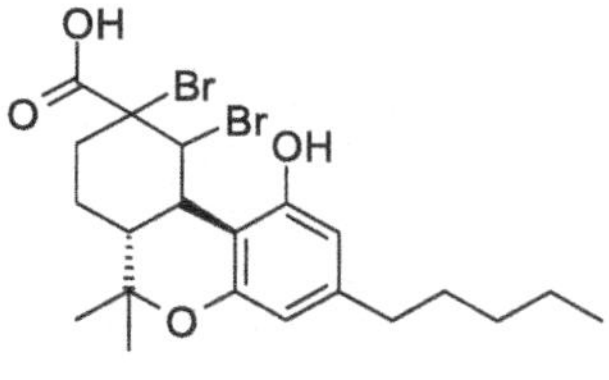

(7.5)

Fig. 7 Deuterated and brominated cannabinoids as internal analytical standards

While the compounds described above contain fundamentally the cannabinoidic structure, there are also compounds with radical changes but which still show high affinity to CB-receptors. Exchange of oxygen with nitrogen in the pyran ring leads to a phenanthridine structure, which can be found in levonantradol (**8.4** in Fig. 8). A compound with total loss of the heterocyclic ring is CP-55,940 (**8.5**). It can be comprehended as a disubstituted cyclohexanole and was synthesized by Pfizer in 1974. This compound was never marketed because of its high psychoactivity, but it is often used for CB-receptor binding studies [79]. Another group of multicore chemical compounds based on the indol structure as a central module in these molecules also shows affinity

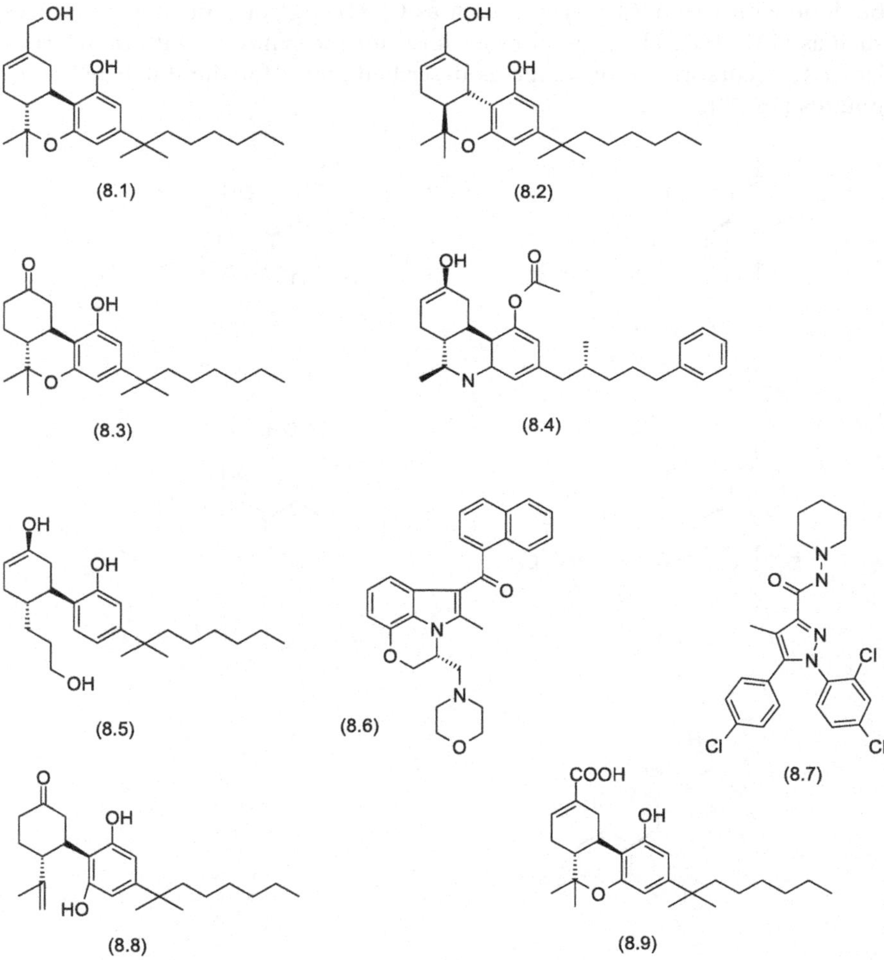

Fig. 8 Synthetic derivatives of Δ9-THC

to CB-receptors. The prototype of this class of aminoalkylindole cannabinoids is the substance named WIN-55,212-2 (**8.6**), which is quite similar to pravadoline, an anti-inflammatory drug [80].

4 Analytics

4.1 Detection of Cannabinoids in Plant Material

The chemical composition of *C. sativa* is very complex and about 500 compounds in this plant are known. A complete list can be found in [81] with some additional supplementations [2, 82]. The complex mixture of about 120 mono- and sesquiterpenes is responsible for the characteristic smell of *C. sativa*. One of these terpenoic compounds, carophyllene oxide, is used as leading substance for hashish detection dogs to find *C. sativa* material [83]. It is a widespread error that dogs that are addicted to drugs are employed for drug detection. Δ9-THC is an odorless substance and cannot be sniffed by dogs.

The aim of the analysis of cannabinoids in plants is to discriminate between the phenotypes (drug-type/fiber-type). Quantification of cannabinoids in plant material is needed if it will be used in medicinal applications, e.g., in *C. sativa* extracts. The ratio between Δ9-THC and CBN can be used for the determination of the age of stored marijuana samples [84].

4.1.1 Analytical Methods for Detection of Δ9-THC and Other Cannabinoids in Plants

Many methods for determination of cannabinoids in plant material have been developed. Commonly HPLC or GC is used, often in combination with mass spectrometry. Molecular techniques are also available to detect these compounds and will be discussed in this section.

4.1.1.1 Sample Preparation

Usually the first step is an extraction of the desired compounds from plant material. This extraction can be done by different solvents, e.g., methanol [85], *n*-hexane [86], petroleum ether or solvent mixtures such as methanol/chloroform [87]. The use of a second liquid–liquid extraction (LLE) with 0.1 M NaOH after extraction with a non-polar solvent like *n*-hexane makes a separate analysis of acidic cannabinoids possible, which can be found as their salts in the water phase [86]. These methods are useful for analysis of plant compartments like flowers or leaves, whereas for seeds a solid phase extraction (SPE) is preferred because of their very low content of cannabinoids [88]. The extracts are commonly used directly for analysis. For analysis of acidic cannabinoids, as they normally appear in plant material, using GC-based methods a previous derivatization of the analytes is usually necessary.

4.1.1.2
Gas Chromatographic Methods (GC)

GC is commonly used for the analysis of cannabinoids, mostly in combination with mass spectrometry (GC-MS). Despite the fact that a lot of different cannabinoids are known almost all of them can be separated by using silica-fused non-polar columns. It is not possible to use GC-based methods for profiling of *C. sativa* samples. The high temperatures that are used in GC cause the decarboxylation of acidic cannabinoids. To detect an acidic cannabinoid such as THCA together with its neutral form such as Δ9-THC, a derivatization is required. This procedure increases the stability of the compounds whereas their volutility is maintained. The most often used reagents for derivatization of cannabinoids in herbal samples are compounds that introduce trimethylsilyl groups (TMS) into the analytes, for example *N*,*O*-bis(trimethylsilyl)trifuoroacetamide (BSTFA), *N*-methyl-*N*-(trimethylsilyl)trifluoroacetamide (MSTFA), or *N*-methyl-*N*-(*tert*-butyldimethylsilyl)trifluoroacetamide (MTBSTFA). Furthermore, mixtures of these compounds with catalysts, e.g., trimethylchlorosilane (TMCS), are used for a quantitative derivatization [89]. While the employment of established detectors such as the flame ionization detector (FID) or electron capture detector (ECD) can only give information about the quantity of a compound, the usage of mass spectrometry (MS) provides additional information about the structures of detected compounds because of their characteristic fragmentation. For the quantification of cannabinoids three-, six-, or even tenfold deuterated compounds such as shown in **7.1**, **7.2** and **7.3** (Fig. 7) are often used as internal standards. The fragmentation of cannabinoids in mass spectrometry is extensively explained by Harvey and the interested reader can find more information about this topic in [90]. A table of about 50 cannabinoids containing free, derivated, and deuterated compounds with their typical mass fragmentations has been published by Raharjo and Verpoorte [89].

4.1.1.3
Liquid Chromatographic Methods (HPLC)

In comparison to GC, an advantage in using HPLC is that there is no decomposition of the acidic forms of cannabinoids. Commonly reversed-phased (RP) materials are used as the stationary phase. Mostly the octadecyl-type

(C-18) is employed. Furthermore, the employment of a guard cartridge containing the same material as used as for the stationary phase is normally recommended. Typical mobile phases are mixtures of methanol and water or acetonitrile and water, acidified with phosphoric acid or formic acid. While for the separation of the main cannabinoids (Δ9-THC, CBD and CBN) an isocratic method is sufficient; the separation of all cannabinoids makes a gradient elution necessary [87]. The use of a photodiode array detector (PDA) is recommended for identification of herbal cannabinoids because of their characteristic UV spectra. If a PDA is used for the detection of cannabinoids Δ8-THC can be employed as an internal standard [91]. According to the law of Lambert–Beer a quantification of cannabinoids based on the strength of the absorption signal is possible. An excellent summary of the most important cannabinoids with their UV spectra and other specific analytical data can be found in [92]. As described in the section on GC-based methods, the employment of mass spectrometry gives the opportunity to identify the structures combined with a better limit of detection (LOD), whereas the use of a UV detector lacks this sensitivity. Another possibility structural identification gives the coupling of HPLC with NMR. The interpretation of [^{1}H]-signals that are specific for different substances can also be used for quantification [93].

4.1.1.4
Immunologically Based Techniques

The enzyme-linked immunosorbent assay (ELISA) technique is often used in laboratories for detection of proteins, but it is also possible to detect small organic molecules by this technique. This assay is based on antibodies that bind with high affinity to certain molecular structures. Testing of cannabinoids by antibodies has been under investigation since the 1970s. The first detections were performed with radiolabeled antibodies made by injection of conjugates from THC, its hemisuccinate, and bovine serum albumin [94]. It was found that the antibody was able to detect cannabinoids and its metabolites from urine and plasma collected from rabbits administered with intravenous cannabinoids. In 1990, Elshoy et al. proved their antibodies to be specific for cannabinoids and related metabolites [95]. Furthermore, they tested against human cannabinoid metabolites excreted via urine and showed that the antibodies against plant cannabinoids were also highly selective and did not bind to any of the non-cannabinoid phenolics. In the early days these studies were all performed with polyclonal antibodies, later monoclonal antibodies were tested and documented the same results [96, 97]. These antibodies may also be used for research. For instance, labeled antibodies have been used against the THC structures to show that THC structures accumulate in the glandular trichoma. Moreover, with this technique it was possible to detect the specific place of accumulation within the trichoma [19]. This indicates that detection by antibodies has an added value over other detection methods such as HPLC and GC. It is thus possible to use these tests either with enzyme, fluorescent or radioactive labels to detect cannabinoids and their metabolites.

4.1.1.5 Molecular Markers and PCR

These detection mechanisms are not able to detect the small organic structure of the cannabinoids. These techniques are designed to make a selection between plant material on a genetic basis. For instance, by the use of only three polynucleotides (primers) and by the use of PCR, discrimination of the major chemotypes (as discussed in Sect. 2.2) was possible. Within the groups selected PCR allowed 100% identification of the chemotypes without any cross reactivity [47]. Furthermore, by a simple PCR technique (two primers used) a separation could be made between drug-type and fiber-type plants [44]. However, it must be stated that a very small number of plants was used and even then polymorphism on the THCA synthase gene was found. The PCR technique cannot be used to detect cannabinoids itself, but maybe it will be of value in plant breeding and cultivation. Furthermore, it may find its place in the detection of illegal *C. sativa* (drug-type) within a population of legal (fiber-type) cultivated plants. However, for this technique, it would be convenient to have the genome of the *C. sativa* plant sequences. This would bring specific information of the differences between male and female plants and could make the design of markers for specific traits easier.

4.2 Detection of Δ9-THC and its Human Metabolites in Forensic Samples

Δ9-THC and its main metabolites are detected and quantified in forensic samples. Determination of these compounds in human beings is needed to make decision on abuse of Δ9-THC-containing drugs by individuals. A careful interpretation of the results is very important to avoid fallacies with regard to the behavior of individuals. The *Cannabis* influence factor (CIF), for example, is an useful tool for distinguishing between acute and chronic intake of Δ9-THC [98].

4.2.1 Metabolism of Δ9-THC by Humane Cytochrome P450 Enzymes

Like other xenobiotics, cannabinoids also undergo extensive metabolism in the human body to increase their hydrophilic properties for a facilitated elimination. The metabolism of Δ9-THC has been very well investigated. More than 100 metabolites of Δ9-THC are known [99] and a good overview of the most important human metabolites is given in [100]. Metabolism takes place mainly in hepatic microsomes, but also in intestines, brain,

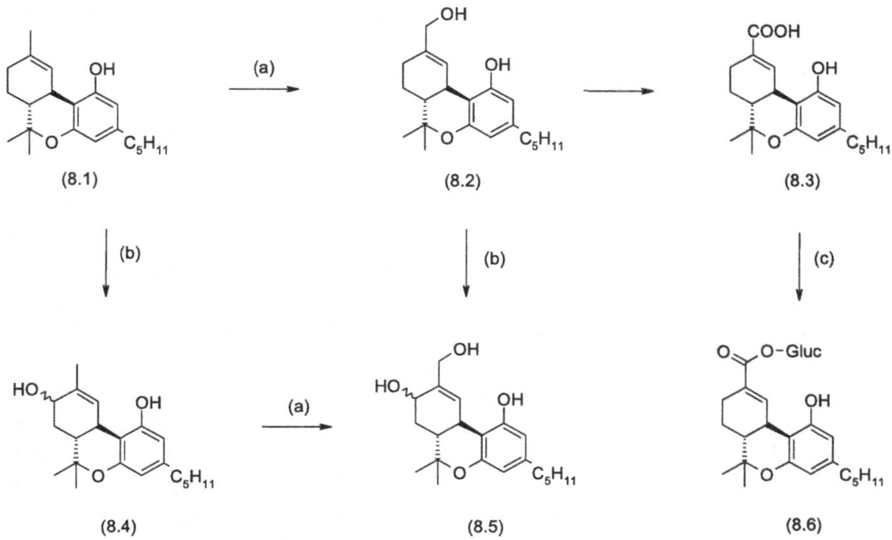

Fig. 9 Main metabolic pathways of Δ9-THC in humans: *a* CYP 2C9, *b* CYP 3A4, *c* UGT

heart, lung, and nearly all tissues of the body. Main metabolites of Δ9-THC are mono-, di- and trihydroxylated compounds, which become carboxylated and glucuronidated further. The metabolism pathway of Δ9-THC and its most important metabolites are shown in Fig. 9. Mostly responsible for metabolism of Δ9-THC in the primary pathway in humans is the cytochrome P450 isoenzyme CYP 2C9 [101]. Hydroxylation of Δ9-THC (**9.1**) at C-11 leads to 11-hydroxy-Δ9-THC (11-OH-THC, **9.2**), which undergoes further oxidation to 11-nor-9-carboxy-Δ9-THC (THC-COOH, **9.3**). 11-OH-THC shows similar psychotropic properties to Δ9-THC whereas THC-COOH is a non-psychotropic compound [102]. CYP 3A4 is the second major cytochrome P450 isoenzyme that is involved in metabolism of Δ9-THC – mainly with hydroxylation at C-8 to 8-OH-Δ9-THC, (**9.4**) [101]. The epoxidation of Δ9-THC at C-9 and C-10 is also described, in addition to oxidation of the alkyl side chain and a following cleavage [8]. Monohydroxylated Δ9-THC can be hydroxylated again, which leads to 8,11-dihydroxy-Δ9-THC, (**9.5**), for example. Metabolites that are formed by CYP 3A4 represent a minority in comparison to those of CYP 2C9. The glucuronide of THC-COOH, (**9.6**), which is formed in the secondary pathway is a human metabolite of Δ9-THC.

4.2.2
Analytical Methods for Detection of Δ9-THC and it Metabolites

As described for the analysis of the plant, GC, HPLC, and immunoassays are commonly used for the analysis of body fluids. Although the general procedures are quite similar to those used in the analysis of *C. sativa* (see Sect. 4.1.1) some differences must be pointed out.

4.2.2.1
Sample Preparation

The typical procedure for analysis of cannabinoids from plasma, urine or oral fluids includes preliminary steps such as a SPE for enhancement of the analytes and for minimizing interfering effects of the matrices. Because the metabolites in humans are often conjugated, an anterior hydrolysis of these conjugates either with chemicals like sodium hydroxide or with enzymes [103] is recommended.

Pretreatment of hair samples also includes an extraction, usually with an alkaline sodium hydroxide solution, followed by cleaning up with LLE with *n*-hexane/ethyl acetate. Instead of LLE, the employment of SPE is also possible. Furthermore, the solid phase microextraction (SPME) in combination with head-space analysis is usable [104–106]. In the case of using hair samples, possible external contamination (e.g., by passive smoking of *Cannabis*) has to be considered as false positive result. False positive results can be avoided by washing of the hair samples previous to extraction [107]. Storage of collected samples is another important fact that can cause false results in their content of Δ9-THC and metabolites [108–110].

4.2.2.2
Gas Chromatographic Methods (GC)

The preferred detection method for cannabinoids in forensic samples is GC-MS with or without preceding derivatization. As described before in the analysis of plant materials, the employment of silica-fused columns is recommend in the analysis of human body fluids. While in analysis of *Cannabis* TMS-reagents are mostly employed for derivatization, in the case of human body material fluoric compounds such as pentafluoropropionic anhydride (PFPA) or 2,2,3,3,3-pentafluoro-1-propanol (PFPOH) as derivatization reagents are used [89]. Halogenation of the analytes in these ways allows the use of an electron capture detector (ECD) to find the desired compounds. In comparison with other detectors such as the flame ionization detector (FID), the detection sensitivity of cannabinoids can be increased by using an ECD. This is important because the amount of these compounds is very low in human forensic samples. However, as mentioned above, these detectors are commonly not used in routine analyses of forensic samples. Among PFPA and PFPOH, acylation reagents such as trifluoroacetic anhydride (TFAA) and *N*-methyl-bis(trifluoroacetamide) (MBTFA) are also used for analysis of cannabinoids in human materials [111–114]. Trideuterated THC-COOH (**7.4**) is the most commonly used internal standard for the analysis of metabolites with GC-MS. Baptista et al. [103] have shown that the limit of quantification (LOQ) for the important metabolite THC-COOH is much more better if negative chemical ionization (NCI) is used instead of electron ionization (EI).

4.2.2.3
Liquid Chromatographic Methods (HPLC)

Whilst for the analysis of plant material for cannabinoids both GC and HPLC are commonly used, in analytical procedures the employment of GC-based methods prevails for human forensic samples. Nonetheless, the usage of HPLC becomes more and more of interest in this field especially in combination with MS [115–120]. Besides the usage of deuterated samples as internal standards Fisher et al. [121] describe the use of a dibrominated THC-COOH (see **7.5**). The usage of Thermospray-MS and electrochemical detection provide good performance and can replace the still-used conventional UV detector. Another advantage in the employment of HPLC rather than GC could be the integration of SPE cartridges, which are needed for sample preparation in the HPLC-system.

4.2.2.4
Immunoassays

Most of the tests that were developed for detection of cannabinoids in plants have shown that antibodies are specific for the cannabinoid structure. Because of this specifity these tests can be extensively applied for the detection of cannabinoids and metabolites in human body fluids such as plasma, urine, and oral fluids. Many different kits based on these methods were developed and they are commercially available, for example Oratect, Branan or Uplink, and OraSure. We must consider, however, that no humans have the same metabolite profile in their blood and that cross-reactivity may always occur [122, 123]. Nevertheless, these tests offer a simple way of excluding most of the suspicious samples, but the results still have to be confirmed with a second method such as GC-MS [124, 125].

5
Medicinal use of *Cannabis* and Cannabinoids

5.1
Historical Aspects

Human use of *C. sativa* goes back over 10 000 years and the medicinal use can be definitely found in ancient Chinese writings from 1000 BC [126]. Modern medicinal use was mainly introduced by William B. O'Shaugnessy who was one of the first physicians who systematically explored its therapeutic potential [127]. Studying the literature of the 19th century it is impressive how efficiently most indications, which are now under intensive research, where already depicted by observation and simple trial and error.

5.2 Modern Use

5.2.1 Natural Cannabinoids

A serious problem in the early Western medicinal use of *C. sativa*, mainly as a tincture, was its highly variable activity and inconsistent results. Medicinal preparations have to handle several particularities due to the structure of the active ingredients of *C. sativa*. The identity of the main active constituent of *C. sativa*, Δ9-tetrahydrocannabinol (INN dronabinol) remained unknown until 1964 [128]; standardized *C. sativa* preparations were not available. The plant itself is found in several different chemotypes, which added to the unpredictable nature of early medicinal preparations.

Cannabinoids are highly lipophilic compounds making bioavailability very dependent on the formulation and the mode of administration. Cannabinoid occurrence in the plant is predominantly in the form of the carboxylic acids, which are pharmacologically totally different and rather unstable, decarboxylating over time to their active neutral form. The carboxylic acids, although not active at the CB receptor, nevertheless add to the overall effect as they possess antibiotic and anti-inflammatory effects.

Last but not least the identification of THC as the main active constituent of *C. sativa* was preceded by an almost total ban on the plant as a narcotic drug, practically ending medicinal research.

So, the 20th century actually led to an almost total disappearance of *C. sativa* for medicinal purposes. The only source for THC, which became the focus of scientific research, was the rather tedious extraction and purification from confiscated hashish or marihuana. In 1972 the first commercially viable total synthesis of Δ9-THC was established and it became the first cannabinoid available as a modern medicine in the form of soft gel capsules (the active ingredient being called dronabinol from tetrahy**dro**cann**abinol**) under the trade name Marinol for the prevention of nausea and vomiting during cancer chemotherapy.

Interestingly this indication resulted from the observation of marihuana-smoking patients rather than from pharmacological research.

In contrast to the *C. sativa* tincture, Marinol soft gel capsules possess clear advantages. Firstly, they contain a single component in an accurate dosage. Secondly, it uses sesame oil as the carrier, making resorption significantly more reliable and also stabilizing the rather sensitive THC molecule.

The indication "prevention of nausea and vomiting during cancer chemotherapy" came from experiences of marihuana-smoking patients, not from pharmacological research [129].

The second indication, being licensed for THC several years later, came from an observation that had been known for a long time for *C. sativa*, namely its appetite-stimulating effects. This sometimes very impressive effect (popularly known as "munchies") was regarded as a side effect until it became apparent that loss of appetite and weight (the "AIDS wasting syndrome") was one of the determining factors influencing mortality of HIV patients [130].

Pharmacological research and the non-prescriptional use of *C. sativa* by patients gave way to new indications. Now well established are the efficacies for the following indications:

- Nausea and vomiting [129]
- Appetite stimulation [131, 132]
- Spasticity [133, 134]
- Tourette syndrome [135]
- Neuropathic pain [136]
- Multiple sclerosis [137]
- Mood elevation
- Glaucoma [138]
- Pruritus
- Asthma
- Epilepsia
- Migraine

After the discovery of specific endocannabinoid receptors, the amount of scientific literature quickly rose and not only new potential indications were established, but also the mechanisms for the already known effects were clarified. Although the most prominent effect of *C. sativa* is clearly related to THC and its activity at the CB1 receptor, most other natural cannabinoids are not active there. Today two other natural cannabinoids CBD and THCV are the focus of medicinal research.

CBD was first isolated from *C. sativa* in 1940 [139]. Unlike the resinous air-sensitive THC, CBD is a crystalline stable substance. Its plant precursor, the carboxylic acid CBDA can be isolated from fiber hemp by extraction and shows potent antibiotic activity. Upon heating it decarboxylates to CBD.

CBD has no activity at the CB1 or CB2 receptor. It is well known that CBD influences the activity of THC if co-administered [140]. Another effect of CBD is the inhibition of cytochrome oxidase [141], which inversely to its antagonistic activity strongly potentiates THC effects above a certain threshhold. CBD is also active as a mild antipsychotic [142] and was proposed as a treatment for anxiety and panic attacks. The mechanism is not fully understood, but it might be caused by an interference with the endocannabinoid system. It is now also under research for the treatment of diabetes and obesity [143].

5.2.2 Synthetic Cannabinoids

Until today only a few synthetic cannabinoids have made their way into clinical use.

5.2.2.1 Nabilone

In contrast to THC (an oxygen-sensitive resin), nabilone (**8.3**) is a crystalline stable substance. It is about five to ten times more potent than THC [144]. It was developed by Lilly and marketed as Cesamed in several countries, mainly for the prevention of nausea and vomiting during chemotherapy. Recently it was approved in the USA for the treatment of neuropathic pain.

5.2.2.2 Levonantradol

Levonantradol (**8.4**) was synthesized with the intention to introduce a basic amino function into the heterocycle in the hope of obtaining water-soluble salts. Although the solubility of the hydrochloride is not good it was possible to get stable aqueous micellar solutions with the aid of emulsifiers [145] and the compound made its way as an injectable into clinical trials, but never was approved.

5.2.2.3 CP-55,940

CP-55,940 (**8.5**) was developed during the search for novel analgesics [146]. Although it is more potent than morphine it was never approved. Nevertheless, in its tritium-labeled form it became a very important tool for research and helped in the first identification of the cannabinoid receptor.

5.2.2.4 WIN-55,212-2

In the search for new anti-inflammatory drugs structurally derived from indomethacine [147], Pravadoline showed psychotropic side effects in clinical trials. It became apparent that these effects are mediated through the cannabinoid receptor. Optimization of the structure finally led to WIN-55,212-2 (**8.6**), which has a higher affinity to the CB1 receptor than THC [148] and became an important research tool. The side effects of substances possessing agonistic activity on the CB1 receptor (mainly psychotropic effects similar to those of cannabis) limited its clinical use and changed the focus of research to the development of compounds without this drawback.

5.2.2.5
Rimonabant

Rimonabant or SR-141716A (**8.7**) is an antagonist at the CB1 receptor [149] and got approval for the treatment of obesity and as an aid in the cessation of cigarette smoking. It is now marketed in Europe under the tradename Acomplia. Interestingly the naturally occurring THCV (the propyl homolog of THC) also acts as an antagonist on the CB1 receptor and might become a competitor for rimonabant.

5.2.2.6
PRS-211,096

PRS-211,096 (**8.8**) is a CB2-selective agonist, thus avoiding the psychotropic side effects related to CB1. It is currently in clinical trial for the treatment of multiple sclerosis.

5.2.2.7
HU-211

HU-210 is (**8.1**) among the most potent cannabinoids known. Its enantiomer HU-211 (**8.2**) does not bind to the cannabinoid receptor and lacks psychotropic side effects (as long as optical purity is guaranteed). In animal models it shows analgesic and antiemetic activity. It also shows neuroprotective effects after brain injury and was tested in humans as anti-trauma agent, where it did not meet the expectations in a clinical phase III trial.

5.2.2.8
Ajulemic Acid

Ajulemic acid (CT3, **8.9**) is the dimethylheptyl homolog of the main metabolite of Δ8-THC. It has no psychotropic activity, but has analgesic and anti-inflammatory effects.

5.3
Drug Delivery

The classical way of application of Δ9-THC from *C. sativa* is smoking of dried *Cannabis* flowers or leaves by patients in traditional medicine. Smoking of

Cannabis as an illegal drug is popular, but not only these drug users but also regular patients suffering from various diseases as discussed above use this form of unprescribed self-medication.

Besides the inhalative use, the development of a drug formulation for Δ9-THC has to address other bioavailability questions. A major problem is the lipophilicity and poor solubility in water, limiting oral uptake when given orally. Because of this, other parenteral routes of application are under investigation like pulmonal uptake by vaporization, sublingual or intranasal administration, and application by injection of Δ9-THC incorporated in liposomes.

Marinol and Sativex are given orally to the patient but, as indicated, the poor solubility of Δ9-THC is responsible for its slow onset and release from drug carriers like soft gelatine capsules [150]. Quite frequently a large variety in the bioavailability and a significant first pass effect can be observed in animal tests and patients. One solution to the solubility problem is the development of new Δ9-THC derivatives with improved solubility (e.g., dexanabinol, which is a hemisuccinate prodrug). However, this strategy is mostly not desirable because of the high risk involved in the cost and time-consuming drug approval process to gain all toxicological and clinical data.

The main strategies in pharmaceutical technology to improve solubility are the reduction of particle size and the increase of particle surface according to the Kelvin equitation. These two strategies have been applied for Δ9-THC production by solid dispersion technology and production of nanosuspensions. Van Drooge et al., created a solid dispersion of inulin in which Δ9-THC was incorporated [151]. Applying freeze drying techniques for evaporation of a mixture of water and tertiary butyl alcohol, which acts as dissolving medium for Δ9-THC and inulin, forms amorphous Δ9-THC in a fast-dissolving solid inulin matrix. The main advantage of the technique is to protect Δ9-THC from degeneration and to optimize the dissolution rate from tablets [151]. A second and easy way to increase the solubility can be achieved by reduction of the particle size. In unpublished work by the author's group, nanosuspensions of Δ9-THC have been achieved indicating a first significant improvement on the physical properties. The main drawbacks of the technique is the poor stability of the highly energetic suspensions and the risk of forming cluster and microparticles without sufficient stabilization of the nanosuspension. Perlin et al., applied Δ9-THC incorporated in gelatine capsules and administered these orally to rhesus monkeys at a dose of 2.5-mg/kg doses and compared the plasma levels with parenteral intravenous and intramuscularly injections [152]. The authors concluded that intramuscularly injection is favorable because of a bioavailability of 89% ± 16% (i.m.) versus 26% ± 14% (p.o.). Interestingly Perlin et al. mentioned that rectal administration was not successful and no significant blood levels were detected [152]. More recently Munjal et al., developed a transmucosal system based on polyethylene oxide (PEO) polymers, which are commonly used for the production of suppositories [150]. In this study the heat-labile Δ9-THC hemisuccinate was used to produce suppositories varying in PEO composition by the hot-melt technique (120 °C). Temperature led to a degradation of between 13.5% and 49.4% depending on the composition, but incorporation of vitamin E succinate reduced processing degradation to 9.2% and gave a shelf half-life of 8 months. No data have been published yet to characterize the bioavailability or pharmacological effect.

To achieve reliable elevated plasma levels and to overcome the first pass effect, alternative parenteral administration systems have been developed. The most obvious route is vaporization of the *Cannabis* plant material or the Δ9-THC directly. Hazekamp et al., conducted an intensive study using the Volcano device [153]. The main principle is evaporation of Δ9-THC from *Cannabis* plant material by a hot air flow. Evaporated compounds are collected in a detectable plastic balloon, which can be removed and fitted with a mouthpiece for inhalation. The main advantage of the Volcano vaporizer is that Δ9-THC is vaporized below the point of combustion, avoiding the production of lung-irritating toxins. Other advantages for the self-medicating patient is the ease of self-titration, fast drug release, and fast reaching of therapeutic blood levels. To compare with alternative smoking procedures, the Δ9-THC recovery was 54% for the Volcano and 39% for the water pipe.

Pulmonal application can be still unpleasant for non-smokers, which is why other administration routes like sublingual or intranasal uptake are also of interest. Valiveti et al., investigated nasal application for Δ9-THC and WIN-55,121-2 mesylate in rats [154]. The latter is a synthetic cannabinoid with a short half life time and a highly variable bioavailability. Both drugs were formulated in ethanol and propylene glycol and were successfully administered. In comparison with i.v. applied reference drugs, a tenfold higher nasal dose (10 mg/kg Δ9-THC) showed similar AUC values with a slightly increased half-life time.

A second alternative is sublingual application, as introduced by Mannila et al., based on cyclodextrin matrices [155]. Cyclodextrins are a group of cyclic oligosaccharides that have been shown to improve aqueous solubility, dissolution rate, and bioavailability of various lipophilic drugs such as testosterone or prostaglandin E, to give two examples. Cyclodextrins have also been successfully studied in a few sublingual and buccal formulations, e.g., hydroxypropyl-β-cyclodextrin (HP-β-CD) led to the effective absorption of sublingual testosterone.

In this study, complexation of Δ9-THC and cannabidiol (prepared by freeze drying) with randomly methylated β-cyclodextrin and hydroxypropyl-β-cyclodextrin (HP-β-CD) was studied by the phase-solubility method. The aqueous solubility of CBD and THC increased as a function of CD concentration, and the dissolution increased for THC and CBD cyclodextrin complexes significantly in contrast to plain THC and CBD. These results demonstrate that cyclodextrins increased both the aqueous solubility and dissolution rate of these cannabinoids, making the development of novel sublingual formulation possible, which has been shown by in vivo studies in New Zealand rabbits.

References

1. Shulgin AT (1968) J Psychedelic Drugs 2:14
2. Elsohly MA, Slade D (2005) Life Sci 78:539
3. Pate DW (2004) In: Grotenhermen F (ed) Cannabis un Cannabinoide. Hans Huber, Bern, p 33
4. Garrett ER, Hunt CA (1974) J Pharm Sci 63:1056
5. Uliss DB, Handrick GR, Dalzell HC, Razdan RK (1978) Tetrahedron 34:1885
6. Gaoni Y, Mechoulam R (1971) J Am Chem Soc 93:217
7. Kriwacki RW, Makriyannis A (1989) Mol Pharmacol 35:495
8. Mechoulam R, Devane WA, Glaser R (1999) In: Nahas GG, Sutin KM, Harvey DJ, Agurell S (eds) Marihuana and medicine. Humana, New Jersey, p 65
9. Gaoni Y, Mechoulam R (1964) J Am Chem Soc 86:1646
10. Miller IJ, McCallum NK, Kirk CM, Peake BM (1982) Cell Mol Life Sci (CMLS) 38:230
11. Turner CE, Elsohly MA (1979) J Heterocycl Chem 16:1667
12. Ranalli P, Venturi G (2004) Euphytica 140:1
13. Collin C, Davies P, Mutiboko IK, Ratcliffe S (2007) Eur J Neurol 14:290
14. Ligresti A, Moriello AS, Starowicz K, Matias I, Pisanti S, De Petrocellis L, Laezza C, Portella G, Bifulco M, Di Marzo V (2006) J Pharmacol Exp Ther 318:1375
15. Williamson EM, Evans FJ (2000) Drugs 60:1303
16. Charles T, Hammond PGM (1973) Am J Botany 60:542
17. Turner JC, Hemphill JK, Mahlberg PG (1981) Bull Narc 33:63
18. Turner JC, Hemphill JK, Mahlberg PG (1981) Bull Narc 33:59
19. Eun-Soo Kim PGM (1997) Am J Botany 84:336
20. Jocelyn C, Turner JKH, Mahlberg PG (1978) Am J Botany 65:1103
21. Sirikantaramas S, Taura F, Tanaka Y, Ishikawa Y, Morimoto S, Shoyama Y (2005) Plant Cell Physiol 46:1578
22. Lange BM, Rujan T, Martin W, Croteau R (2000) Proc Natl Acad Sci USA 97:13172
23. Eisenreich W, Bacher A, Arigoni D, Rohdich F (2004) Cell Mol Life Sci 61:1401
24. Rohdich F, Kis K, Bacher A, Eisenreich W (2001) Curr Opin Chem Biol 5:535
25. Julsing MK, Koulman A, Woerdenbag HJ, Quax WJ, Kayser O (2006) Biomol Eng 23:265
26. Raharjo TJ, Chang W-T, Choi YH, Peltenburg-Looman AMG, Verpoorte R (2004) Plant Sci 166:381
27. Raharjo TJ, Chang WT, Verberne MC, Peltenburg-Looman AM, Linthorst HJ, Verpoorte R (2004) Plant Physiol Biochem 42:291
28. Fellermeier M, Eisenreich W, Bacher A, Zenk MH (2001) Eur J Biochem 268:1596
29. Lichtenthaler HK, Schwender J, Disch A, Rohmer M (1997) FEBS Lett 400:271
30. Lichtenthaler HK (1999) Annu Rev Plant Physiol Plant Mol Biol 50:47
31. Austin MB, Noel JP (2003) Nat Prod Rep 20:79
32. Fellermeier M, Zenk MH (1998) FEBS Lett 427:283
33. Shoyama Y, Hirano H, Nishioka I (1984) Phytochemistry 23:1909
34. Funa N, Ozawa H, Hirata A, Horinouchi S (2006) Proc Natl Acad Sci USA 103:6356
35. Kajima M, Piraux M (1982) Phytochemistry 21:67

36. Horper W, Marner F-J (1996) Phytochemistry 41:451
37. Kuzuyama T, Noel JP, Richard SB (2005) Nature 435:983
38. Sirikantaramas S, Morimoto S, Shoyama Y, Ishikawa Y, Wada Y, Shoyama Y, Taura F (2004) J Biol Chem 279:39767
39. Taura F, Morimoto S, Shoyama Y (1996) J Biol Chem 271:17411
40. Morimoto S, Komatsu K, Taura F, Shoyama Y (1998) Phytochemistry 49:1525
41. Taura F, Morimoto S, Shoyama Y, Mechoulam R (1995) J Am Chem Soc 117:9766
42. Taura U (2004) Research report, Graduate School of Pharmceutical Sciences, Kyushu University. Available at http://www.nisr.or.jp/englishHP/report2004/NISR04taura.pdf, last visited: 29 June 2007
43. Yoshikai K, Morimoto S, Shoyama Y (2001) Japanese Patent 2000–78979
44. Kojoma M, Seki H, Yoshida S, Muranaka T (2006) Forensic Sci Int 159:132
45. Cherniak L (1982) The great books of *Cannabis*, vol I, Book II. Damele, Oakland, CA
46. Hillig KW (2005) Gene Res Crop Evolut 52:1573
47. Pacifico D, Miselli F, Micheler M, Carboni A, Ranalli P, Mandolino G (2006) Mole Breed 17:257
48. Renner SS, Ricklefs RE (1995) Am J Bot 82:596
49. Sakamoto K, Abe T, Matsuyama T, Yoshida S, Ohmido N, Fukui K, Satoh S (2005) Genome 48:931
50. Sakamoto K, Akiyama Y, Fukui K, Kamada H, Satoh S (1998) Cytologica 63:459
51. Mandolino GAC, Bagatta MVM, Moliterni C, Ranalli P (2002) Euphytica 126:211
52. Mandolino G, Bagatta M, Carboni A, Ranalli P, Meijer E (2003) J Industrial Hemp 8:51
53. de Meijer EPM, Bagatta M, Carboni A, Crucitti P, Moliterni VMC, Ranalli P, Mandolino G (2003) Genetics 163:335
54. de Meijer EPM, Hammond KM (2005) Euphytica 145:189
55. Sharma GK (1975) Bull Torrey Bot Club 102:199
56. Jansen M, Gaba V, Greenberg BM (1998) Trends Plant Sci 3:131
57. Lydon J (1986) Effects of ultraviolet-B radiation on the growth, physiology and cannabinoid production of *Cannabis sativa* L. Dissertation, Maryland University, USA
58. Lydon J, Teramura AH, Coffman CB (1987) Photochem Photobiol 46:201
59. Pijlman FTA, Rigter SM, Hoek J, Goldschmidt HMJ, Niesink RJM (2005) Addiction Biol 10:171
60. Heitrich A, Binder M (1982) Cell Mol Life Sci 38:898
61. Feeney M, Punja ZK (2003) In Vitro Cell Develop Biol Plant 39:578
62. Braemer R, Paris M (1987) Plant Cell Reports 6:150
63. Koziel MG, Beland GL, Bowman C, Carozzi NB, Crenshaw R, Crossland L, Dawson J, Desai N, Hill M, Kadwell S, Launis K, Lewis K, Maddox D, McPherson K, Meghji MR, Merlin E, Rhodes R, Warren GW, Wright M, Evola SV (1993) Nat Biotech 11:194
64. Herbers K (2003) J Plant Physiol 160:821
65. Mason HS, Lam DM, Arntzen CJ (1992) Proc Natl Acad Sci USA 89:11745
66. Gaoni Y, Mechoulam R (1964) J Am Chem Soc 86:1646
67. Mechoulam R, Braun P, Gaoni Y (1972) J Am Chem Soc 94:6159
68. Mechoulam R, Braun P, Gaoni Y (1967) J Am Chem Soc 89:4552
69. Razdan RK, Handrick GR, Dalzell HC (1975) Experientia 31:16
70. Petrzilka T, Haefliger W, Sikemeier C (1969) Helv Chim Acta 52:1102
71. Razdan RK, Dalzell HC, Handrick GR (1974) J Am Chem Soc 96:5860
72. Mechoulam R, Lander N, University A, Zahalka J (1990) Tetrahedron: Asymmetry 1:315
73. Fahrenholtz KE, Lurie M, Kierstead RW (1967) J Am Chem Soc 89:5934

74. Archer RA, Blanchard WB, Day WA, Johnson DW, Lavagnino ER, Ryan CW, Baldwin JE (1977) J Org Chem 42:2277
75. Kachensky DF, Hui AHF (1997) J Org Chem 62:7065
76. Szirmai M (1995) Total synthesis and analysis of major human urinary metabolites of d1-tetrahydrocannabinol, the principal psychoactive component of *Cannabis sativa* L. Dissertation, Uppsala University, Sweden
77. Siegel C, Gordon PM, Razdan RK (1989) J Org Chem 54:5428
78. Seltzman HH, Begum KM, Wyrick DC (1991) J Labelled Compds Radiopharm 29:1009
79. Melvin LS, Milne GM, Johnson MR, Subramaniam B, Wilken GH, Howlett AC (1993) Mol Pharmacol 44:1008
80. Eissenstat MA, Bell MR, D'Ambra TE, Alexander EJ, Daum SJ, Ackerman JH, Gruett MD, Kumar V, Estep KG, Olefirowicz EM, et al. (1995) J Med Chem 38:3094
81. Turner CE, Elsohly MA, Boeren EG (1980) J Nat Prod 43:169
82. Ross SA, El-Sohly MA (1995) Zagazig J Pharm Sci 4:1
83. Mediavilla V, Steinemann S (1997) J Inter Hemp Assoc 4:80
84. Ross SA, ElSohly MA (1998) Bull Narcotics
85. Bacigalupo MA, Ius A, Meroni G, Grassi G, Moschella A (1999) J Agric Food Chem 47:2743
86. Lercker G, Bocci F, Frega N, Bortolomeazzi R (1992) Farmaco 47:367
87. Lehmann T, Brenneisen R (1995) J Liquid Chrom 18:689
88. Ross SA, Mehmedic Z, Murphy TP, Elsohly MA (2000) J Anal Toxicol 24:715
89. Raharjo TJ, Verpoorte R (2004) Phytochem Anal 15:79
90. Harvey DJ (1987) Mass Spectrom Rev 6:135
91. Rustichelli C, Ferioli V, Baraldi M, Zanoli P, Gamberini G (1998) Chromatographia 48:215
92. Hazekamp A, Peltenburg A, Verpoorte R, Giroud C (2005) J Liquid Chrom Rel Technol 28:2361
93. Hazekamp A, Choi YH, Verpoorte R (2004) Chem Pharm Bull 52:718
94. Teale JD, Forman EJ, King LJ, Piall EM, Marks V (1975) J Pharm Pharm 27:465
95. ElSohly MA, Jones AB, ElSohly HN (1990) J Anal Toxicol 14:277
96. Tanaka H, Goto Y, Shoyama Y (1996) J Immunoassay 17:321
97. Tanaka H, Shoyama Y (1999) Forensic Sci Int 106:135
98. Musshoff F, Madea B (2006) Ther Drug Monit 28:155
99. Brenneisen R (2004) In: Grotenhermen F (ed) Cannabis und Cannabinoide. Hans Huber, Bern, p 89
100. Agurell S, Halldin M, Lindgren JE, Ohlsson A, Widman M, Gillespie H, Hollister L (1986) Pharmacol Rev 38:21
101. Watanabe K, Yamaori S, Funahashi T, Kimura T, Yamamoto I (2007) Life Sci 80:1415
102. Grotenhermen F (2003) Clin Pharmacokinet 42:327
103. Baptista MJ, Monsanto PV, Pinho Marques EG, Bermejo A, Avila S, Castanheira AM, Margalho C, Barroso M, Vieira DN (2002) Forensic Sci Int 128:66
104. Sporkert F, Pragst F (2000) Forensic Sci Int 107:129
105. Musshoff F, Lachenmeier DW, Kroener L, Madea B (2003) Forensic Sci Int 133:32
106. de Oliveira CDR, Yonamine M, de Moraes Moreau RL (2007) J Sep Sci 30:128
107. Boumba VA, Ziavrou KS, Vougiouklakis T (2006) Int J Toxicol 25:143
108. Skopp G, Potsch L (2004) J Anal Toxicol 28:35
109. Jamerson MH, McCue JJ, Klette KL (2005) J Anal Toxicol 29:627
110. Moore C, Vincent M, Rana S, Coulter C, Agrawal A, Soares J (2006) Forensic Sci Int 164:126

111. Kim JY, Suh SI, In MK, Paeng KJ, Chung BC (2005) Arch Pharm Res 28:1086
112. Nadulski T, Sporkert F, Schnelle M, Stadelmann AM, Roser P, Schefter T, Pragst F (2005) J Anal Toxicol 29:782
113. Jamerson MH, Welton RM, Morris-Kukoski CL, Klette KL (2005) J Anal Toxicol 29:664
114. Moore C, Rana S, Coulter C, Feyerherm F, Prest H (2006) J Anal Toxicol 30:171
115. Franski R, Tezyk A, Wachowiak R, Schroeder G (2004) J Mass Spectrom 39:458
116. Concheiro M, de Castro A, Quintela O, Cruz A, Lopez-Rivadulla M (2004) J Chromatogr B Analyt Technol Biomed Life Sci 810:319
117. Maralikova B, Weinmann W (2004) J Mass Spectrom 39:526
118. Teixeira H, Proenca P, Verstraete A, Corte-Real F, Vieira DN (2005) Forensic Sci Int 150:205
119. Kolmonen M, Leinonen A, Pelander A, Ojanpera I (2007) Anal Chim Acta 585:94
120. Laloup M, Ramirez Fernandez MM, Wood M, De Boeck G, Henquet C, Maes V, Samyn N (2005) J Chromatogr A 1082:15
121. Fisher DH, Broudy MI, Fisher LM (1996) Biomed Chromatogr 10:161
122. Walsh JM, Crouch DJ, Danaceau JP, Cangianelli L, Liddicoat L, Adkins R (2007) J Anal Toxicol 31:44
123. Clarke J, Wilson JF (2005) Forensic Sci Int 150:161
124. Moody DE, Fang WB, Andrenyak DM, Monti KM, Jones C (2006) J Anal Toxicol 30:50
125. Cirimele V, Villain M, Mura P, Bernard M, Kintz P (2006) Forensic Sci Int 161:180
126. Li HL (1974) Econ Bot 28:437
127. O'Shaughnessy WB (1838) Trans Med Phys Soc Bengal, p 71
128. Gaoni Y, Mechoulam R (1964) J Am Chem Soc 86:1646
129. Sallan SE, Zinberg NE, Frei E 3rd (1975) N Engl J Med 293:795
130. Palenicek JP, Graham NM, He YD, Hoover DA, Oishi JS, Kingsley L, Saah AJ (1995) J Acquir Immune Defic Syndr Hum Retrovirol 10:366
131. Beal JE, Olson R, Laubenstein L, Morales JO, Bellman P, Yangco B, Lefkowitz L, Plasse TF, Shepard KV (1995) J Pain Symptom Manage 10:89
132. Beal JE, Olson R, Lefkowitz L, Laubenstein L, Bellman P, Yangco B, Morales JO, Murphy R, Powderly W, Plasse TF, Mosdell KW, Shepard KV (1997) J Pain Symptom Manage 14:7
133. Brenneisen R, Egli A, Elsohly MA, Henn V, Spiess Y (1996) Int J Clin Pharmacol Ther 34:446
134. Maurer M, Henn V, Dittrich A, Hofmann A (1990) Eur Arch Psychiatry Clin Neurosci 240:1
135. Muller-Vahl KR, Schneider U, Kolbe H, Emrich HM (1999) Am J Psychiatry 156:495
136. Finnegan-Ling D, Musty RE (1994) Symposium on the cannabinoids, Burlington, Vermont. International Cannabinoid Research Society, p 53
137. Petro DJ, Ellenberger C Jr (1981) J Clin Pharmacol 21:413S
138. Merritt JC, McKinnon S, Armstrong JR, Hatem G, Reid LA (1980) Ann Ophthalmol 12:947
139. Jacob A, Todd AR (1940) J Chem Soc, p 649
140. Karniol IG, Carlini EA (1973) Psychopharmacologia 33:53
141. Jones G, Pertwee RG (1972) Br J Pharmacol 45:375
142. Zuardi AW, Morais SL, Guimaraes FS, Mechoulam R (1995) J Clin Psychiatry 56:485
143. Weiss L, Zeira M, Reich S, Har-Noy M, Mechoulam R, Slavin S, Gallily R (2006) Autoimmunity 39:143
144. Einhorn LH, Nagy C, Furnas B, Williams SD (1981) J Clin Pharmacol 21:64S

145. Aguiar AJ, Rasadi B (1983) US Patent 4406888. Pfizer, New York
146. Johnson MR, Melvin LS Jr (1983) US Patent 4371720. Pfizer, New York
147. Haubrich DR, Ward SJ, Baizman E, Bell MR, Bradford J, Ferrari R, Miller M, Perrone M, Pierson AK, Saelens JK, et al. (1990) J Pharmacol Exp Ther 255:511
148. D'Ambra TE, Estep KG, Bell MR, Eissenstat MA, Josef KA, Ward SJ, Haycock DA, Baizman ER, Casiano FM, Beblin NC, et al. (1992) J Med Chem 35:124
149. Rinaldi-Carmona M, Barth F, Millan J, Derocq JM, Casellas P, Congy C, Oustric D, Sarran M, Bouaboula M, Calandra B, Portier M, Shire D, Breliere JC, Le Fur GL (1998) J Pharmacol Exp Ther 284:644
150. Munjal M, ElSohly MA, Repka MA (2006) AAPS Pharm Sci Tech 7:71
151. van Drooge DJ, Hinrichs WL, Wegman KA, Visser MR, Eissens AC, Frijlink HW (2004) Eur J Pharm Sci 21:511
152. Perlin E, Smith CG, Nichols AI, Almirez R, Flora KP, Cradock JC, Peck CC (1985) J Pharm Sci 74:171
153. Hazekamp A, Ruhaak R, Zuurman L, Van Gerven J, Verpoorte R (2006) J Pharm Sci 95:1308
154. Valiveti S, Agu RU, Hammell DC, Paudel KS, Earles DC, Wermeling DP, Stinchcomb AL (2007) Eur J Pharm Biopharm 65:247
155. Mannila J, Jarvinen T, Jarvinen K, Tarvainen M, Jarho P (2005) Eur J Pharm Sci 26:71

Top Heterocycl Chem (2007) 10: 43–73
DOI 10.1007/7081_2007_060

Published online: 9 May 2007

Quantitative Structure–Activity Relationships of Heterocyclic Topoisomerase I and II Inhibitors

Corwin Hansch · Rajeshwar P. Verma (✉)

Department of Chemistry, Pomona College, 645 North College Avenue, Claremont, CA 91711, USA
rverma@pomona.edu

Abstract Deoxyribonucleic acid (DNA) topoisomerases are ubiquitous enzymes that are involved in diverse cellular processes, such as replication, recombination, transcription, and chromosome segregation. These enzymes solve topological problems related to DNA double helical structure by breaking and rejoining DNA strands. There are two major classes of topoisomerases: topoisomerase I (topo I), which breaks and reseals one strand of DNA, and topoisomerase II (topo II), which alters DNA topology by catalyzing the passage of an intact DNA double helix through a transient double stranded break made in a second helix. A variety of heterocyclic antitumor agents currently used in chemotherapy or evaluated in clinical trials are known to inhibit either DNA topo I or II. The clinical use of these inhibitors is limited due to sever toxic effects on normal cells. Therefore, there has been increasing interest in discovering and developing novel heterocyclic molecules that inhibit topo I or II or both, and which have the ability to spare normal cells. Interest in the application of the quantitative structure–activity relationship (QSAR) has steadily increased in recent decades because it has repeatedly proven itself to be a low-cost, high-return investment. Potential use of QSAR models for screening of chemical databases or virtual libraries before their synthesis appears equally attractive to chemical manufacturers, pharmaceutical companies, and government agencies. We hope it may also be useful in the design and development of new heterocyclic topo I or II inhibitors. In this chapter, an attempt has been made to collect the inhibitory data on different series of heterocyclic compounds against topos I and II, and to discuss it in terms of QSAR to understand the chemical–biological interactions.

Keywords Heterocycles · Hydrophobicity · Molar refractivity · QSAR · Topoisomerase

Abbreviations

Clog *P*	Calculated hydrophobicity of the whole molecule (calculated logarithm of partition coefficient (*P*) in *n*-octanol/water)
π	Calculated hydrophobicity of the substituent
CMR	Calculated molar refractivity of the whole molecule
DNA	Deoxyribonucleic acid
MR	Calculated molar refractivity of the substituent
MgVol	Calculated molar volume of the whole molecule (McGowan volume)
MW	Molecular weight
NVE	Number of valence electrons
log 1/*C*	Inverse logarithms of the biological activity
LOO	Leave-one-out
MRA	Multiple regression analysis
QSAR	Quantitative structure–activity relationship
REC	Relative effective concentration
topo	Topoisomerase
CPT	Camptothecin
CPT-11	Irinotecan (camptostar)
SN-38	7-Ethyl-10-hydroxy-camptothecin
TPT	Topotecan
EGCG	Epigallocatechin-3-gallate
CoMFA	Comparative molecular field analysis
CoMSIA	Comparative molecular similarity index analysis
HBD	Hydrogen-bond donor

1 Introduction

Deoxyribonucleic acid (DNA) topoisomerases (topos) are ubiquitous enzymes that can manipulate DNA by changing the number of topological links between two strands of the same or different DNA molecules [1]. These enzymes are involved in many cellular processes, such as replication, recombination, transcription, and chromosome segregation. There are two major classes of topoisomerases: topoisomerase I (topo I), which breaks and reseals one strand of DNA, and topoisomerase II (topo II), which alters DNA topology by catalyzing the passage of an intact DNA double helix through a transient double-stranded break made in a second helix [2–4].

DNA topo I is an essential human enzyme and can be trapped by anticancer drugs as it cleaves DNA. Moreover, topo I can be trapped by endogenous alterations to DNA (mismatches, abasic sites, nicks, and adducts) and apoptotic alterations to chromatin. Camptothecin is a natural product of which topo I is the only cellular target. Recently, two anticancer camptothecin derivatives, topotecan (for ovarian and lung cancers) and irinotecan (for colorectal cancer), have been approved by the US Food and Drug Administration [5]. These drugs bind to a transient topo I–DNA covalent complex and inhibit the resealing of a single-strand break that the enzyme creates to relieve superhelical tension in the duplex DNA [6]. Various non-camptothecin anticancer agents like indolocarbazoles, phenanthridines, and indenoisoquinolines can inhibit topo I through inhibition of the catalytic site. These inhibitors of topo I are expected to be active in cancers that are currently resistant to camptothecins, and to have a greater therapeutic value [5, 7]. The crystal structure of topo I shows the enzyme encircling the DNA tightly like a clamp, which accounts for the fact that topo I controls the processive relaxation of supercoiled DNA [8]. Topo I enzymes have been subdivided into two classes, 1A and 1B, based on differences in their amino acid sequences and reaction mechanisms [2, 9].

DNA topo II is a homodimeric nuclear enzyme that inhibits antineoplastic agents such as etoposide and doxorubicin, which are among the most effective anticancer drugs currently available for the treatment of human cancers. These agents are shown to induce accumulation of the DNA–topo II cleavable complex that causes tumor cell death [10]. Topo II inhibitors prevent the enzyme from re-ligating cleaved DNA and generate DNA double-strand breaks (DSBs), which constitute a major threat to genome integrity [11]. In mammalian cells, DSBs can be repaired by two major pathways: (i) the non-homologous DNA end-joining (NHEJ) pathway, which joins, precisely or not, broken DNA ends containing little or no homology, and (ii) the homologous recombination (HR) pathway, which requires a homologous sequence provided by either a sister chromatid or a homologous chromosome [12–14]. Many anticancer topo II inhibitors act as a result of interactions with both the enzyme and DNA; for example, hypericin interacts with DNA at the N_7 sites of the purine residue [15].

Topo inhibitors are found to be the most efficient inducers of apoptosis. The main pathways leading from topo-mediated DNA damage to cell death involve activation of caspases in the cytoplasm by pro-apoptotic molecules released from mitochondria. In some cells, the apoptotic response also involves the death receptor Fas (APO-1/CD95). The engagement of these apoptotic effect or pathways is tightly controlled by upstream regulatory pathways that respond to DNA lesions – induced by topo inhibitors in cells undergoing apoptosis [16]. It has also been suggested that the both classes of DNA topos can control pre-mRNA splicing of the caspase-2 transcript [17].

Green tea, especially its major polyphenolic constituents, epigallocatechin-3-gallate (EGCG), has received much attention over the past few years as a potential cancer chemopreventive agent [18, 19]. The use of green tea is found to be beneficial because its major polyphenolic constituent (EGCG) impacts only the growth inhibitory responses to cancer cells but not to normal cells, resulting in a dose-dependent inhibition of cell growth, G0/G1 phase arrest of the cell cycle, and DNA damage leading to induction of apoptosis [20, 21]. Based on this knowledge, Berger et al. [22] evaluated the possibility of the involvement of topo in the antiproliferative response of EGCG and showed that EGCG inhibits topo I, but not topo II in several human colon carcinoma cell lines. Thus, the use of green tea with other topo inhibitors could be an improved strategy for the treatment of colon cancer.

A variety of heterocyclic antitumor agents currently used in chemotherapy or evaluated in clinical trials are known to inhibit either DNA topo I or II. The clinical use of these inhibitors is limited due to sever toxic effects on normal cells as well as several side effects such as myelosuppression, nausea, hair loss, congestive heart failure, and even increasing the risk of secondary malignancies leading to early death [22]. Therefore, there has been increasing interest in discovering and developing novel heterocyclic molecules that inhibit topo I or II or both, and which have the ability to spare normal cells as well as having minimal/or no side-effects.

Interest in the application of the quantitative structure–activity relationship (QSAR) has steadily increased in recent decades because it has repeatedly proven itself to be a low-cost, high-return investment. Potential use of QSAR models for screening of chemical databases or virtual libraries before their synthesis appears equally attractive to chemical manufacturers, pharmaceutical companies, and government agencies. The quality of a QSAR model depends strictly on the type and quality of the data, and not on the hypotheses, and is valid only for the compound structures analogous to those used to build the model. QSAR models can stand alone to augment other computational approaches or can be examined in tandem with equations of a similar mechanistic genre to establish their authenticity and reliability [23]. We hope it may also be useful in the design and development of new heterocyclic topo I or II inhibitors. In this chapter, an attempt has been made to collect the inhibitory data on different series of heterocyclic compounds against topos I and II, and to discuss it in terms of QSAR to understand the chemical–biological interactions.

2 Materials and Methods

All the data/equations have been collected from the literature (see individual QSAR for respective references). C is the molar concentration of a compound and log $1/C$ is the dependent variable that defines the biological parameter for the QSAR equations. Physicochemical descriptors are autoloaded, and multiregression analyses (MRA) used to derive the QSAR are executed with the C-QSAR program [24]. Selection of descriptors is made on the basis of permutation and correlation matrix among the descriptors (to avoid collinearity problems). Details of the C-QSAR program, the search engine, the choice of parameters and their use in the development of QSAR models have already been discussed [25]. The parameters used in this chapter have also been discussed in detail along with their application [26]. Briefly, Clog P is the calculated partition coefficient in n-octanol/water and is a measure of hydrophobicity, and π is the hydrophobic parameter for substituents. σ, σ^+ and σ^- are Hammett electronic parameters that apply to substituent effects on aromatic systems.

$B1$, $B5$ and L are Verloop's sterimol parameters for substituents [27]. $B1$ is a measure of the minimum width of a substituent, $B5$ is an attempt to define maximum width of the substituent, and L is the substituent length. CMR is the calculated molar refractivity for the whole molecule. MR is calculated from the Lorentz–Lorenz equation and is described as follows: $[(n^2 - 1)/(n^2 + 2)](\mathrm{MW}/\delta)$, where n is the refractive index, MW is the molecular weight, and δ is the density of the substance. MR is dependent on volume and polarizability. It can be used for a substituent or for the whole molecule. A new polarizability parameter, NVE, was developed, which is shown to be effective at delineating various chemico-biological interactions [28–31]. NVE represents the total number of valence electrons and is calculated by simply summing up the valence electrons in a molecule, that is, H = 1, C = 4, Si = 4, N = 5, P = 5, O = 6, S = 6 and halogens = 7. It may also be represented as: $\mathrm{NVE} = \mathrm{n}_\sigma + \mathrm{n}_\pi + \mathrm{n}_n$, where n_σ is the number of electrons in σ-orbital, n_π is the number of electrons in π-orbitals, and n_n is the number of lone pair electrons. MgVol is the molar volume for the whole molecule [32]. The indicator variable I is assigned the value of 1 or 0 for special features with special effects that cannot be parameterized and has been explained wherever used.

In QSAR equations, n is the number of data points, r is the correlation coefficient between observed values of the dependent and the values predicted from the equation, r^2 is the square of the correlation coefficient and represents the goodness of fit, q^2 is the cross-validated r^2 (a measure of the quality of the QSAR model), and s is the standard deviation. The cross-validated r^2 (q^2) is obtained by using leave-one-out (LOO) procedure [33]. Q is the quality factor (quality ratio), where $Q = r/s$. Chance correlation, due to the excessive number of parameters (which increases the r and s values also), can, thus, be detected by the examination of Q value. F is the Fischer statistics (Fischer ratio), $F = fr^2 / [(1 - r^2)m]$, where f is the number of degree of freedom, $f = n - (m + 1)$, n = number of data points, and m = number of variables. The modeling was taken to be optimal when Q reached a maximum together with F, even if slightly non-optimal F values have normally been accepted. A significant decrease in F with the introduction of one additional variable (with increasing Q and decreasing s) could mean that the new descriptor is not as good as expected, that is, its introduction has endangered the statistical quality of the combination. However, the statistical quality could be improved by the introduction of a more useful descriptor [34–36]. Compounds were assigned to be outliers on the basis of their deviation between observed and calculated activities from the equation (>2s) [37–39]. Each regression equation includes 95% confidence limits for each term in parentheses.

3 QSAR Studies

3.1 Topoisomerase I Inhibitors

3.1.1 Benzimidazoles

Hoechst 33342 [2′-(4-ethoxyphenyl)-5-(4-methyl-1-piperazinyl)-2,5′-bi-1*H*-benzimidazole; **Ho342; I**], a bisbenzimidazole dye, binds to adenine/thymine-rich regions in the minor groove of DNA. This dye induces apoptosis and inhibits topo I activity in vivo. It has been suggested that the destruction of immunoreactive topo I and topo I–DNA complexes or cleavable complexes results in inhibition of topo I activity, a key step in the Hoechst 33342-induced apoptotic process [40].

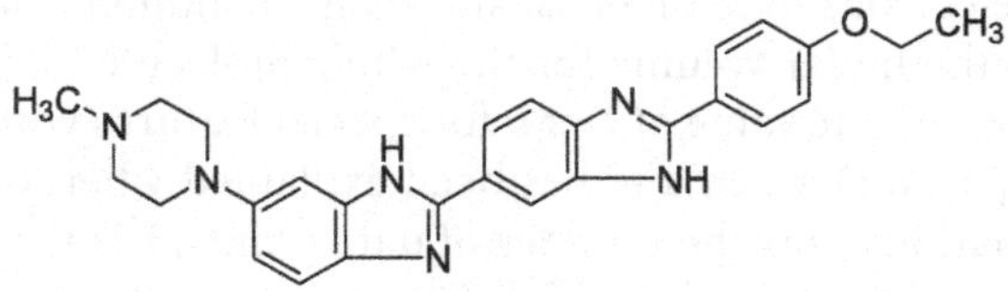

I (Ho342)

Sun et al. [41] synthesized a series of benzimidazoles (**II**) and evaluated their topo I-mediated cleavage as well as cytotoxicity against four cancer cell lines. Topo I cleavage values are reported as REC, the relative effective concentration (i.e., concentrations relative to that of Hoechst 33342, whose value

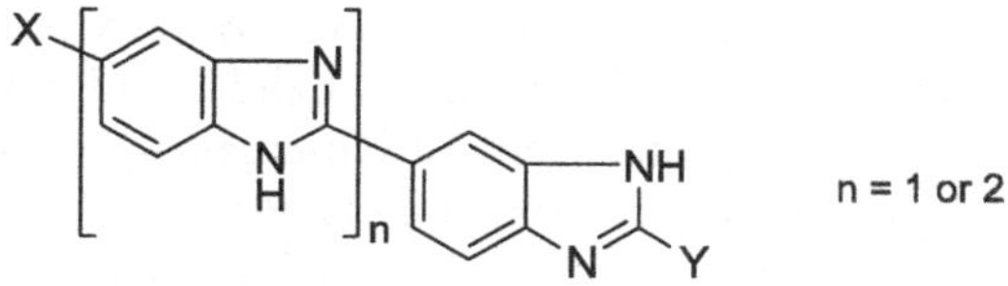

II

is arbitrarily assumed as 1) that is able to produce the same cleavage on the plasmid DNA in the presence of calf thymus topo I. From the topo I cleavage values of benzimidazoles (**II**), Eq. 1 was derived [42]:

$$\log \text{REC} = 7.56(\pm 1.21)\text{C}\log P - 0.83(\pm 0.13)\text{C}\log P^2 - 17.32(\pm 2.74) \quad (1)$$
$$n = 9, \quad r^2 = 0.976, \quad s = 0.222,$$
$$q^2 = 0.844, \quad Q = 4.450, \quad F_{2,6} = 122.000$$
$$\text{optimum C}\log P = 4.55(4.44 - 4.66)$$

This is a parabolic correlation in terms of Clog *P*, which suggests that the topo I-mediated cleavage of benzimidazoles (**II**) first increases with an increase in the hydrophobicity up to an optimum Clog *P* of 4.55 and then decreases.

3.1.2 Camptothecins

In 1985, it was reported by Hsiang et al. [43] that the cytotoxic activity of 20-(*S*)-camptothecin (CPT; **III**) was attributed to a novel mechanism of action involving the nuclear enzyme topo I, and this discovery of unique mechanism of action revived the interest in CPT and its analogues as anticancer agents. CPT stabilizes the covalent, reversible topo I–DNA complex leading to the inhibition of DNA synthesis in mammalian cells and interferes with the topo I breakage–reunion reaction [44]. Clinical trials and structure–activity relationships have demonstrated the requirement of the α-hydroxy group, the

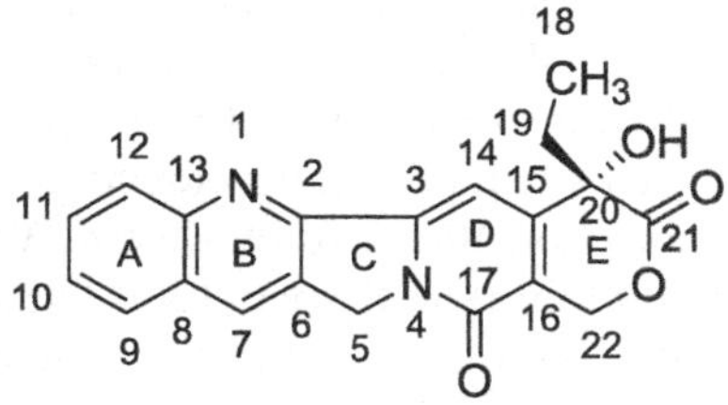

III

pyridone moiety, and the pentacyclic ring system for maximum activity. Substitution at positions 9 and 10 of the A ring by halides and other electron-rich groups (e.g., amino, hydroxy, etc.) generally increases the topo I inhibition. The addition of a 10,11-methylenedioxy moiety at the A ring substantially increases the activity. Substitutions at position 7 have been found to be more potent, and increase water solubility depending on the nature of the substituent [45].

Wall et al. [46] synthesized a series of X-camptothecin analogues (**IV**; where, X = 9- or 10- or 9, 10- or 10, 11- or 9, 10, 11- substitutions) and evaluated their inhibitory activity against DNA topo I. Using topo I inhibition data of these X-camptothecin analogues (**IV**), Eq. 2 was derived [47]:

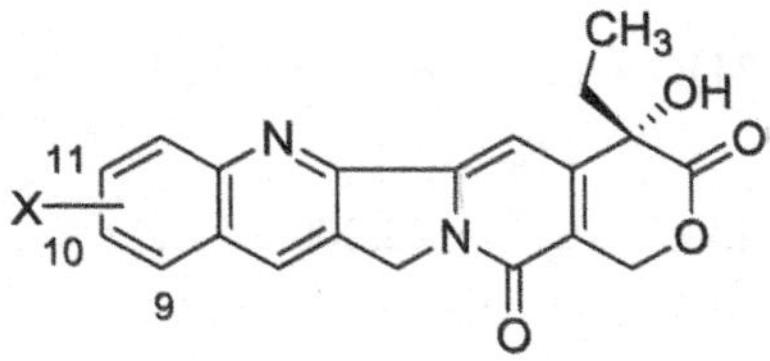

IV

$$\log 1/C = 0.43(\pm 0.29)\mathrm{C}\log P - 0.43(\pm 0.22)\sigma_X^+ + 1.11(\pm 0.35)I - 0.89(\pm 0.40)\mathrm{MR}_9 + 6.37(\pm 0.19) \quad (2)$$

$n = 17, \quad r^2 = 0.862, \quad s = 0.226,$

$q^2 = 0.681, \quad Q = 4.108, \quad F_{4,12} = 18.739$

In Eq. 2, σ_X^+ is the sum of σ^+ values for X = 9-, 10-, and 11-substituents, while MR_9 is the molar refractivity only for the 9-X-substituent. The negative coefficient with σ_X^+ (– 0.43) implies that highly electron-releasing substituents at positions 9, 10, and 11 may strengthen the inhibitory activity of these compounds against topo I. The negative coefficient of MR_9 (– 0.89) suggests an unfavorable steric interaction at this position. The indicator variable I is assigned the value of 1 and 0 for the presence and absence of X = 10-OCH_2O-11 group. Its positive coefficient suggests that the presence of a 10,11-methylenedioxy moiety at the A ring increases the activity. A positive hydrophobic effect for the whole molecule also appears in this equation. However, this equation does not allow any clue for adequate distinction between the σ^+ responsiveness of topo I inhibition for 9-, 10-, and 11-substituents. Considering these drawbacks of Eq. 2, we developed Eq. 3 using the same data of Wall et al. [46] but only for 10-X-camptothecin (**V**), which gave a good correlation between the inhibitory activity of topo I and the hydrophobic parameters of X-substituents. Biological and physicochemical parameters used to derive QSAR Eq. 3 are shown in Table 1.

Table 1 Biological and physicochemical parameters used to derive QSAR Eq. 3

No.	X	Obsd.	log 1/C (Eq. 3) Pred.	Δ	π_X
1 [a]	OH	6.96	5.74	1.22	-0.67
2	Br	6.89	6.89	0.00	0.86
3 [a]	NH_2	6.85	5.32	1.53	-1.23
4	Cl	6.85	6.78	0.07	0.71
5	CH_3	6.52	6.66	-0.14	0.56
6	F	6.43	6.35	0.08	0.14
7	NO_2	6.19	6.03	0.16	-0.28
8	H	6.17	6.24	-0.07	0.00
9	COOH	6.00	6.01	-0.01	-0.32
10	CN	5.72	5.82	-0.10	-0.57

[a] Not included in the derivation of QSAR Eq. 3

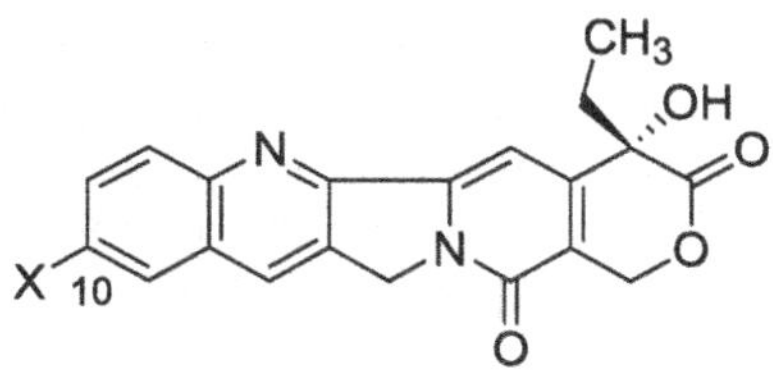

V

$$\log 1/C = 0.75(\pm 0.20)\pi_X + 6.24(\pm 0.10) \quad (3)$$
$$n = 8, \quad r^2 = 0.936, \quad s = 0.110, \quad q^2 = 0.890,$$
$$Q = 8.795, \quad F_{1,6} = 87.750 \quad \text{outliers: } X = OH; NH_2$$

π_X represents the hydrophobicity of the substituents at position 10. Its positive coefficient (+0.75) suggests that the presence of highly hydrophobic substituents at position 10 increases the activity. The outlier (X = OH) is much more active than expected by 11 times the standard deviation. This may be due to the formation of a phenoxyl radical that interacts with DNA [48]. The other derivative (X = NH_2) is also considered as an outlier due to being much more active than expected by 14 times the standard deviation. This anomalous behavior may be attributed to its nature as an aniline. This could result in hydrogen abstraction, or involve microsomal N-oxidation [48, 49].

In an effort to improve the water solubility of camptothecin, Rahier et al. [50] synthesized four 20-*O*-phosphate derivatives (**VI**). These compounds are freely water-soluble, stable to physiological pH, and stabilize the human topo I–DNA covalent binary complex with the same sequence-selectivity as camptothecin itself. All four compounds inhibited the growth

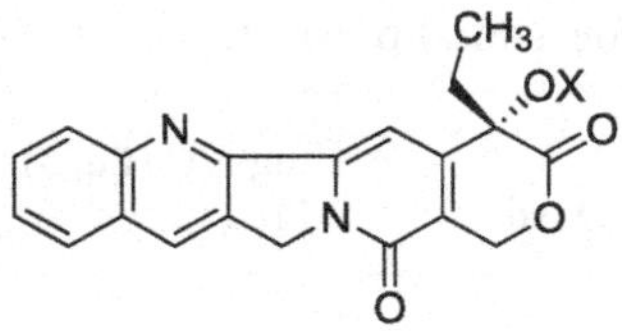

VI

of yeast expressing human topo I in an enzyme-dependent fashion. From the topo I-dependent cytotoxicity data of these four compounds along with camptothecin (**III**) in *S. cerevisiae*, Eq. 4 was developed (Table 2):

$$\log 1/C = 1.77(\pm 0.86)\mathrm{C}\log P + 4.39(\pm 0.41) \qquad (4)$$
$$n = 4, \quad r^2 = 0.975, \quad s = 0.156, \quad q^2 = 0.718,$$
$$Q = 6.327, \quad F_{1,2} = 78.000, \quad \text{outlier:} \quad \mathrm{X = P(=O)(OH)Ph}$$

Hydrophobicity is found to be the single most important parameter for this data set, which shows that at all the parts where substituents have been entered, hydrophobic contacts have been made. The linear Clog *P* model suggests that the highly hydrophobic molecules will be more active. Although this is a very small data set it is the best model and explains 97.5% of the variance in log 1/*C*.

Vladu et al. [51] synthesized a series of 7-X-10-Y-camptothecins (**VII**) and evaluated their ability to trap human DNA topo I in cleavable complexes. We used these data to develop Eq. 5 (Table 3):

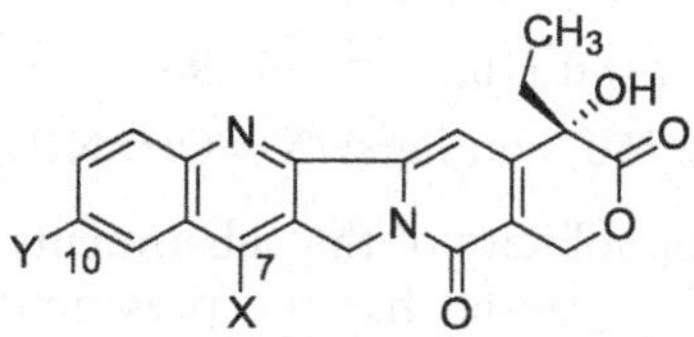

VII

$$\log 1/C = -0.30(\pm 0.19)\pi_X + 0.90(\pm 0.28)I + 5.89(\pm 0.28) \qquad (5)$$
$$n = 14, \quad r^2 = 0.855, \quad s = 0.230,$$
$$q^2 = 0.742, \quad Q = 4.022, \quad F_{2,11} = 32.431$$

π_X is the hydrophobic parameter of the substituents at position 7. Its negative coefficient indicates that the increase in hydrophobicity of the substituents at position 7 would be detrimental to their activity. The indicator variable *I* is assigned the value of 1 and 0 for the presence and absence of OH at pos-

Table 2 Biological and physicochemical parameters used to derive QSAR Eq. 4

No	X	log 1/*C* (Eq. 4) Obsd.	Pred.	Δ	Clog *P*
1	H (CPT; **III**)	6.05	5.97	0.08	0.90
2	$P(=O)(OH)_2$	4.70	4.88	−0.18	0.28
3	$P(=O)(OH)(OCH_3)$	4.48	4.47	0.01	0.05
4	$P(=O)(OH)(CH_3)$	4.24	4.14	0.10	−0.14
5[a]	P(=O)(OH)(Ph)	4.43	6.50	−2.07	1.20

[a] Not included in the derivation of QSAR Eq. 4

Table 3 Biological, physicochemical, and structural parameters used to derive QSAR Eq. 5

No	X	Y	log 1/*C* (Eq. 5) Obsd.	Pred.	Δ	π_X	*I*
1	CH_3	H	5.97	5.72	0.25	0.56	0
2	CH_2CH_3	H	5.35	5.58	−0.23	1.02	0
3	$(CH_2)_2CH_3$	H	5.14	5.42	−0.28	1.55	0
4	$(CH_2)_3CH_3$	H	5.14	5.25	−0.11	2.13	0
5	H	OH	6.46	6.78	−0.32	0.00	1
6	CH_3	OH	6.89	6.61	0.28	0.56	1
7	CH_2CH_3	OH	6.49	6.48	0.01	1.02	1
8	$(CH_2)_2CH_3$	OH	6.22	6.32	−0.10	1.55	1
9	$(CH_2)_3CH_3$	OH	6.28	6.14	0.14	2.13	1
10	H	OCH_3	6.14	5.89	0.25	0.00	0
11	CH_3	OCH_3	5.62	5.72	−0.10	0.56	0
12	CH_2CH_3	OCH_3	5.55	5.58	−0.03	1.02	0
13	$(CH_2)_2CH_3$	OCH_3	5.35	5.42	−0.07	1.55	0
14	$(CH_2)_3CH_3$	OCH_3	5.55	5.25	0.30	2.13	0

ition 10. Its positive coefficient suggests that the presence of an OH group at position 10 increases the activity.

3.1.3 Isoaurostatins

Suzuki et al. [52] synthesized a series of isoaurostatin derivatives (**VIII**) and evaluated their inhibitory activities as well as structure–activity relationships against topo I and II. They predicted from their results that the addition of hydroxyl groups on aromatic rings increases the activity. From the inhibitory data of these compounds (**VIII**) against topo I, Eq. 6 was derived (Table 4):

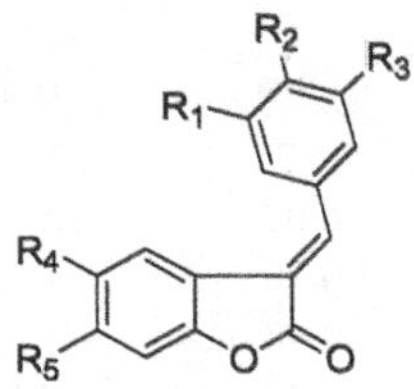

VIII

$$\log 1/C = -1.45(\pm 0.48)\mathrm{C}\log P + 7.53(\pm 1.22) \quad (6)$$
$$n = 11, \quad r^2 = 0.840, \quad s = 0.406, \quad q^2 = 0.771, \quad Q = 2.259,$$
$$F_{1,9} = 47.250 \quad \text{outliers:} \quad R_1 = R_3 = R_4 = R_5 = \mathrm{H},$$
$$R_2 = \mathrm{NO_2}; \quad R_1 = R_3 = R_4 = R_5 = \mathrm{H}, \quad R_2 = \mathrm{F}$$

The negative Clog$P$ term shows that highly hydrophilic molecules for this data set would present better inhibitory activities against topo I. Two compounds ($R_1 = R_3 = R_4 = R_5 = H$, $R_2 = NO_2$ and $R_1 = R_3 = R_4 = R_5 = H$, $R_2 = F$) in Table 4 for the development of QSAR Eq. 6 were deemed to be outliers on the basis of their deviation (>2s). The outlier ($R_1 = R_3 = R_4 = R_5 = H$, $R_2 = NO_2$) is much more active than expected, by three times the standard deviation. This may be due to the formation of nitro anion radicals that interact with DNA [48]. The other derivative ($R_1 = R_3 = R_4 = R_5 = H$, $R_2 = F$) is

Table 4 Biological and physicochemical parameters used to derive QSAR Eq. 6

No.	R_1	R_2	R_3	R_4	R_5	log 1/C (Eq. 6) Obsd.	Pred.	Δ	Clog$P$
1	OH	H	H	H	H	3.58	3.16	0.42	3.00
2	H	OH	H	H	H	3.60	3.16	0.44	3.00
3	OH	OH	H	H	H	3.80	4.03	−0.23	2.41
4	OH	OH	OH	H	H	5.40	5.00	0.40	1.74
5	H	H	H	OH	H	3.02	2.90	0.12	3.18
6	H	OH	H	OH	H	3.26	3.87	−0.61	2.52
7	OH	OH	H	OH	H	5.22	4.74	0.48	1.92
8	OH	OH	OH	OH	H	5.52	5.71	−0.19	1.25
9	OH	OCH_3	H	H	H	3.10	3.38	−0.28	2.85
10	OCH_3	OH	H	H	H	3.20	3.38	−0.18	2.85
11 [a]	H	NO_2	H	H	H	3.92	2.57	1.35	3.41
12 [a]	H	F	H	H	H	3.41	1.99	1.42	3.81
13	H	OH	H	H	OH	3.51	3.87	−0.36	2.52

[a] Not included in the derivation of QSAR Eq. 6

also an outlier as it is more active than expected by three times the standard deviation. The anomalous behavior of this compound, due to the presence of fluorine, is not very clear.

3.1.4 Naphthyridinones

A series of 6-substituted 8,9-dimethoxy-2,3-methylenedioxy-6*H*-dibenzo[*c*,*h*][2,6] naphthyridin-5-ones (**IX**) was synthesized and evaluated for topo I-targeting activity as well as for cytotoxicity against different cell lines by Zhu et al. [53]. From the topo I-mediated DNA cleavage data of these compounds (**IX**), we developed Eq. 7 (Table 5):

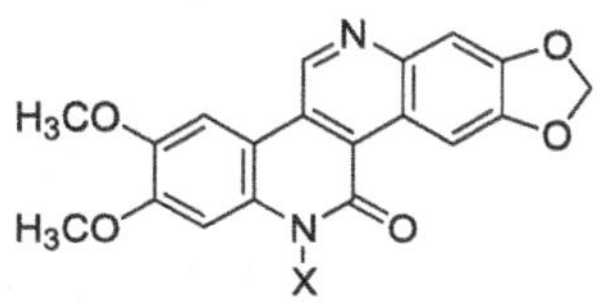

IX

$$\log \mathrm{REC} = -1.16(\pm 0.46)\pi_X + 0.86(\pm 0.23)\pi_X^2 + 1.41(\pm 0.36)I - 0.93(\pm 0.25) \quad (7)$$

$n = 10,\quad r^2 = 0.962,\quad s = 0.172,\quad q^2 = 0.896,\quad Q = 5.703,\quad F_{3,6} = 50.632$

Inversion point for $\pi_X = 0.67(0.53 - 0.78)$

outlier: $X = CH_2CH_2N(-CH_2CH_2OCH_2CH_2-)$

Table 5 Biological, physicochemical, and structural parameters used to derive QSAR Eq. 7

No.	X	log REC (Eq. 7) Obsd.	Pred.	Δ	π_X	*I*
1	$CH_2CH_2N(CH_3)_2$	0.08	0.11	-0.03	0.52	1
2	$CH_2CH(CH_3)N(CH_3)_2$	0.08	0.11	-0.03	0.83	1
3	$CH_2CH_2N(C_2H_5)_2$	-0.70	-0.62	-0.08	1.58	0
4	$CH2CH_2N[-CH_2CH_2CH_2CH_2-]$	-1.15	-1.27	0.12	0.92	0
5	$CH_2CH_2N[-CH_2(CH_2)_3CH_2-]$	-1.00	-0.77	-0.23	1.48	0
6	$CH_2CH_2N[-(CH_2)_2CH(CH_3)(CH_2)_2-]$	0.48	0.18	0.30	2.00	0
7[a]	$CH_2CH_2N[-CH_2CH_2OCH_2CH_2-]$	0.95	-1.18	2.13	0.27	0
8	$CH_2CH_2CH_2N(CH_3)_2$	0.15	0.09	0.06	0.67	1
9	$CH_2CH_2N(CH_3)CH_2C_6H_5$	1.11	1.21	-0.10	2.39	0
10	$CH_2CH_2NH_2$	-0.46	-0.42	-0.04	-0.35	0
11	$CH_2CH_2NHCH_3$	-0.82	-0.86	0.04	-0.06	0

[a] Not included in the derivation of QSAR Eq. 7

This is an inverted parabolic relation in terms of π_X (calculated hydrophobic parameter of the substituents), which suggests that activity of these compounds first decreases as the hydrophobicity of substituents increases and after a certain point (inversion point; $\pi_X = 0.67$), activity begins to increase. This may correspond to an allosteric reaction [54]. The indicator variable I is assigned the value of 1 and 0 for the presence and absence of $N(CH_3)_2$ substituent at the X position. Its positive coefficient suggests that the presence of a $N(CH_3)_2$ substituent at X position, increases the activity. REC is the relative effective concentration i.e., concentration relative to topotecan, whose value is arbitrarily assumed as 1, that is able to produce the same cleavage on the plasmid DNA in the presence of human topo I.

3.1.5
Phenanthridines

Zhu et al. [55] also synthesized a series of ester/amide derivatives of 2,3-dimethoxy-8,9-methylenedioxy-benzo[*i*]phenanthridine-12-carboxylic acid (**X**) and evaluated their topo I-targeting activity as well as cytotoxicity against different cell lines. Topo I cleavage values are given as REC, the relative ef-

Table 6 Biological, physicochemical, and structural parameters used to derive QSAR Eq. 8

No.	X	log REC (Eq. 8) Obsd.	Pred.	Δ	π_X	I
1	OCH_2CH_3	1.00	0.89	0.11	0.80	0
2	$OCH_2CH_2N(CH_3)_2$	0.70	0.64	0.06	0.36	0
3 [a]	$OCH(CH_3)CH_2N(CH_3)_2$	0.00	0.82	−0.82	0.67	0
4	$OC(CH_3)_2CH_2N(CH_3)_2$	0.90	1.04	−0.14	1.07	0
5	$OCH_2CH_2CH_2N(CH_3)_2$	0.81	0.84	−0.03	0.71	0
6	$NHCH_2CH_2N(CH_3)_2$	−0.30	−0.25	−0.05	−0.15	1
7	$NHCH(CH_3)CH_2N(CH_3)_2$	−0.22	−0.07	−0.15	0.16	1
8	$NHCH_2CH(CH_3)N(CH_3)_2$	−0.40	−0.07	−0.33	0.16	1
9 [a]	$N(CH_3)CH_2CH_2N(CH_3)_2$	1.00	−0.37	1.37	−0.37	1
10	$NHCH_2CH_2N(C_2H_5)_2$	−0.15	0.36	−0.51	0.91	1
11	$NHCH_2CH_2NCH_3(CH_2C_6H_5)$	1.08	0.81	0.27	1.71	1
12	$NHCH_2CH_2N(CH_2C_6H_5)_2$	1.78	1.79	−0.01	3.43	1
13	$N(\text{-}CH_2CH_2CH_2CH_2\text{-})$	−0.22	−0.40	0.17	−0.41	1
14	$N(\text{-}CH_2CH_2CH_2CH_2CH_2\text{-})$	0.18	−0.08	0.25	0.15	1
15	$N(\text{-}CH_2CH_2N(CH_3)CH_2CH_2\text{-})$	−0.52	−0.41	−0.11	−0.43	1
16 [a]	$NHCH_2CH_2CH_2N(CH_3)_2$	0.70	−0.09	0.79	0.13	1
17	$N[CH_2CH_2N(CH_3)_2]_2$	0.48	0.01	0.47	0.31	1

[a] Not included in the derivation of QSAR Eq. 8

fective concentration (i.e., concentration relative to topotecan, whose value is arbitrarily assumed as 1) that is able to produce the same cleavage on the plasmid DNA in the presence of human topo I. We used these data to formulate Eq. 8 (Table 6):

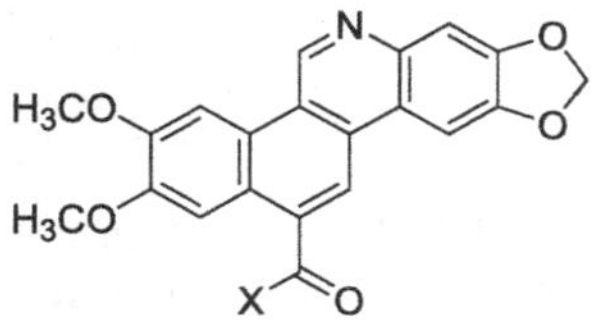

X

$$\log \mathrm{REC} = 0.57(\pm 0.17)\pi_X - 0.60(\pm 0.36)I + 0.43(\pm 0.33) \quad (8)$$

$n = 14$, $r^2 = 0.869$, $s = 0.274$, $q^2 = 0.827$, $Q = 3.401$,
$F_{2,11} = 36.485$ outliers: $X = OCH(CH_3)CH_2N(CH_3)_2$;
$N(CH_3)CH_2CH_2N(CH_3)_2$;
$NHCH_2CH_2CH_2N(CH_3)_2$

The indicator variable I is assigned the value of 1 for the presence of amide derivatives and 0 for the esters. Its negative coefficient suggests that esters would be preferred over amides for this data set. π_X is the calculated hydrophobic parameter of the X-substituents. Its positive coefficient suggests that the highly hydrophobic X-substituents would be preferred.

3.1.6 Terpenes

Zhang et al. [56] isolated two new daphne diterpene esters yuanhuajine (**XIc**) and yuanhuagine (**XId**), together with three known daphne diterpene esters yuanhuacine (**XIa**), yuanhuadine (**XIb**), and yuanhuapine (**XIe**) from *Daphne genkwa*, a traditional Chinese medicine. It was mainly used for dispelling retained water, abortion, and mammary cancer. In order to explore the structure–activity relationships, three more derivatives (**XIf**, **XIg**, and **XIh**) were synthesized by this research group and the inhibitory activities of all these compounds (Fig. 1) against DNA topo I evaluated. On the basis of their results, Zhang et al. [56] suggested that the orthoester group of daphne diterpene esters is necessary for the inhibitory activity against DNA topo I. From the inhibitory data of these compounds (**XIa–XIh**) against topo I, Eq. 9 was derived (Table 7):

$$\log 1/C = 0.31(\pm 0.15)\mathrm{C}\log P - 1.17(\pm 0.49)\mathrm{MgVol} + 8.28(\pm 1.68) \quad (9)$$

$n = 7$, $r^2 = 0.918$, $s = 0.096$, $q^2 = 0.782$, $Q = 9.979$,
$F_{2,4} = 22.390$ outlier: **XIg**

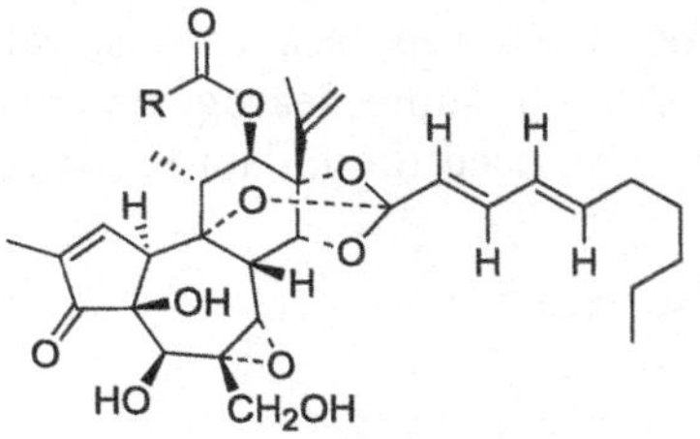

XIa; R = C_6H_5 (Yuanhuacine)
XIb; R = CH_3 (Yuanhuadine)

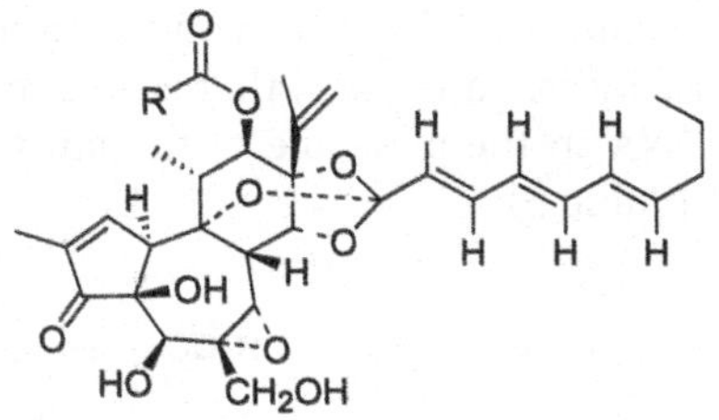

XIc; R = C_6H_5 (Yuanhuajine)
XId; R = CH_3 (Yuanhuagine)

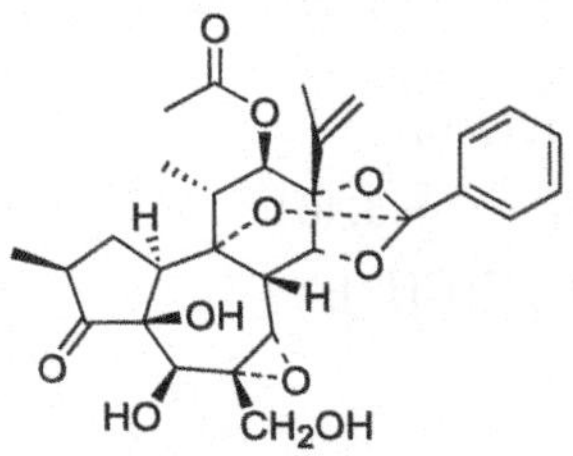

XIe; (Yuanhuapine)

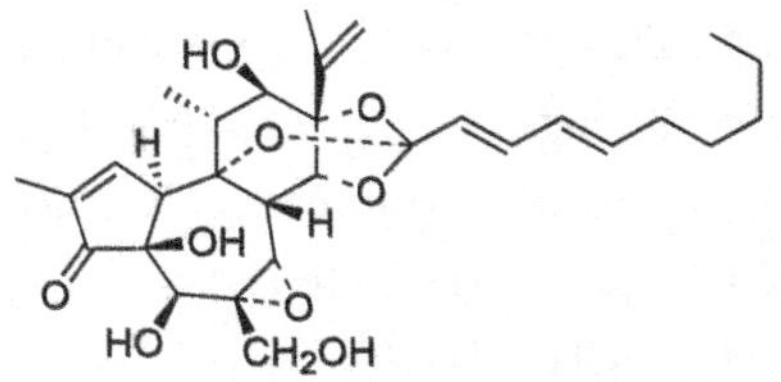

XIf

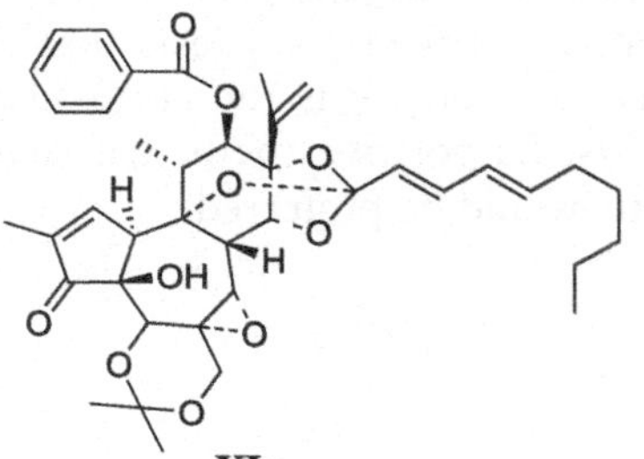

XIg

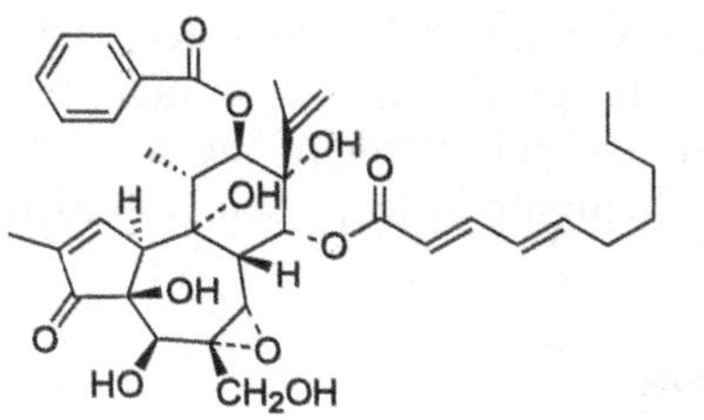

XIh

Fig. 1 Structure of compounds **XIa–XIh** used in the derivation of QSAR Eq. 9

The negative sign of the molar volume (MgVol) suggests that the steric interactions are unfavorable to the activity.

3.1.7
Miscellaneous Compounds

Green tea, especially its major polyphenolic constituent, epigallocatechin-3-gallate (EGCG), has received much attention over the past few years as a potential cancer chemopreventive agent because EGCG imparts growth inhibitory responses to cancer cells but not to normal cells. Based on this fact, Berger et al. [22] compared the inhibition of DNA topo I activity of EGCG with that of several known topo I inhibitors (CPT, CPT-11, SN-38, and TPT) in three human colon carcinoma cells. From their data, we developed Eqs. 10–12 (Table 8 and 9)

Table 7 Biological and physicochemical parameters used to derive QSAR Eq. 9

No.	Compd.	log 1/C (Eq. 9) Obsd.	Pred.	Δ	Clog *P*	MgVol
1	XIa	4.40	4.41	−0.01	5.51	4.76
2	XIb	4.28	4.42	−0.14	3.79	4.29
3	XIc	4.42	4.32	0.10	5.06	4.72
4	XId	4.30	4.33	−0.03	3.33	4.25
5	XIe	4.27	4.25	0.02	1.21	3.76
6	XIf	4.55	4.49	0.06	2.89	4.00
7[a]	XIg	4.95	4.36	0.59	6.54	5.08
8	XIh	3.69	3.70	−0.01	4.12	4.99

[a] Not included in the derivation of QSAR Eq. 9

Table 8 Biological and physicochemical parameters used to derive QSAR Eqs. 10 and 11

No	Compd.	log 1/C (Eq. 10) Obsd.	Pred.	Δ	log 1/C (Eq. 11) Obsd.	Pred.	Δ	MgVol
1[a]	CPT	5.32	6.01	−0.69	5.31	5.99	−0.68	2.43
2	CPT-11	3.17	3.17	0.00	3.15	3.15	0.00	4.32
3	SN-38	5.49	5.50	−0.01	5.50	5.48	0.02	2.77
4	TPT	5.26	5.14	0.12	5.28	5.12	0.16	3.01
5	EGCG	5.07	5.17	−0.10	4.97	5.15	−0.18	2.99

[a] Not included in the derivation of QSAR Eqs. 10 and 11

Table 9 Biological and physicochemical parameters used to derive QSAR Eq. 12

No.	Compd.	log 1/C (Eq. 12) Obsd.	Pred.	Δ	MgVol
1	CPT	5.43	5.66	−0.23	2.43
2	CPT-11	3.19	3.23	0.04	4.32
3	SN-38	5.55	5.22	0.33	2.77
4	TPT	5.02	4.91	0.11	3.01
5	EGCG	4.78	4.94	−0.16	2.99

Inhibition of DNA topo I activity in HCT 116 (human colon carcinoma) cells by CPT, CPT-11, SN-38, TPT, and EGCG (Table 8):

$$\log 1/C = -1.50(\pm 0.39)\text{MgVol} + 9.66(\pm 1.30) \tag{10}$$

$n = 4,\quad r^2 = 0.993,\quad s = 0.111,\quad q^2 = 0.972,\quad Q = 8.973,$

$F_{1,2} = 283.714$ outlier: CPT

Inhibition of DNA topo I activity in VACO 241 (human colon carcinoma) cells by CPT, CPT-11, SN-38, TPT, and EGCG (Table 8):

$$\log 1/C = -1.51(\pm 0.61)\text{MgVol} + 9.66(\pm 2.03) \quad (11)$$

$n = 4$, $r^2 = 0.983$, $s = 0.174$, $q^2 = 0.964$, $Q = 5.695$, $F_{1,2} = 115.647$

outlier: CPT

Inhibition of DNA topo I activity in SW 480 (human colon carcinoma) cells by CPT, CPT-11, SN-38, TPT, and EGCG (Table 9):

$$\log 1/C = -1.28(\pm 0.57)\text{MgVol} + 8.77(\pm 1.80) \quad (12)$$

$n = 5$, $r^2 = 0.945$, $s = 0.256$, $q^2 = 0.810$, $Q = 3.797$, $F_{1,3} = 51.545$

The above three equations (Eqs. 10–12) for the different cell lines are very similar to each other, which suggests that the inhibition against DNA topo I is probably one of the most important antitumor mechanisms for these compounds (CPT, CPT-11, SN-38, TPT, and EGCG) against the three human colon carcinoma (HCT 116, VACO 241, and SW 480) cell lines. In these equations, the number of data points (four or five) is small, but the correlations are statistically significant.

3.2 Topoisomerase II Inhibitors

3.2.1 Anthrapyrazoles

Losoxantrone (**XII**; $R_1 = R_4 = H$, $R_2 = R_5 = OH$, $R_3 = NH(CH_2)_3NH(CH_2)_2OH$) and piroxantrone (**XII**; $R_4 = H$, $R_1 = R_2 = R_5 = OH$, $R_3 = NH(CH_2)_3NH_2$) are anthrapyrazole antitumor agents that were developed as the non-cardiotoxic alternatives to daunorubicin and doxorubicin [57]. Both losoxantrone and piroxantrone exert their antitumor activity by acting as DNA topo II poisons [58, 59]. Liang et al. [60] considered a series of anthrapyrazole analogues (**XII**) and evaluated their topo II inhibitory activities, DNA binding, and growth inhibition of K562 cells to understand their QSAR and structure-based 3D-QSAR. The QSAR correlation and 3D-QSAR analyses showed the importance of anthrapyrazole–DNA van der Waals interactions, while 3D-QSAR (CoMFA and CoMSIA) showed hydrogen-bond donor (HBD) interactions and electrostatic interactions with the protonated amino side-chains of the anthrapyrazoles. Using the DNA topo II inhibitory data of Liang et al. [60] for anthrapyrazole analogues (**XII**), we developed Eq. 13 (Table 10):

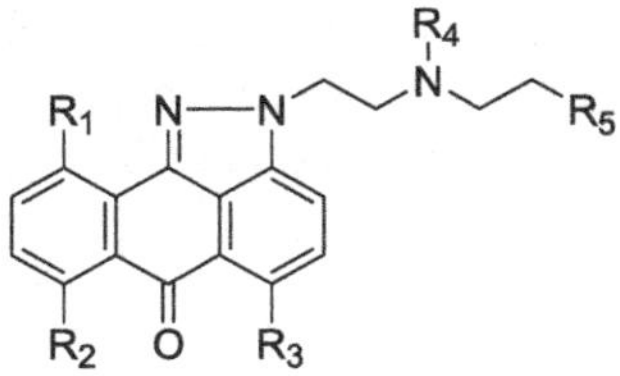

XII

$$\log 1/C = -0.48(\pm 0.21)\mathrm{C}\log P + 40.33(\pm 8.65)(\beta \times 10^{\mathrm{C}\log P} + 1) + 6.38(\pm 0.71) \quad (13)$$

$n = 12$, $r^2 = 0.949$, $s = 0.159$, $q^2 = 0.843$, $Q = 6.126$, $F_{3,8} = 49.621$

Inversion point for $\mathrm{C}\log P = 3.90$

$\log \beta = -5.82$

outliers: $R_1 = R_2 = H$, $R_3 = Cl$, $R_4 = Me$, $R_5 = OH$;
$R_1 = R_3 = H$, $R_2 = Cl$, $R_4 = Me$, $R_5 = OH$

This is an inverted bilinear relation in terms of Clog*P* (calculated hydrophobic parameter of the whole molecule), which suggests that activity of

Table 10 Biological and physicochemical parameters used to derive QSAR Eq. 13

No.	R_1	R_2	R_3	R_4	R_5	log 1/*C* (Eq. 13) Obsd.	Pred.	Δ	Clog *P*
1[a]	H	H	Cl	Me	OH	5.54	4.71	0.83	3.97
2	H	H	Cl	Me	Cl	6.60	6.34	0.26	4.98
3	H	H	Cl	H	OH	4.89	4.83	0.06	3.32
4	H	H	Cl	H	Cl	4.92	4.90	0.02	4.39
5	H	Cl	H	Me	Cl	6.06	6.33	-0.27	4.97
6	H	Cl	H	H	OH	4.74	4.83	-0.09	3.32
7	H	Cl	H	H	Cl	4.85	4.90	-0.05	4.39
8[a]	H	Cl	H	Me	OH	5.14	4.71	0.43	3.97
9	H	H	$NH(CH_2)_2NMe_2$	H	OH	4.64	4.80	-0.16	3.41
10	H	H	$NH(CH_2)_2NMe_2$	Me	OH	4.80	4.72	0.08	4.06
11	H	$NH(CH_2)_2NMe_2$	H	H	OH	4.96	4.83	0.13	3.32
12	H	$NH(CH_2)_2NMe_2$	H	Me	OH	4.72	4.71	0.01	3.96
13[b]	H	OH	$NH(CH_2)_3NH(CH_2)_2OH$	H	OH	5.14	5.13	0.01	2.61
14[c]	OH	OH	$NH(CH_2)_3NH_2$	H	OH	5.34	5.33	0.01	2.19

[a] Not included in the derivation of QSAR Eq. 13
[b] Losoxantrone
[c] Piroxantrone

these compounds first decreases linearly as the hydrophobicity of the whole molecules increases, and that after a certain point (inversion point; Clog P = 3.90) activity begins to increase gradually. This may correspond to an allosteric reaction [54, 61].

3.2.2 Benzimidazoles

The effects of bis-benzimidazoles (**XIII**) on *Pneumocystis carinii* topo II were evaluated by Dykstra et al. [62]. From their data, Eq. 14 was developed [47]:

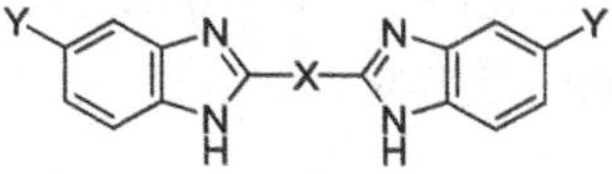

XIII

$$\log 1/C = 3.28(\pm 0.60)\text{MgVol} - 4.59(\pm 1.76) \quad (14)$$
$$n = 7, \quad r^2 = 0.975, \quad s = 0.180, \quad q^2 = 0.963, \quad Q = 5.486, \quad F_{1,5} = 195.000$$

The steric hindrance as MgVol (molar volume) was found to be the most significant variable for this data set.

3.2.3 Benzonaphthofurandiones

The inhibitory activity (% inhibition) of benzonaphthofurandiones (**XIV**) was evaluated by Rhee et al. [63] against topo II using a decatenation assay. From their data, Eq. 15 was developed (Table 11):

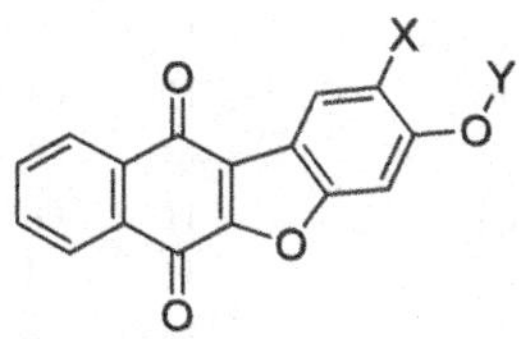

XIV

$$\log \% = -0.30(\pm 0.15)\text{CMR} + 4.93(\pm 1.53) \quad (15)$$
$$n = 7, \quad r^2 = 0.844, \quad s = 0.070, \quad q^2 = 0.699, \quad Q = 13.129, \quad F_{1,5} = 27.051$$
outlier: $X = Cl, Y = (CH_2)_2N(CHMe_2)_2$

CMR represents the overall calculated molar refractivity. Its negative sign bring out a steric effect. It is interesting to note here that there is a high mutual correlation between Clog P and CMR (r= 0.966). Thus, it is very hard to predict for this data set if it is a negative hydrophobic or a polarizability effect.

Table 11 Biological and physicochemical parameters used to derive QSAR Eq. 15

No.	X	Y	log % (Eq. 15) Obsd.	Pred.	Δ	CMR	Clog *P*
1	H	$(CH_2)_2NEt_2$	1.96	1.87	0.09	10.23	5.17
2	H	$(CH_2)_2N(CHMe_2)_2$	1.56	1.59	−0.03	11.16	5.78
3	H	$CH(Me)CH_2NMe_2$	1.92	2.01	−0.09	9.76	4.42
4	H	$CH_2CH(Me)NMe_2$	2.00	2.01	−0.01	9.76	4.42
5	Cl	$(CH_2)_2NEt_2$	1.70	1.72	−0.02	10.72	5.67
6 [a]	Cl	$(CH_2)_2N(CHMe_2)_2$	1.77	1.44	0.33	11.65	6.28
7	Cl	$CH(Me)CH_2NMe_2$	1.94	1.86	0.08	10.26	4.92
8	Cl	$CH_2CH(Me)NMe_2$	1.83	1.86	−0.03	10.26	4.92

[a] Not included in the derivation of QSAR Eq. 15

We derived Eq. 15a with Clog *P* and finally preferred Eq. 15 on the basis of its statistics, which are better than those of Eq. 15a:

$$\log\% = -0.25(\pm 0.18)C\log P + 3.12(\pm 0.92) \quad (15a)$$

$$n = 7, \quad r^2 = 0.720, \quad s = 0.094, \quad q^2 = 0.442, \quad Q = 9.032, \quad F_{1,5} = 12.857$$

outlier: $X = Cl, Y = (CH_2)_2N(CHMe_2)_2$

3.2.4 Desoxypodophyllotoxins

Etoposide (**XV**) is a semisynthetic gylcoside derivative of podophyllotoxin, which is one of the most extensively used anticancer drugs in the treatment of various types of tumors [64, 65]. The anticancer activity of this drug is mainly due to its ability to inhibit an ubiquitous and essential enzyme: human DNA topo II [66, 67]. Despite its extensive use in the treatment of cancers, it has several limitations, such as poor water solubility, drug resistance, metabolic inactivation, myelosuppression, and toxicity [68]. In order to overcome these

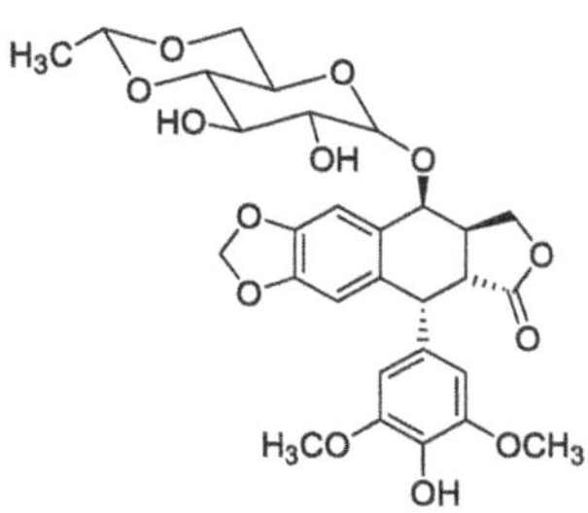

Etoposide (XV)

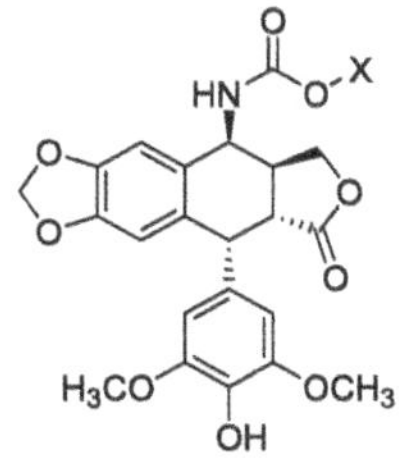

XVI

limitations, and to develop more active and more potent compounds, Duca et al. [69] synthesized a novel series of desoxypodophyllotoxins (**XVI**) and evaluated their inhibitory activities (% inhibition) against DNA topo II. From their data, we developed Eq. 16 (Table 12):

$$\log\% = -0.24(\pm 0.11)\mathrm{MR_X} + 0.56(\pm 0.23)I + 1.83(\pm 0.20) \quad (16)$$

$$n = 11, \quad r^2 = 0.815, \quad s = 0.108, \quad q^2 = 0.668, \quad Q = 8.361, \quad F_{2,8} = 17.622$$

$$\text{outliers:} \quad X = (CH_2)_3CH_3; \quad CH_2Ph(2,4-Cl_2)$$

$$\pi_X \text{ vs. } \mathrm{MR_X}; \quad r = 0.470$$

MR_X is the calculated molar refractivity of X-substituents, whereas **I** is an indicator variable taking the value of 1 and 0 for the presence and absence of a phenyl ring in the X-substituents. The negative sign of MR_X brings out a steric effect for the X-substituents that do not appear to reach a hydrophobic surface; for π_X (calculated hydrophobicity of X-substituents), $r = 0.470$. The indicator variable (I) with positive coefficient suggests that the presence of a phenyl ring in the X-substituents would be favorable.

Table 12 Biological, physicochemical, and structural parameters used to derive QSAR Eq. 16

No.	X	log % (Eq. 16) Obsd.	Pred.	Δ	MR_X	I
1	CH_3	1.70	1.72	−0.02	0.46	0
2	CH_2CH_3	1.64	1.61	0.03	0.93	0
3 [a]	$(CH_2)_3CH_3$	1.60	1.39	0.22	1.86	0
4	$(CH_2)_3Cl$	1.49	1.38	0.11	1.88	0
5	$CH_2CH{=}CH_2$	1.40	1.50	−0.11	1.37	0
6	$CH_2C{\equiv}CH$	1.46	1.55	−0.08	1.19	0
7	$(CH_2)_2OCH_3$	1.62	1.46	0.16	1.54	0
8	$CH_2CH(\text{-}O(CH_2)_3CH_2\text{-})$	1.0	1.17	−0.17	2.76	0
9	$(CH_2)_2N(\text{-}COCH_2CH_2CO\text{-})$	1.18	1.10	0.08	3.05	0
10	CH_2Ph	1.70	1.68	0.02	2.97	1
11	$CH_2Ph(4\text{-}F)$	1.65	1.68	−0.02	2.99	1
12	$CH_2CH_2Ph(4\text{-}F)$	1.57	1.56	0.01	3.45	1
13 [a]	$CH_2Ph(2,4\text{-}Cl_2)$	.60	1.44	−0.84	3.96	1

[a] Not included in the derivation of QSAR Eq. 16

3.2.5 Isoaurostatins

From the inhibitory data of Suzuki et al. [52] for a series of isoaurostatin derivatives (**VIII**) against topo II, Eq. 17 was derived (Table 13):

$$\log 1/C = -0.25(\pm 0.15)\mathrm{C}\log P + 4.18(\pm 0.35) \quad (17)$$
$$n = 5, \quad r^2 = 0.907, \quad s = 0.065, \quad q^2 = 0.549, \quad Q = 14.646, \quad F_{1,3} = 29.258$$
outlier: $R_1 = R_2 = R_3 = OH$, $R_4 = R_5 = H$

The negative Clog *P* term shows that the highly hydrophilic molecules for this data set would present better inhibitory activities against topo II.

Table 13 Biological and physicochemical parameters used to derive QSAR Eq. 17

No.	R_1	R_2	R_3	R_4	R_5	log 1/*C* (Eq. 17) Obsd.	Pred.	Δ	Clog *P*
1	H	OH	H	H	H	3.49	3.43	0.06	3.00
2	OH	OH	H	H	H	3.57	3.58	-0.01	2.41
3 [a]	OH	OH	OH	H	H	4.20	3.74	0.46	1.74
4	OH	OH	H	OH	H	3.62	3.70	-0.08	1.92
5	OH	OH	OH	OH	H	3.92	3.86	0.06	1.25
6	OCH_3	OH	H	H	H	3.45	3.47	-0.02	2.85

[a] Not included in the derivation of QSAR Eq. 17

3.2.6 Quinolines

A series of 5*H*-indolo[2,3-*b*]quinoline derivatives (**XVII**) was synthesized by Kaczmarek et al. [70] as novel DNA topo II inhibitors. Using their data for topo II-induced DNA cleavage by this compound series (**XVII**), Eq. 18 was derived (Table 14):

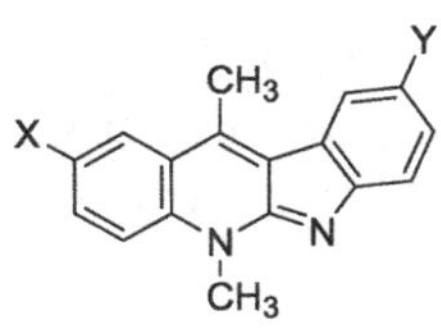

XVII

$$\log 1/C = 0.67(\pm 0.26)\mathrm{CMR} + 2.27(\pm 0.05) \quad (18)$$
$$n = 6, \quad r^2 = 0.928, \quad s = 0.096, \quad q^2 = 0.733, \quad Q = 10.031, \quad F_{1,4} = 51.556$$
outlier: $X = Y = CH_3$

Table 14 Biological and physicochemical parameters used to derive QSAR Eq. 18

No	X	Y	log 1/C (Eq. 18) Obsd.	Pred.	Δ	CMR
1	H	OCH_3	5.85	5.79	0.06	8.52
2	CH_3	OCH_3	6.00	6.11	-0.11	8.98
3	OCH_3	CH_3	6.10	6.11	-0.01	8.98
4	OCH_3	OCH_3	6.22	6.21	0.01	9.13
5	OCH_3	H	5.92	5.79	0.13	8.52
6	H	H	5.30	5.38	-0.08	7.90
7 [a]	CH_3	CH_3	6.40	6.00	0.40	8.83

[a] Not included in the derivation of QSAR Eq. 18

The positive sign of the coefficient associated with the CMR term indicates that an increase in overall molar refractivity should result in stronger topo II-induced DNA cleavage. Lipophilicity and electronic factors are not found to play a definite role.

3.2.7 Quinolones

Eissenstat et al. [71] synthesized a series of quinolone derivatives (**XVIII**) and evaluated their inhibitory activities against topo II. From their data Eq. 19 was derived (Table 15):

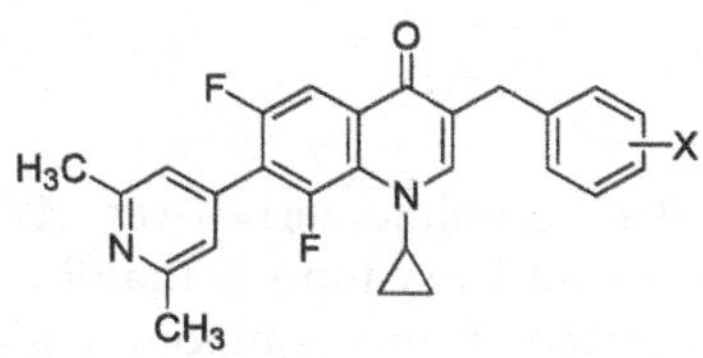

XVIII

$$\log 1/C = 0.84(\pm 0.53)B1_{2,6} + 1.27(\pm 0.28)I_{OH} + 3.30(\pm 1.20) \quad (19)$$

$$n = 19, \quad r^2 = 0.887, \quad s = 0.274, \quad q^2 = 0.834, \quad Q = 3.438,$$

$$F_{2,16} = 62.796 \quad \text{outlier:} \quad X = 2-\text{COOH}$$

$$C\log P \text{ vs. } B1_{2,6}; \quad r = 0.589$$

$$\text{CMR vs. } B1_{2,6}; \quad r = 0.347$$

$B1_{2,6}$ represents the sum of the minimum width of the *ortho* X-substituents in the benzyl ring. I_{OH} is an indicator variable taking the value of 1 and 0 for the presence and absence of an OH group in X-substituents. Its positive coefficient suggests that the presence of an OH group in X-substituents would

Table 15 Biological, physicochemical, and structural parameters used to derive QSAR Eq. 19

No.	X	log 1/*C* (Eq. 19) Obsd.	Pred.	Δ	$B1_{2,6}$	I_{OH}
1	H	5.41	4.99	0.42	2.00	0
2	4-Cl	5.11	4.99	0.12	2.00	0
3	4-OCH_3	4.85	4.99	−0.14	2.00	0
4	4-OH	5.92	6.26	−0.34	2.00	1
5	4-NH_2	5.14	4.99	0.15	2.00	0
6	3-OCH_3	4.85	4.99	−0.14	2.00	0
7	3-OH	5.96	6.26	−0.30	2.00	1
8	2-OCH_3	4.85	5.29	−0.44	2.35	0
9	2-OH	6.48	6.55	−0.07	2.35	1
10	2-CH_2OH	5.17	5.43	−0.26	2.52	0
11 [a]	2-COOH	3.70	5.50	−1.80	2.60	0
12	2-NH_2	5.68	5.29	0.39	2.35	0
13	2-$NHCOCH_3$	5.29	5.29	0.00	2.35	0
14	2-$NHCOCF_3$	5.33	5.66	−0.33	2.79	0
15	2-$NHSO_2CH_3$	5.48	5.29	0.19	2.35	0
16	2,3-$(OH)_2$	6.44	6.55	−0.11	2.35	1
17	2,4-$(OH)_2$	6.80	6.55	0.25	2.35	1
18	2,5-$(OH)_2$	6.80	6.55	0.25	2.35	1
19	2,6-$(OH)_2$	7.02	6.85	0.17	2.70	1
20	2,4,6-$(OH)_3$	7.01	6.85	0.16	2.70	1

[a] Not included in the derivation of QSAR Eq. 19

be favorable. The indicator variable (I_{OH}) is of critical importance in describing the inhibition of DNA topo II, which alone accounts for 80.5% of the variance in the data. It seems that as the minimum width of the substituent at *ortho* positions of the benzyl ring increases, followed by the presence of at least one OH group at the X position, inhibition of DNA topo II also increases. For example, compounds **19** and **20** (Table 15) are the most active analogues ($\log 1/C \approx 7.02$) with $B1_{2,6} = 2.70$, but compound **14** is less active with the highest value of $B1_{2,6}$ in the set (2.79). This is probably due to the presence of an OH group in compounds **19** and **20** and its absence in compound **14**.

A series of 1-cyclopropyl-6,8-difluoro-1,4-dihydro-7-(2,6-dimethyl-4-pyridinyl)-4-substituted-quinolones (**XIX**) was synthesized by Kuo et al. [72] and their inhibitory activities against topo II evaluated. Using their data, we developed Eq. 20 (Table 16):

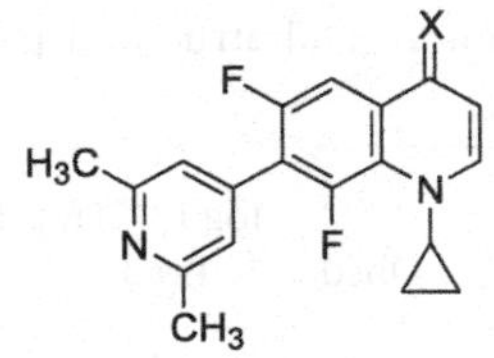

XIX

Table 16 Biological and physicochemical parameters used to derive QSAR Eq. 20

No	X	log 1/*C* (Eq. 20) Obsd.	Pred.	Δ	MR_X
1	O	4.77	4.91	−0.14	−0.06
2 [a]	S	4.12	5.38	−1.26	0.79
3	NNH_2	5.77	5.37	0.40	0.76
4	$NNHCH_3$	5.72	5.51	0.21	1.23
5	$NN(CH_3)_2$	5.64	5.58	0.06	1.69
6	$NNH(CH_2)_2OH$	5.49	5.58	−0.09	1.84
7 [a]	NNH-2-pyridinyl	6.01	5.31	0.70	3.06
8	NNH-3-quinolinyl	4.04	4.01	0.03	4.75
9 [a]	NNH-9-acridinyl	4.68	1.66	3.02	6.44
10 [b]	NNPhth	4.80	4.64	0.16	4.10
11	$NNHCONH_2$	5.51	5.58	−0.07	1.63
12	N-cy-C_3H_5	5.62	5.56	0.06	1.50
13	$N(CH_2)_3N(CH_3)_2$	5.26	5.36	−0.10	2.93
14	N-2-pyridinyl	5.44	5.53	−0.09	2.39
15	N-2-OH-Ph	5.80	5.43	0.37	2.76
16	N-4-NH_2-Ph	5.12	5.35	−0.23	2.97
17	NH	4.92	5.02	−0.10	0.09
18 [a]	NOH	5.85	5.27	0.58	0.55
19	$NOCH_3$	5.59	5.46	0.13	1.01
20	NS-4-Cl-Ph	5.03	4.80	0.23	3.90
21	NS-2-pyridinyl	5.42	5.24	0.18	3.20
22	NSO_2-4-CH_3-Ph	4.35	4.77	−0.42	3.94
23	$C(CN)_2$	5.20	5.56	−0.36	1.50
24	$C(CN)CO_2CH_2CH_3$	5.32	5.48	−0.16	2.60
25	CHCN	5.40	5.46	−0.06	1.02

[a] Not included in the derivation of QSAR Eq. 20

[b] NNPhth = 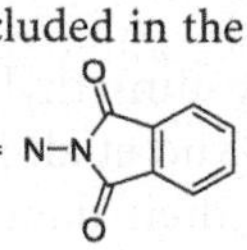

$$\log 1/C = 0.68(\pm 0.28)\mathrm{MR_X} - 0.19(\pm 0.06)\mathrm{MR_X^2} + 4.96(\pm 0.29) \qquad (20)$$

$n = 21,\quad r^2 = 0.787,\quad s = 0.226,\quad q^2 = 0.722,\quad Q = 3.925,$

$F_{2,18} = 33.254$ optimum $\mathrm{MR_X} = 1.84(1.52 -- 2.07)$

outliers: X = S; NNH – 2 – pyridinyl; NNH – 9 – acridinyl; NOH

π_X vs. $\mathrm{MR_X}$; $r = 0.639$

This is a parabolic relation in terms of $\mathrm{MR_X}$ (calculated molar refractivity of X-substituents), which suggests that the inhibitory activities of quinolones (**XIX**) against topo II first increases with an increase in the molar refractivity of X-substituents up to an optimum $\mathrm{MR_X}$ of 1.84 and then decreases.

4 Validation of the QSAR Models

QSAR model validation is an essential task in developing a statistically valid and predictive model, because the real utility of a QSAR model is in its ability to predict accurately the modeled property for new compounds. The following approaches have been used for the validation of QSAR Eqs. 1–20:

- **Fraction of the variance:** The fraction of the variance of an MRA model is expressed by r^2. It is believed that the closer the value of r^2 to unity, the better the QSAR model. The values of r^2 for these QSAR models are from 0.787 to 0.993, which suggests that these QSAR models explain 78.7–99.3% of the variance of the data. According to the literature, the predictive QSAR model must have $r^2 > 0.6$ [73, 74].
- **Cross-validation test:** The values of q^2 for these QSAR models are from 0.549 to 0.972. The high values of q^2 validate the QSAR models. From the literature, it must be greater than 0.50 [73, 74].
- **Standard deviation (*s*):** *s* is the standard deviation about the regression line. The smaller the value of *s* the better the QSAR model. The values of *s* for these QSAR models are from 0.065 to 0.406.
- **Quality factor or quality ratio (*Q*):** The high values of *Q* (2.259–14.646) for these QSAR models suggest that the high predictive power for these models as well as no over-fitting.
- **Fischer statistics (*F*):** Fischer statistics (*F*) is the ratio between explained and unexplained variance for a given number of degree of freedom. The larger the *F* value the greater the probability that the QSAR equation is significant. The *F* values obtained for these QSAR models are from 17.622 to 283.714, which are statistically significant at the 95% level.
- All the QSAR models (except Eqs. 7 and 9) also fulfill the thumb rule condition that (number of data points)/(number of descriptors) ≥ 4.

5
Overview

An analysis of QSAR results on the inhibition of six different heterocyclic compound series (benzimidazoles, camptothecins, isoaurostatins, naphthyridinones, phenanthridines, and terpenes) as well as some miscellaneous heterocyclic compounds against topo I reveals a number of interesting points. The most important parameter for these correlations is hydrophobicity, which is one of the most important determinants for the activity. Out of 12 QSAR, nine contain a correlation between inhibitory activity and hydrophobicity. A positive linear correlation is found in five equations (Eqs. 2–4, 8, and 9). The coefficient with the hydrophobic parameter varies considerably, from a low value of 0.31 (Eq. 9) to a high value of 1.77 (Eq. 4). These data suggest that the inhibitory activity against topo I might be improved by increasing compound/substituent hydrophobicity. A negative linear correlation is found in two equations (Eqs. 5 and 6), and the coefficient ranges from – 1.45 (Eq. 6) to – 0.30 (Eq. 5). Less hydrophobic congeners in these compound families might display enhanced activity. Parabolic correlations with hydrophobic parameter are found in Eqs. 1 and 7. One of these (Eq. 7) reflects the situation where activity declines with increasing hydrophobicity of the substituents and then changes direction and increases. This may correspond to an allosteric reaction. The other (Eq. 1) situation shows that activity is optimal for a particular value, or range of values, of log *P*. The optimal log *P* for this equation is 4.55 (Eq. 1, benzimidazoles; **II**). The second important parameter is molar volume, which is present in four equations (Eqs. 9–12) with negative coefficient. In three cases, this parameter correlates all of the observed variation in activity, but they do not seem to play as important a role as hydrophobicity for the data sets that we have examined.

On considering topo II, only two QSAR (Eqs. 13 and 17) out of eight have a hydrophobic term. Equation 13 is an inverted bilinear correlation with hydrophobic parameters, which reflects the situation where activity declines linearly with increasing hydrophobicity of the compounds and then changes direction and increases gradually. This may correspond to an allosteric reaction. Equation 17 is a linear correlation with a hydrophobic parameter with a negative coefficient (– 0.25), which suggests that less hydrophobic congeners in this compound family might display enhanced activity. The most important parameter for these correlations is molar refractivity, which is one of the most important determinants for the inhibitory activity of heterocyclic compound series (under consideration) against topo II. Out of eight QSAR, four contain a correlation between inhibitory activity and molar refractivity. A positive linear correlation is found in one equation (Eq. 18). The coefficient with molar refractivity is 0.67, which suggests that the inhibitory activity against topo II might be improved by increasing compound molar refractivity/polarizability. A negative linear correlation is found in two equations (Eqs. 15 and 16), and the coefficient ranges from – 0.30 (Eq. 15) to – 0.24 (Eq. 16). The congeners of these compound families having less molar refractivity/polarizability might display enhanced activity. Parabolic correlation with molar refractivity of the substituents is found in one (Eq. 20), which suggests that activity is optimal for a particular value, or range of values, of MR_X. The optimal MR_X for this equation is 1.84 (Eq. 20, quinolones; **XIX**). The other important steric parameters are molar volume and Verloop's sterimol parameters, which are present in Eqs. 14 and 19, respectively, with positive coefficients, indicating the importance of the steric factor.

6 Conclusion

In this chapter, an attempt has been made to present a total number of 20 QSAR models (12 QSAR models for topo I inhibitors and eight QSAR models for topo II inhibitors) on 11 different heterocyclic compound series (anthrapyrazoles, benzimidazoles, benzonaphthofurandiones, camptothecins, desoxypodophyllotoxins, isoaurostatins, naphthyridinones, phenanthridines, quinolines, quinolones, and terpenes) as well as on some miscellaneous heterocyclic compounds for their inhibition against topo I and II. They have been found to be well-correlated with a number of physicochemical and structural parameters. The conclusion, from the analysis of these 20 QSAR, has been drawn that the inhibition of topo I is largely dependent on the hydrophobicity of the compounds/substituents. On the other hand, steric parameters (molar refractivity, molar volume, and Verloop's sterimol parameters) are important for topo II inhibition.

References

1. Champoux JJ (2001) Annu Rev Biochem 70:369
2. Wang JC (1996) Annu Rev Biochem 65:635
3. Nitiss JL (1998) Biochim Biophys Acta 1400:63
4. Corbett AH, Osheroff N (1993) Chem Res Toxicol 6:585
5. Pommier Y (2006) Nature Rev Cancer 6:789
6. Staker BL, Feese MD, Cushman M, Pommier Y, Zembower D, Stewart L, Burgin AB (2005) J Med Chem 48:2336
7. Meng LH, Liao ZY, Pommier Y (2003) Curr Top Med Chem 3:305
8. Koster DA, Croquette V, Dekker T, Shuman S, Dekker NH (2005) Nature 434:671
9. Berger JM, Fass D, Wang JC, Harrison SC (1998) Proc Natl Acad Sci USA 95:7876
10. Kamata J, Okada T, Kotake Y, Niijima J, Nakamura K, Uenaka T, Yamaguchi A, Tsukahara K, Nagasu T, Koyanagi N, Kitoh K, Yoshimatsu K, Yoshino H, Sugumi H (2004) Chem Pharm Bull 52:1071
11. Froelich-Ammon SJ, Osheroff N (1995) J Biol Chem 270:21429
12. van Gent DC, Hoeijmakers JH, Kanaar R (2001) Nat Rev Genet 2:196

13. Haber JE (2000) Trends Genet 16:259
14. Coiteux V, Onclercq-Delic R, Fenaux P, Amor-Guéret M (2007) Leukemia Res 31:353
15. Burden DA, Osheroff N (1998) Biochim Biophys Acta 1400:139
16. Sordet O, Khan QA, Konh KW, Pommier Y (2003) Curr Med Chem Anti-Cancer Agents 3:271
17. Solier S, Lansiaux A, Logette E, Wu J, Soret J, Tazi J, Bailly C, Desoche L, Solary E, Corcos L (2004) Mol Cancer Res 2:53
18. Isemura M, Saeki K, Kimura T, Hayakawa S, Minami T, Sazuka M (2000) Biofactors 13:81
19. Mukhtar H, Ahmad N (1999) Toxicol Sci 52:111
20. Ahmad N, Gupta S, Mukhtar H (2000) Arch Biochim Biophys 376:338
21. Ahmad N, Feyes DK, Nieminen AL, Agarwal R, Mukhtar H (1997) J Natl Cancer Inst 89:1881
22. Berger SJ, Gupta S, Belfi CA, Goski DM, Mukhtar H (2001) Biochem Biophys Res Commun 288:101
23. Selassie CD, Mekapati SB, Verma RP (2002) Curr Top Med Chem 2:1357
24. BioByte Corp (2006) C-QSAR program. Claremont, CA, USA, http://www.biobyte.com, last visited: 13 April 2007
25. Hansch C, Hoekman D, Leo A, Weininger D, Selassie CD (2002) Chem Rev 102:783
26. Hansch C, Leo A (1995) Exploring QSAR: fundamentals and applications in chemistry and biology. Am Chem Soc, Washington, DC
27. Verloop A (1987) The sterimol approach to drug design. Marcel Dekker, New York
28. Hansch C, Steinmetz WE, Leo AJ, Mekapati SB, Kurup A, Hoekman D (2003) J Chem Inf Comput Sci 43:120
29. Hansch C, Kurup A (2003) J Chem Inf Comput Sci 43:1647
30. Verma RP, Kurup A, Hansch C (2005) Bioorg Med Chem 13:237
31. Verma RP, Hansch C (2005) Bioorg Med Chem 13:2355
32. Abraham MH, McGowan JC (1987) Chromatographia 23:243
33. Cramer RD III, Bunce JD, Patterson DE, Frank IE (1988) Quant Struct-Act Relat 7:18
34. Pogliani L (1996) J Phys Chem 100:18065
35. Pogliani L (2000) Chem Rev 100:3827
36. Agrawal V, Singh J, Khadikar PV, Supuran CT (2006) Bioorg Med Chem Lett 16:2044
37. Selassie CD, Kapur S, Verma RP, Rosario M (2005) J Med Chem 48:7234
38. Verma RP, Hansch C (2006) Mol Pharmaceutics 3:441
39. Verma RP, Hansch C (2006) Virology 359:152
40. Zhang X, Kiechle F (2001) Annal Clin Lab Sci 31:187
41. Sun Q, Gatto B, Yu C, Liu A, Liu LF, LaVoie EJ (1995) J Med Chem 38:3638
42. Mekapati SB, Hansch C (2001) Bioorg Med Chem 9:2885
43. Hsiang YH, Hertzberg R, Hecht S, Liu LF (1985) J Biol Chem 260:14873
44. Hsiang YH, Liu LF, Wall ME, Wani MC, Nicholas AW, Manikumar G, Kirschenbaum S, Silber R, Potmesil M (1989) Cancer Res 49:4385
45. Carrigan SW, Fox PC, Wall ME, Wani MC, Bowen JP (1997) J Comput Aided Mol Design 11:71
46. Wall ME, Wani MC, Nicholas AW, Manikumar G, Tele C, Moore L, Truesdale A, Leitner P, Besterman JM (1993) J Med Chem 36:2689
47. Verma RP (2005) Bioorg Med Chem 13:1059
48. Selassie CD, Garg R, Kapur S, Kurup A, Verma RP, Mekapati SB, Hansch C (2002) Chem Rev 102:2585
49. Kapur S, Shusterman A, Verma RP, Hansch C, Selassie CD (2000) Chemosphere 41:1643

50. Rahier NJ, Eisenhauer BM, Gao R, Jones SH, Hecht SM (2003) Org Lett 6:321
51. Vladu B, Woynarowski JM, Manikumar G, Wani MC, Wall ME, von Hoff DD, Wadkins RM (2000) Mol Pharmacol 57:243
52. Suzuki K, Okawara T, Higashijima T, Yokomizo K, Mizushima T, Otsuka M (2005) Bioorg Med Chem Lett 15:2065
53. Zhu S, Ruchelman AL, Zhou N, Liu A, Liu LF, LaVoie EJ (2006) Bioorg Med Chem 14:3131
54. Mekapati SB, Kurup A, Verma RP, Hansch C (2005) Bioorg Med Chem 13:3737
55. Zhu S, Ruchelman AL, Zhou N, Liu AA, Liu LF, LaVoie EJ (2005) Bioorg Med Chem 13:6782
56. Zhang S, Li X, Zhang F, Yang P, Gao X, Song Q (2006) Bioorg Med Chem 14:3888
57. Gogas H, Mansi JL (1996) Cancer Treat Rev 21:541
58. Leteurtre F, Kohlhagen G, Paull KD, Pommier Y (1994) J Natl Cancer Inst 86:1239
59. Capranico G, Palumbo M, Tinelli S, Mabilia M, Pozzan A, Zunino F (1994) J Mol Biol 235:1218
60. Liang H, Wu X, Guziec LJ, Guziec FS Jr, Larson KK, Lang J, Yalowich JC, Hasinoff BB (2006) J Chem Inf Model 46:1827
61. Verma RP (2005) Lett Drug Design Discov 2:205
62. Dykstra CC, McClernon DR, Elwell LP, Tidwell RR (1994) Antimicrob Agents Chemother 38:1890
63. Rhee H-K, Park HJ, Lee SK, Lee C-O, Choo H-YP (2007) Bioorg Med Chem 15:1651
64. Belani CP, Doyle LA, Aisner J (1994) Cancer Chemother Pharmacol 34(Suppl):S118
65. Meresse P, Dechaux E, Monneret C, Bertounesque E (2004) Curr Med Chem 11:2443
66. Fortune JM, Osheroff N (2000) Prog Nucleic Acid Res Mol Biol 64:221
67. Bromberg KD, Burgin AB, Osheroff (2003) J Biol Chem 278:7406
68. Kobayashi K, Ratain MJ (1994) Cancer Chemother Pharmacol 34(Suppl):S64
69. Duca M, Arimondo PB, Léonce S, Pierré A, Pfeiffer B, Monneret C, Dauzonne D (2005) Org Biomol Chem 3:1074
70. Kaczmarek L, Peczynska-Czoch W, Osiadacz J, Mordarski M, Sokalski WA, Boratynski J, Marcinkowska E, Glazman-Kusnierczyk H, Radzikowski C (1999) Bioorg Med Chem 7:2457
71. Eissenstat MA, Kuo G-H, Weaver JD III, Wentland MP, Robinson RG, Klingbeil KM, Danz DW, Corbett TH, Coughlin SA (1995) Bioorg Med Chem Lett 5:1021
72. Kuo G-H, Eissenstat MA, Wentland MP, Robinson RG, Klingbeil KM, Danz DW, Coughlin SA (1995) Bioorg Med Chem Lett 5:399
73. Golbraikh A, Tropsha A (2002) J Mol Graph Modl 20:269
74. Tropsha A, Gramatica P, Gombar VK (2003) QSAR Comb Sci 22:69

Top Heterocycl Chem (2007) 10: 75–97
DOI 10.1007/7081_2007_069

Published online: 20 June 2007

Molecular Modeling of the Biologically Active Alkaloids

Mahmud Tareq Hassan Khan

PhD School of Molecular and Structural Biology, and Department of Pharmacology,
Institute of Medical Biology, Faculty of Medicine,
University of Tromsø, 9037 Tromsø, Norway
mahmud.khan@fagmed.uit.no

Abstract A large number of compounds are nitrogen-containing or alkaloid in nature and have immense biological or pharmacological properties. Several of them are potentially toxic and lethal to mankind. Different kinds of molecular modeling techniques have been employed to clarify and predict several of these types of compounds. This chapter critically reviews a few of these aspects; in addition, discussions on the modeling approaches are briefly considered.

Keywords Alkaloids · Molecular modeling · Molecular dynamics simulation · Telomerase · Tyrosinase inhibitor · Sildenafil · Serotonin antagonist

Abbreviations

2D	Two-dimensional
3D	Three-dimensional
5-HT	5-Hydroxytryptamine or serotonin
CNS	Central nervous system
EC	Enzyme Commission
GABA	γ-Aminobutyric acid

HYBOT	Hydrogen Bond Thermodynamics
LCAP	Long-chain arylpiperazine
MD	Molecular dynamics
NMDA	*N*-Methyl-D-aspartate
PPO	Polyphenol oxidase
QSAR	Quantitative structure–activity relationship
rms	Root mean square
SAR	Structure–activity relationship
TRAP	Telomerase repeat amplification protocol

1 Introduction

Large numbers of drug molecules are derived from or analogues of the active molecules isolated from plants or microorganisms. Nevertheless, a lot of species have not yet been studied; they constitute an important source of novel and diverse structures that could possess interesting pharmacological activities and/or very unusual mechanisms of action, for example in the field of anticancer or antiparasitic drugs (e.g., taxol, artemisinin, camptothecin, etc.). A huge number of new alkaloids are being isolated every year from the crude extracts of natural products, such as higher plants, bacteria, and fungi. Some of them possess interesting antibiotic, cytotoxic, antitumor, antiparasitic, and toxic properties [1]. An immense number of these biologically and pharmacologically active compounds have been synthesized and reported by several authors in the last few decades. Some of the potential molecules have also been studied by employing different molecular modeling approaches to analyze their interactions with the biochemical targets.

In recent years it has been realized that molecular modeling studies of the alkaloidal molecules having different pharmacological activities are highly important in order to explain their mechanisms, at least partially in some cases. This chapter presents and critically reviews some examples of molecular modeling studies of alkaloids, based on their different biological properties or sometimes performed in parallel to explain their biochemical effects.

2 General Three-Dimensional Structural Studies of Pharmacologically Active Molecules

To establish the structure–activity relationships (SARs) of a set of molecules, a knowledge of the 3D structure is of great importance [2]. Thevand et al. recently (2004) reported the 3D structural analysis of tetrandrine using NMR and molecular modeling (the structure is shown in Fig. 1) [2]. They employed

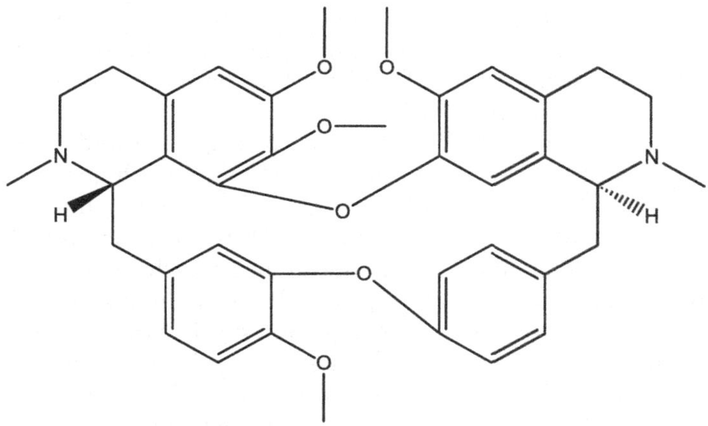

Fig. 1 Molecular structure of tetrandrine [2]

high-resolution 1- and 2D NMR spectra of tetrandrine and molecular modeling to characterize its structure in solution.

Tetrandrine, (6,6′,7,12-tetramethoxy-2,2′-dimethylberbaman), is a bisbenzylisoquinoline alkaloid isolated from the roots of *Stephania tetrandrae* S. Moore [3]. Tetrandrine is used in traditional Chinese medicine to cure angina and silicosis [4]. As a calcium antagonist of the diltiazem class it is effective against hypertension and other cardiovascular disorders [5–8]. For the molecular modeling of tetrandrine, the authors used the computer aided chemistry (CAChe™) program. The X-ray structure was the starting geometry for computation. The AM1 semiempirical method was used to determine the minimum conformation energy in water ($\varepsilon = 78.5$, solvent radius 1 Å) with the EF optimization algorithm [9]. The authors reported complete and unambiguous assignment of all proton and carbon resonance signals. Scalar couplings have been determined and reported, from dihedral angles with the Karplus equation. A comparison of simulated and X-ray conformations of tetrandrine reveals only small differences [2].

3
Telomerase Inhibitors

Telomeres are the nucleoprotein assemblies at the ends of chromosomes that comprise guanine-rich tandem-repeating DNA sequences (the size of a typical telomere is around 6–12 kb in length in humans) together with a number of structural and regulatory proteins. The major function of telomeres is to protect chromosome ends from base-pair loss and end-to-end fusions, thereby safeguarding the integrity of each chromosome [10, 11]. The enzyme telomerase has an important role in tumorigenesis [12] and its inhibition by transfection of dominant (-)hTERT (the catalytic domain of human telomerase) or with antisense oligonucleotides, which has been shown to result in the selective death of cancer cells [13–16].

One approach to telomerase inhibition envisages the folding of the extreme 3′ single-stranded end of telomeric DNA into a higher-order four-stranded quadruplex structure [17–21]. Such a structure cannot hybridize with the complementary single-stranded RNA template in the telomerase complex, the essential first step in the catalytic cycle of telomeric DNA length extension. A number of G-quadruplex ligands have now been described, including substituted anthraquinones [22–24], porphyrins [25], quinoacridines [26–29], phenanthrolines [30], substituted triazines [31], cyclic oligo-oxazoles [32, 33], and acridines [34, 35]. These molecules exhibited their potencies against telomerase as measured by the telomerase repeat amplification protocol (TRAP) assays between low nanomolar and tens of micromolar amounts [11].

Recently, Guyen and coworkers (2004) reported the synthesis and evaluation of telomerase inhibitory and quadruplex DNA binding properties of several rationally designed quindoline analogues, substituted at the 2- and 7-positions. The ability of these compounds to interact with and stabilize an intramolecular G-quadruplex DNA against increases in temperature was evaluated by a fluorescence-based (FRET) melting assay. Finally, the interactions of a number of compounds with a quadruplex DNA molecular structure were simulated by molecular modeling methods [11]. The molecular structures of the compounds (**1** and **2**) used in this study are shown in Fig. 2. The authors also reported the synthesis of compounds **1** and **2** starting from 2-amino-5-bromobenzoic acid. The synthetic pathway is presented in Scheme 1.

In the molecular modeling step, they used two human G-quadruplexes from the crystal structure [36] of d[AGGG(TTAGGG)$_3$] to build a 45-mer molecule, joining the two quadruplexes with a TTA loop. Then they investigated the possible structures of the complexes using the AFFINITY™ docking program of INSIGHT II™ (www.accelrys.com). Both the ligand and the two G-quartets forming the binding site (including the phosphate backbones) allowed movement by employing a flexible docking method. Then molecular dynamics (MD) simulations were carried out using the SANDER module of the AMBER 7 program package (http://amber.scripps.edu/) [37]. The AMBER 1999 force field [38] was used, along with the GAFF force field, to provide any necessary force-field parameters for compound **1** [11]. The binding of **1** and **2** to G-quadruplexes was investigated using MD simulations. For the purpose of the simulations, a structure containing a quadruplex-binding site was created by stacking two 22-mer human G-quadruplexes in a 3′ to 5′ orientation, and joining them with a TTA loop. Previous studies have shown that ligands are most likely to stack on the exterior surfaces of quadruplexes, rather than binding by intercalation between internal G-quartets [11,

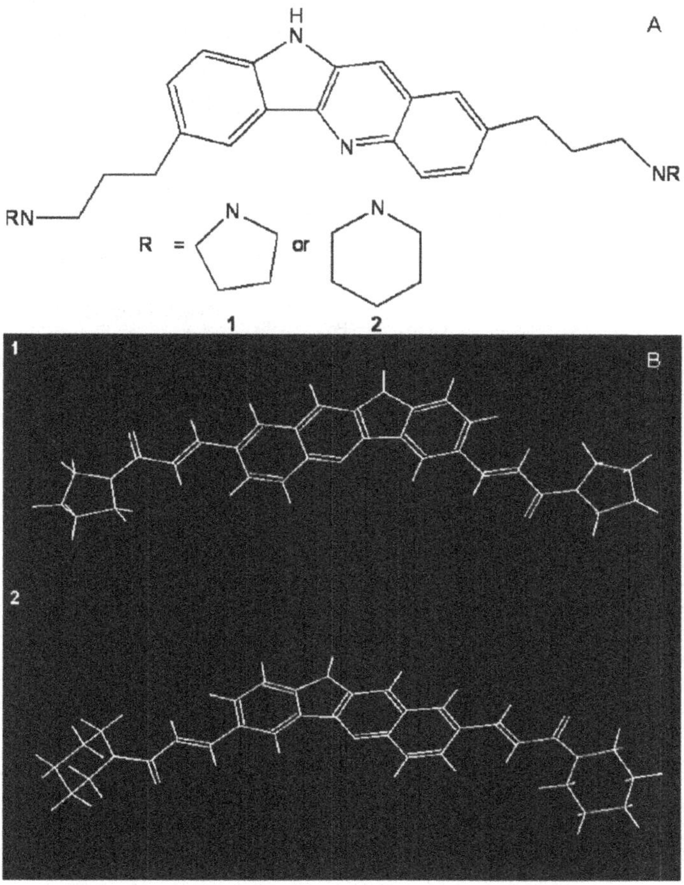

Fig. 2 Molecular structures of compounds 1 and 2. **A** 2D structures of 1 and 2, and **B** their energy-minimized 3D structures, which used the molecular modeling by Guyen et al. (2004) [11]

27, 39, 40]. Several possible ligand–quadruplex complexes have been obtained using the DOCKING module of the INSIGHT II modeling package [11]. The MD simulations were carried out on the lowest-energy **1**–quadruplex complex. The loop residues had very high root mean square (rms) deviations, 3.21 Å on average, compared to 1.89 Å for the atoms in the guanine residues [11].

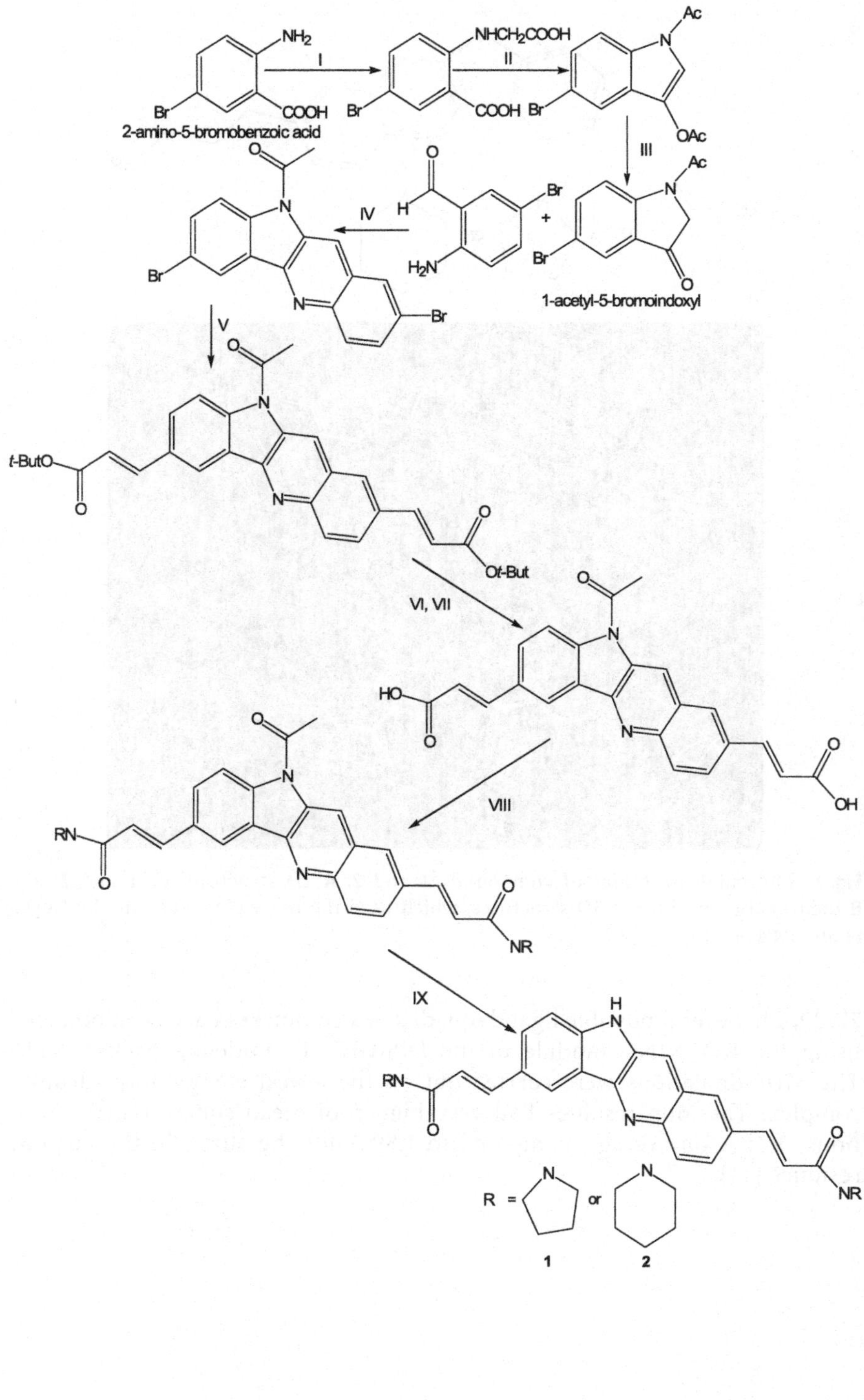
NH2
I
NHCH2COOH
II
Ac
N
Br
COOH
Br
COOH
Br
OAc
2-amino-5-bromobenzoic acid
III
1-acetyl-5-bromoindoxyl
IV
V
t-ButO
Ot-But
VI, VII
HO
OH
VIII
RN
NR
IX
R =
or
1
2

Scheme 1 Synthetic steps for compounds 1 and 2. I: 2-chloroacetic acid, 2 M aq. Na_2CO_3, 80 °C, 20 h, yield 51%; II: acetic anhydride, anhydrous sodium acetate, 60 °C, 5 h, yield 65%; III: sodium sulfite, H_2O, 90 °C, 4 h, yield 82%; IV: toluene/piperidines, reflux 4 h, yield 61%; V: *tert*-butyl acrylate, tri-(*O*-tolyl)phosphine, $Pd(OAc)_2$, Et_3N, DMF, reflux, yield 70–91%; VI: H_2, 10% Pd/C, rt, yield 96–100%; VII: TFA, DCM, rt, yield 75–100%; VIII: (a) isobutylchloroformate, Et_3N, dry DCM, 0 °C, (b) R_2NH, DCM, rt, 62–70%; IX: $LiAlH_4$, dry THF, yield 24–44% [11]

4 Tyrosinase Inhibitors

The enzyme tyrosinase (EC 1.14.18.1), also known as polyphenol oxidase (PPO), is a multifunctional copper-containing enzyme widely distributed in plants and animals. It catalyzes the *ortho* hydroxylation of monophenols and also the oxidation of *o*-diphenols to *o*-quinones. Tyrosinase is known to be a key enzyme for melanin biosynthesis in plants and animals. Therefore, tyrosinase inhibitors should be clinically useful for the treatment of some dermatological disorders associated with melanin hyperpigmentation, and are also important in cosmetics for whitening and depigmentation after sunburn. In addition, tyrosinase is known to be involved in the molting process of insects and adhesion of marine organisms [41]. Also, tyrosinase inhibitors are becoming important constituents of cosmetic products that relate to hyperpigmentation. Therefore, there is a concerted effort to search for naturally occurring tyrosinase inhibitors from plants, since plants constitute a rich source of bioactive chemicals and many of them are largely free from harmful adverse effects [42].

A large number and diverse classes of compounds have been reported to possess tyrosinase inhibitory activities to various extents from the millimolar to nanomolar concentration ranges. From our laboratory we have also reported a large variety of tyrosinase inhibitors and for some of them we have performed molecular modeling as well. Among them there are several alkaloids, and nitrogen-containing molecules are also present.

4.1 Sildenafil Analogues

In 2005 we reported the synthesis of some sildenafil (**3**) analogues and their tyrosinase inhibitory potential [43]. The compounds were synthesized using microwave irradiation in key steps, such as the SNAr reaction on important precursor bromopyrazole [44]. Their molecular structures are shown in Fig. 3.

This study reflected the fact that the inhibition was enhanced with increase of the carbon chain. In the case of compound **7**, the – OH group was replaced with $-CH_2-CH_2-OH$ with a resulting increase in inhibition against tyrosinase. Compound **7** was found to be more potent than the potent refer-

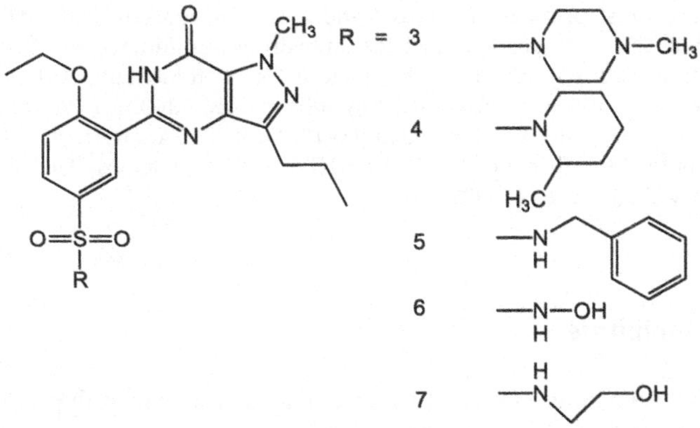

Fig. 3 Molecular structures of the synthesized sildenafil (**3**) and some of its analogues (**4–7**) which exhibited promising tyrosinase inhibitory activities [43]

ence inhibitors L-mimosine and kojic acid [43]. The inhibitory potencies of **3–7** are shown in Table 1 and compared with those of the standard inhibitors L-mimosine and kojic acid [43].

These compounds were further employed for molecular modeling studies. The 2D and 3D hydrogen-bonding descriptors that help to study quantitative structure–property relationships (QSPRs) were also calculated. The energetically most stable conformations of these compounds were analyzed. Their kinetic potential and total energies were also calculated through MD simulation [43]. Their energy-minimized 3D structures are shown in Fig. 4. Here, the energy minimization experiments were carried out using HyperChem® employing the block-diagonal Newton–Raphson algorithm at the rms gradients of 0.1 kcal/(Å mol) for different cycles in vacuo. For rendering and refinement of the 3D structures, the Persistence of Vision™ Ray-Tracer (Pov-Ray®) MS windows XP® version program was used [43].

Table 1 The tyrosinase inhibitory activities of compounds **3–7**, as compared to the standard inhibitors (kojic acid and L-mimosine) [43]

Compound	IC_{50} (in μM)
3	Not active
4	19.95
5	8.69
6	54.43
7	3.54
Kojic acid	16.67
L-Mimosine	3.68

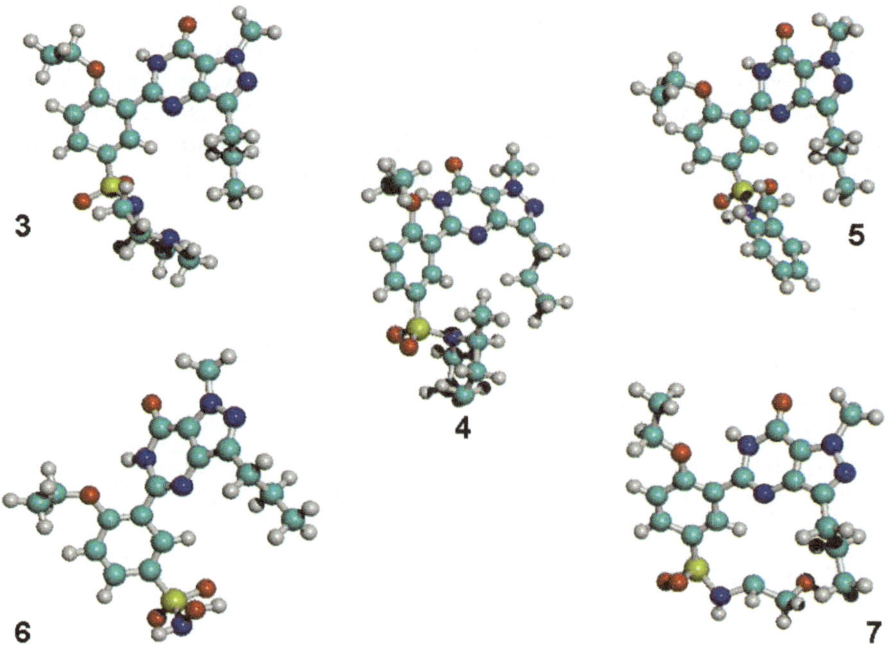

Fig. 4 Energy-minimized 3D structures of compounds **3–7**, which showed potential inhibitory activities against the enzyme tyrosinase [43]

The MD simulation (MM+, molecular mechanics force field, MMFF) experiments were performed using HyperChem® Professional version 7.1 (HyperCube Inc., Gainesville, FL, USA) on MS Windows XP®. Electrostatic bond dipoles were taken for the option of MM+ simulations. In these experiments, the total run time was fixed at 1.0 ps; heating and cooling times were 0.5 ps and the step size was 0.001 ps. The simulations were run in vacuo at a temperature of 300 K, with a starting and final temperature of 0 K and a temperature step of 20 K. For the force field analysis bond angle, torsion, nonbonded, electrostatic, and hydrogen bonded were taken as the components for the MM+ simulations [43, 45, 46]. Hydrogen bonding plays a critical role in the functional biology of living organisms. The hydrogen-bonding properties of the same compounds (**3–7**) have been calculated using the software 2D and 3D Hydrogen Bond Thermodynamics (HYBOT), where different descriptor parameters were calculated [47, 48].

4.2 Diterpenoid and Napelline Type Alkaloids

Sultankhodzhaev et al. (2005) reported tyrosinase inhibition studies on 15 diterpenoid alkaloids with the lycoctonine skeleton, and their semisynthetic derivatives and six napelline-type compounds were discussed [49]. The biological data for the tyrosinase active compounds (here IC_{50} values, in μM) are presented in Table 2. The SARs (structures are shown in Fig. 5) for tyrosinase inhibition are also discussed. These activities were compared with two referenced tyrosinase inhibitors, kojic acid and L-mimosine. The study showed that lappaconitine HBr (**8**) is the most potent member of the series (IC_{50} = 13.30 μM) [49].

Table 2 Tyrosinase inhibitory activities of the diterpenoid and napelline type alkaloids and their derivatives [49]

Compound	IC_{50} (in μM)
Lappaconitine hydrobromide (**8**)	13.30
Methyllycaconitine perchlorate (**9**)	477.84
Aconine (**10**)	220.70
Napelline (**11**)	167.66
1-*O*-Benzoylnapelline (**12**)	33.10

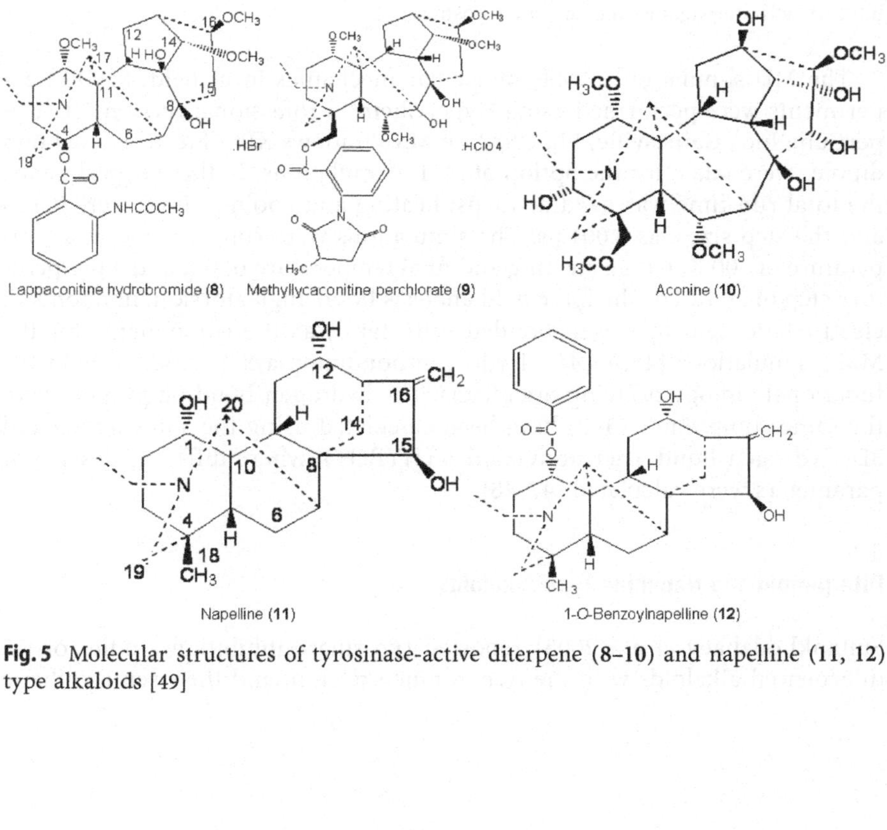

Fig. 5 Molecular structures of tyrosinase-active diterpene (**8–10**) and napelline (**11, 12**) type alkaloids [49]

More recently (2006) we performed and reported quantitative structure–activity relationship (QSAR) modeling of the same compounds based on their atomic linear indices, for finding functions that discriminate between the tyrosinase inhibitor compounds and inactive ones [50]. Discriminant models have been applied and globally good classifications of 93.51 and 92.46% were observed for nonstochastic and stochastic linear indices best models, respectively, in the training set. The external prediction sets had accuracies of 91.67 and 89.44% [50]. In addition to this, these fitted models have also been employed in the screening of new cycloartane compounds isolated from herbal plants. Good behavior was observed between the theoretical and experimental results. These results provide a tool that can be used in the identification of new tyrosinase inhibitor compounds [50].

5 Molecules Against Epileptic Seizures

Epilepsy is a clinical disorder characterized by spontaneous, recurrent seizures arising from excessive electrical activity in certain parts of the brain [51]. Currently available drugs, such as phenytoin, carbamazepine, valproic acid, lamotrigine, and topiramate (for molecular structures see Fig. 6), provide symptomatic seizure suppression in only 60–70% of those receiving treatment [52–54]. These drugs are also associated with unwanted side

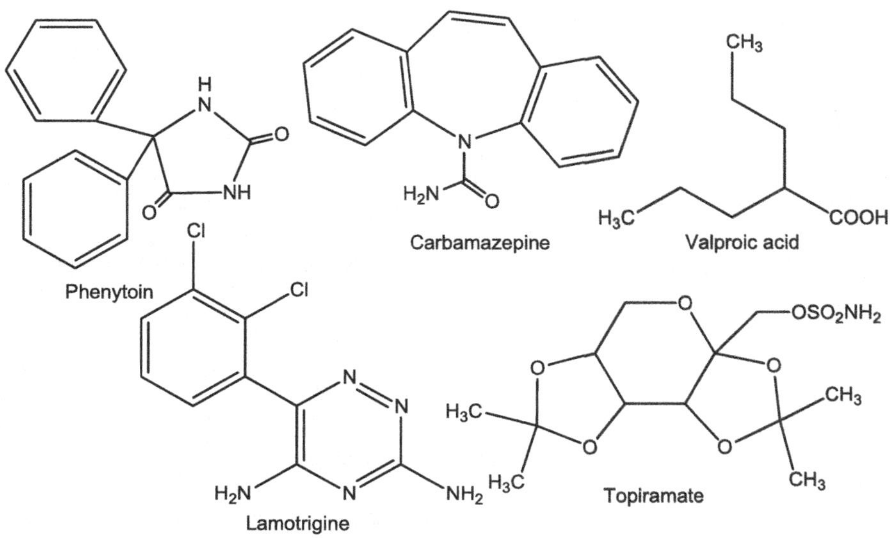

Fig. 6 Molecular structures of currently available antiepileptic drugs. These molecules possess several severe life-threatening adverse effects [51–54]

effects, ranging from cosmetic (gingival hyperplasia) to life threatening (hepatotoxicity, megaloblastic anemia) adverse effects [55–58].

It has been proved by various research projects that *β*-alanine synthesized naturally in the central nervous system (CNS) is released by electrical stimulation, can inhibit neuronal excitability, and has binding sites [59–62]. Conventionally, for antiepileptic drug design the main target is the neuronal voltage gated Na^+ channel protein, resulting in the majority of currently available antiepileptic molecules [51]. Extensive background design research led Tan et al. (2003) to select *β*-alanine as the prototype "lead" for designing new molecules against epilepsy [51]. They also proposed a model three-point pharmacophore, based on the modeling data obtained from the study of antagonists for both the glial *γ*-aminobutyric acid (GABA)-uptake site and the glycine co-agonist site of *N*-methyl-D-aspartate (NMDA) receptor. Three series of 3-aminopropionic acids, containing N, *α*, and *β* substituents, were designed and synthesized to probe the position and size of a lipophilic binding pocket within the proposed pharmacophore. These analogues were also tested in vivo for both their antiepileptic activities and their neurologic toxicities. Among the 14 novel 3-aminopropionic acids presented, eight of them were found to be promising antiepileptic molecules [51].

The three-point pharmacophore models are shown in Fig. 7. The structures of the *β*-alanine analogues and their antiepileptic activities against pilocarpine test are tabulated in Table 3.

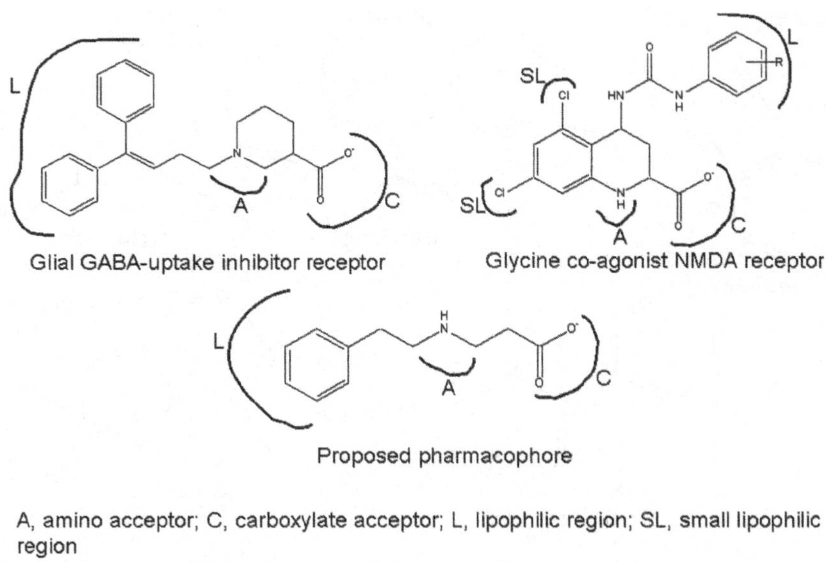

Fig. 7 The three-point pharmacophore models proposed by Tan et al. [51, 87, 88]

Table 3 Molecular structures of β-alanine analogues and their antiepileptic activities against pilocarpine test [51]

Structures of the compounds	Response on pilocarpine test
N-substituted	
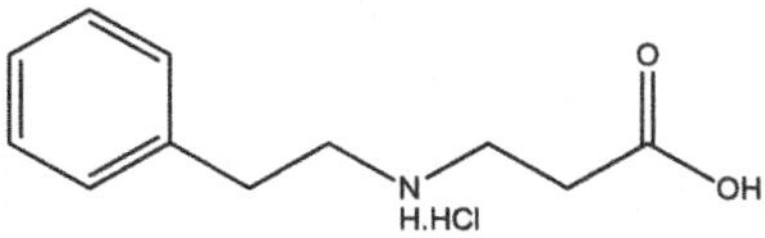	Active
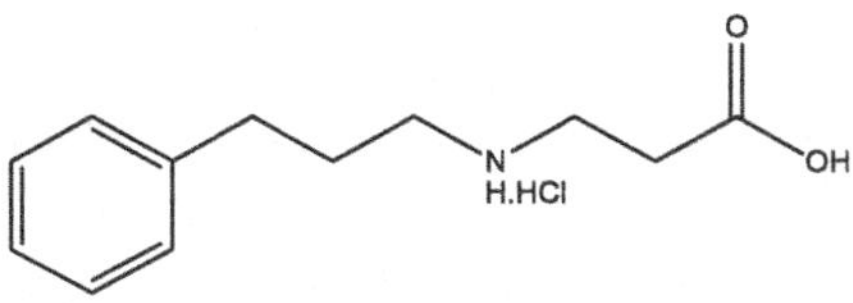	Active
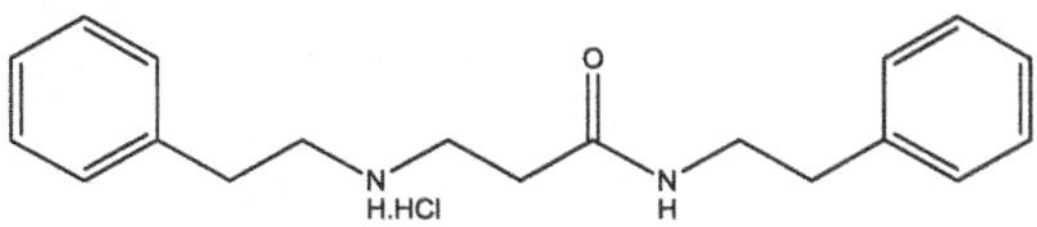	Active
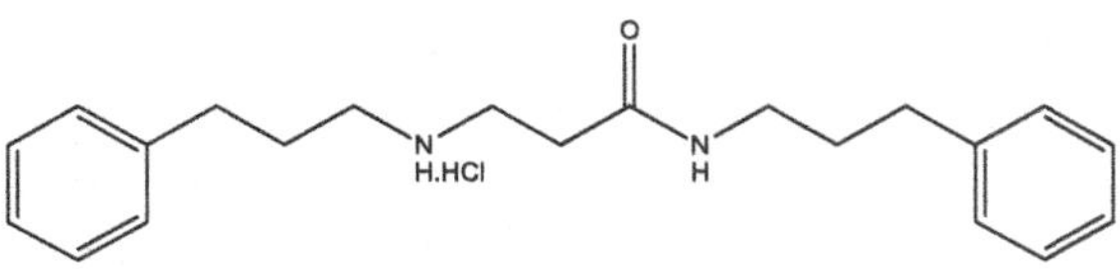	–
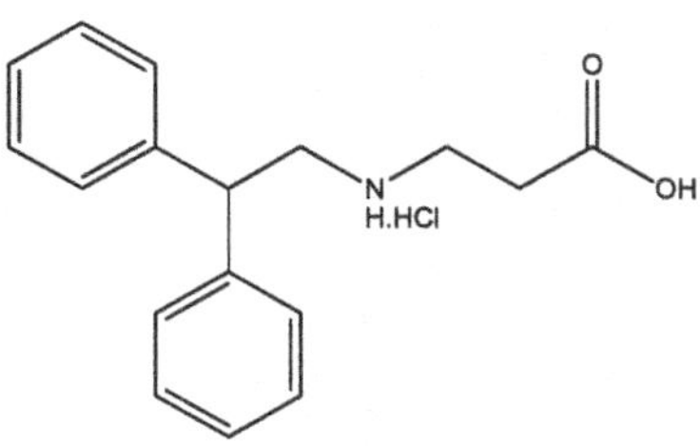	–
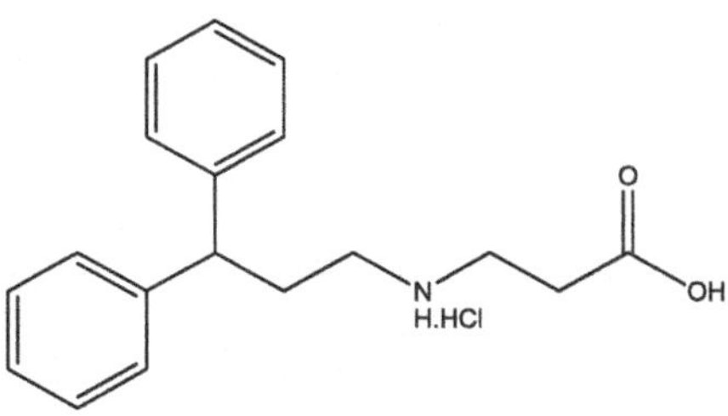	Active

Table 3 (continued)

Structures of the compounds	Response on pilocarpine test
β-substituted	
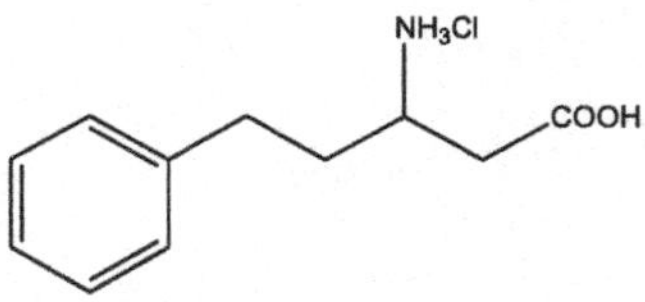	Active
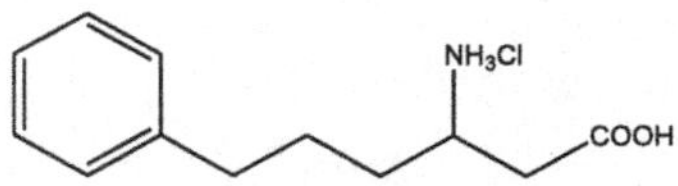	–
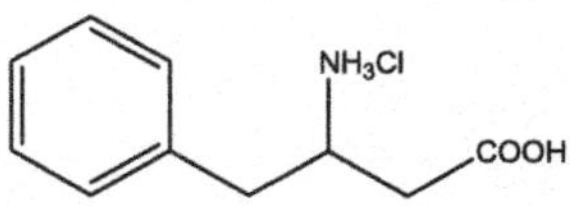	Active
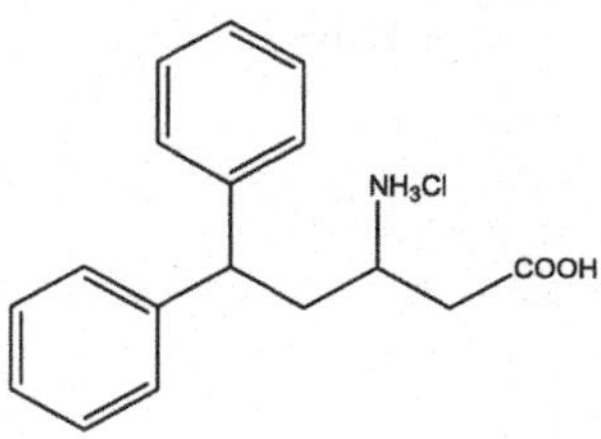	–
α-substituted	
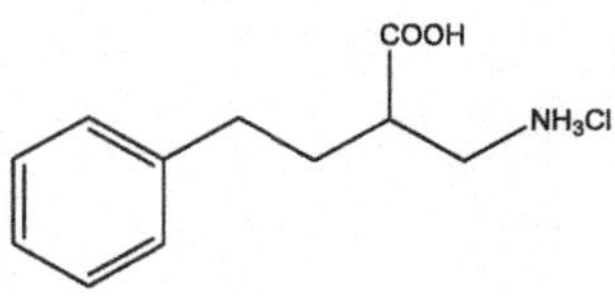	–
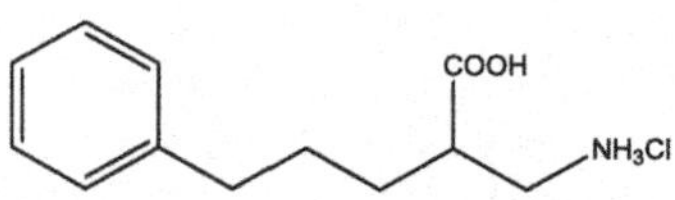	Active
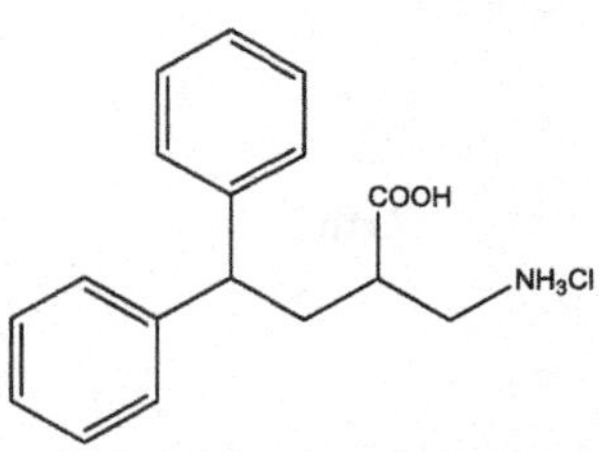	Active

Table 3 (continued)

Structures of the compounds	Response on pilocarpine test
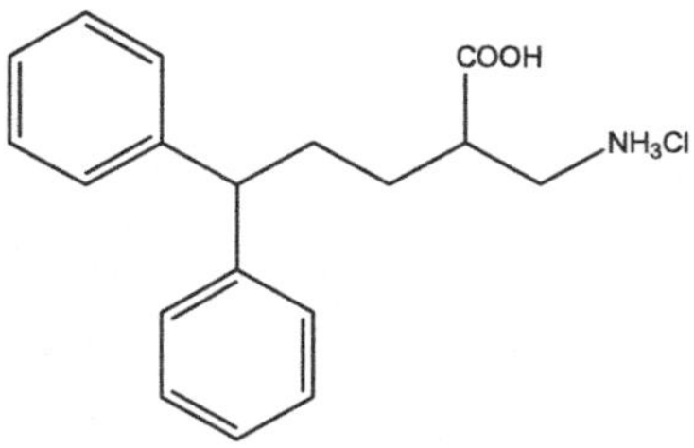	-

6 Serotonin Antagonists

Arylpiperazines have immensely important effects on various and diverse biological targets, in particular on CNS receptors. In the case of serotonin (5-HT) receptors, compounds containing this arylpiperazine moiety represent the largest systematically studied class of 5-HT_{1A} receptor ligands [63]. Structural alterations within long-chain arylpiperazines (LCAPs) occur mainly at the two opposite ends of a molecule and have been described by many authors [64–71].

Paluchowska et al. (2002) reported the synthesis, pharmacological studies, and conformational analysis utilizing classical molecular modeling approaches of some arylpiperazine or 1,2,3,4-tetrahydroisoquinoline derivatives of the known and flexible 5-HT_{1A} receptor ligands with different intrinsic activities at nanomolar levels [63]. The SAR is shown in Table 4. The synthetic steps involved for some of the compounds mentioned in Table 4 are shown in Scheme 2 [63].

Finally, the authors performed molecular modeling of the ligand. All the modeling was performed using the SYBYL™ package version 6.6 (Tripos Associates Inc., St. Louis, USA, www.tripos.com) [63]. For the study of the dihedral angles $\tau_2-\tau_4$ by the Mopac/AM1 method, the structures of compounds **16–24** were minimized over all the bonds and angles except for the respective torsion, which was constrained for values between 0 and 360° with a 10° increment. Investigations of the conformational space of compounds **15** and **16** were carried out by a standard random search method. Here, for the optimization, Tripos force field™ (Tripos Associates Inc., St. Louis, USA, www.tripos.com) was utilized [63].

Table 4 SAR data with molecular structures of 5-HT_{1A} receptor ligands reported by Paluchowska et al. (2002) [63]

	 Series A	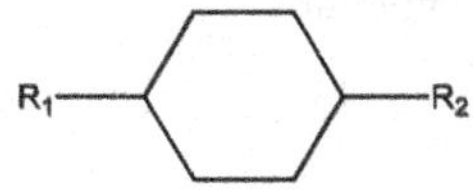 Series B	
Compound	R_1	R_2	K_i value (in nM)
13	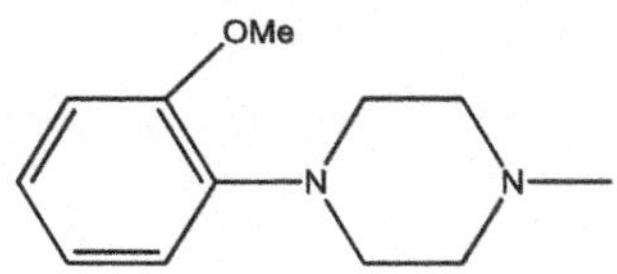	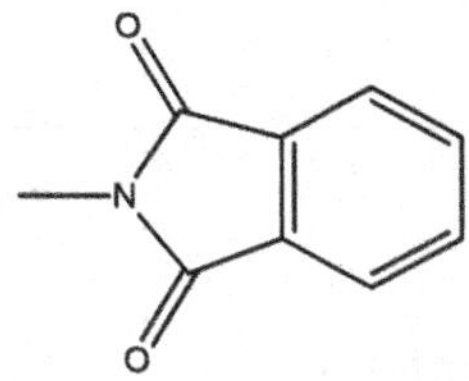	0.55
14	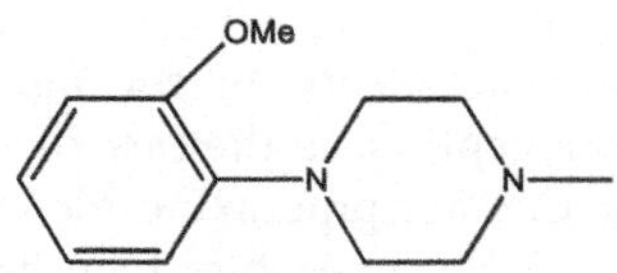	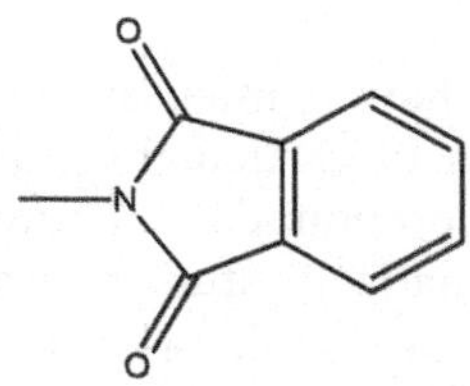	8.0
15	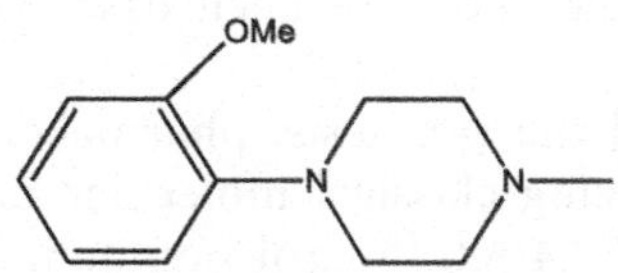	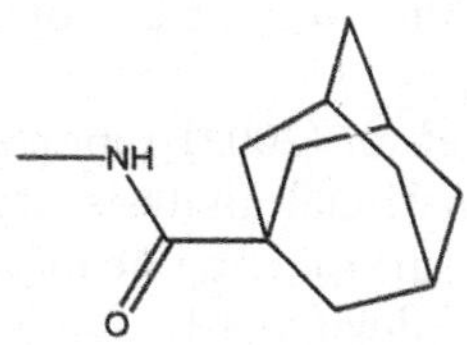	7.0
16	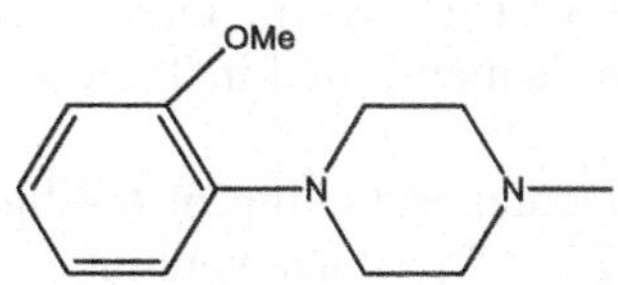	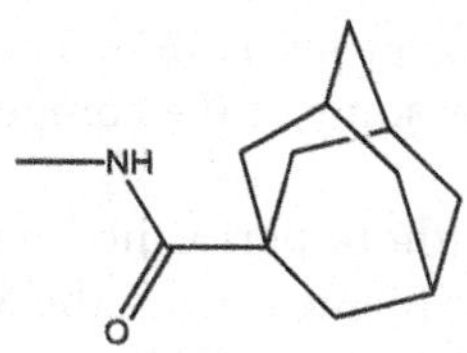	52.0
17	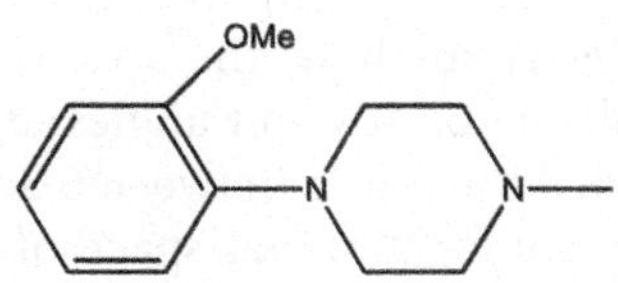	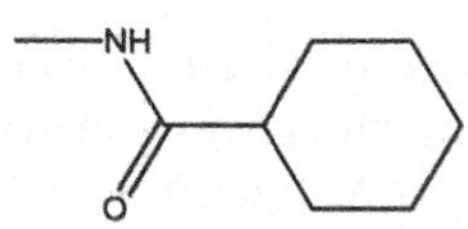	4.0
18		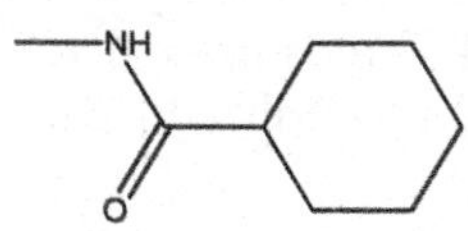	44.0

Table 4 (continued)

	 	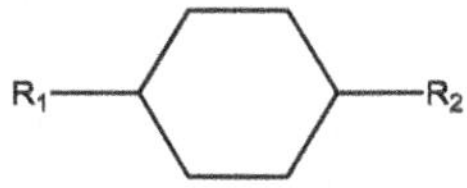 	
	Series A	Series B	
Compound	R_1	R_2	K_i value (in nM)
19	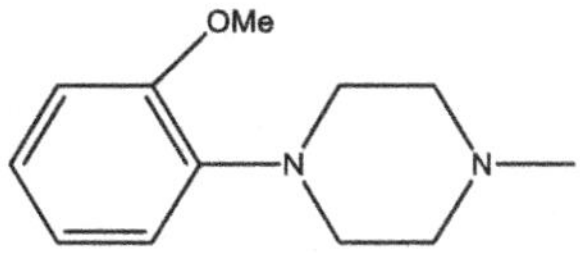 	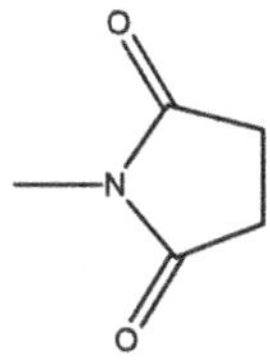 	6.4
20	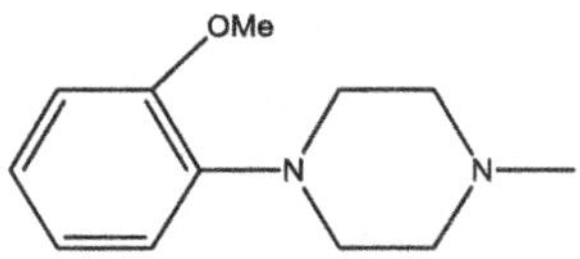 	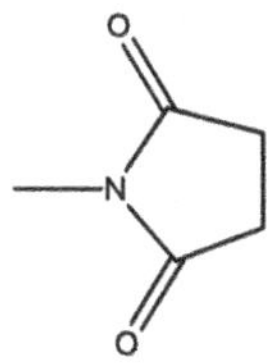 	15.0
21	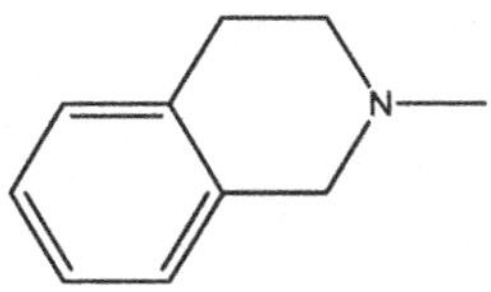 	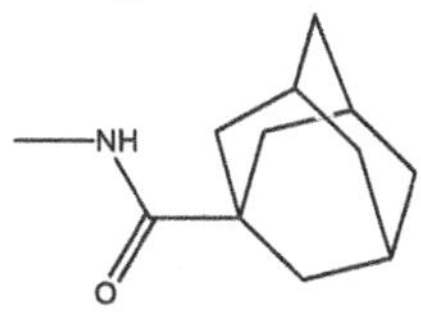 	0.95
22	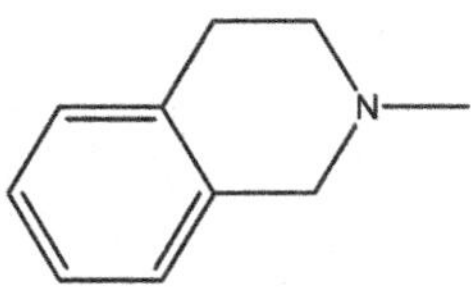 	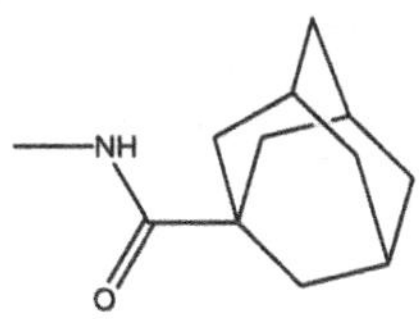 	1354.0
23	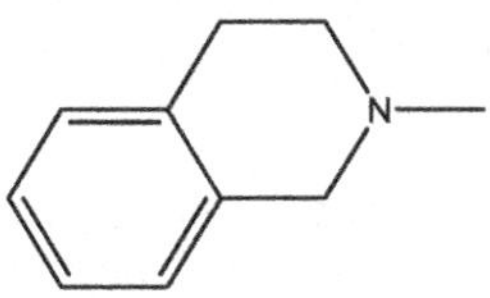 	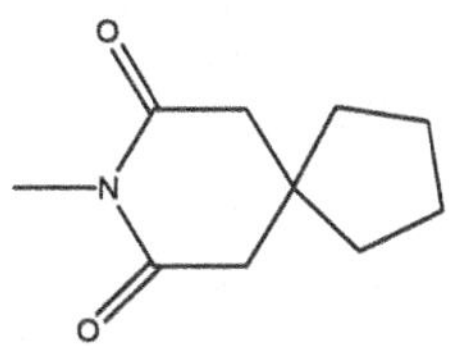 	5.0
24	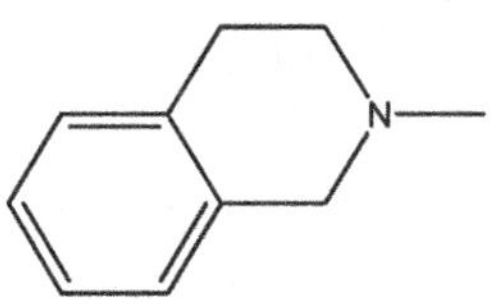 	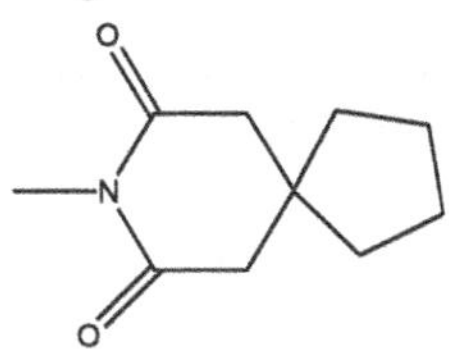 	2356.0

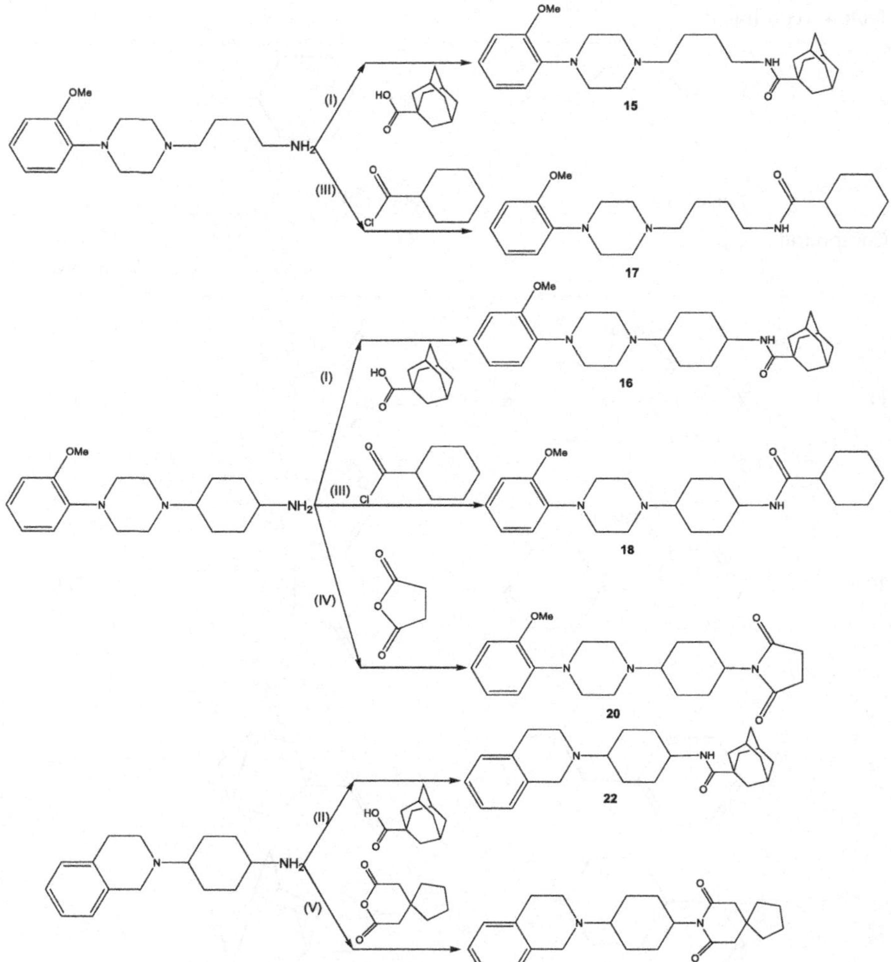

Scheme 2 The steps involved in the synthesis of the 5-HT_{1A} receptor ligands reported by Paluchowska et al. (2002) [63]. I: Ph_3P, NBS, Et_3N, CH_2Cl_2, 0–25 °C; II: BOP, Et_3N, CH_3CN, rt; III: 20% K_2CO_3 – $CHCl_3$, rt; IV: xylene, reflux; V: pyridine, reflux

While all the achievable conformations of the constrained compounds belong to an extended family—as indicated by molecular modeling studies—the hypothesis that such conformations are responsible for the blockade of postsynaptic 5-HT_{1A} receptors has been confirmed by Paluchowska et al. [63].

7
Toxicity Studies

A large number of research and review papers have been published in recent years on the integration of data on physicochemical properties, in vitro derived toxicity data, and physiologically based kinetics and dynamics as a modeling tool in hazard and risk assessment [72–85].

Recently (2004), Bello-Ramírez and Nava-Ocampo reported QSAR modeling studies on the toxicities of *Aconitum* alkaloids [86]. The QSAR modeling was performed on two groups of alkaloids: compounds with an aroyl/aroyloxy group at the R14 position (yunaconitine, bulleyaconitine, aconitine, beiwutine, nagarine, 3-acetyl aconitine, and penduline), and compounds with the aroyloxy group at the R4 position (*N*-deacetyllappaconitine, lappaconitine, ranaconitine, *N*-deacetylfinaconitine, and *N*-deacetylranaconitine) [86].

8
Conclusion

Alkaloids and nitrogen-containing compounds show immense biological activities. Unfortunately some of the alkaloids are potentially toxic, like *Aconitine* type diterpenoid alkaloids [86]. Interestingly, these compounds also possess very promising pharmacological activities in quite low doses. These classes of compounds should be carefully studied to explore their exact dose below lethal concentrations. The molecular modeling approach can be a good solution to study their SARs and QSARs against certain biological targets, such as different clinically important enzymes, proteins, DNA, and RNA, etc., faster and more accurately.

References

1. Quetin-Leclercq J (1994) Potential anticancer and antiparasitic indole alkaloids. J Pharm Belg 49(3):181–192
2. Thevand A et al. (2004) Total assignment and structure in solution of tetrandrine by NMR spectroscopy and molecular modelling. Spectrochim Acta A Mol Biomol Spectrosc 60(8–9):1825–1830
3. Schiff JL (1997) Bisbenzylisoquinoline alkaloids. J Nat Prod 60(9):934–953
4. Pang L, Hoult JR (1997) Cytotoxicity to macrophages of tetrandrine, an antisilicosis alkaloid, accompanied by an overproduction of prostaglandins. Biochem Pharmacol 53(6):773–782
5. Achike FI, Kwan CY (2002) Characterization of a novel tetrandrine-induced contraction in rat tail artery. Acta Pharmacol Sin 23(8):698–704

6. Kwan CY, Achike FI (2002) Tetrandrine and related bis-benzylisoquinoline alkaloids from medicinal herbs: cardiovascular effects and mechanisms of action. Acta Pharmacol Sin 23(12):1057–1068
7. Wu S et al. (2001) Cardiac effects of the extract and active components of *Radix stephaniae tetrandrae*. I. Electrically induced intracellular calcium transient and protein release during the calcium paradox. Life Sci 68(25):2853–2861
8. Yu XC et al. (2001) Cardiac effects of the extract and active components of *Radix stephaniae tetrandrae* II. Myocardial infarct, arrhythmias, coronary arterial flow and heart rate in the isolated perfused rat heart. Life Sci 68(25):2863–2872
9. Dewar MJS et al. (1985) Development and use of quantum mechanical molecular models. 76. AM1: a new general purpose quantum mechanical molecular model. J Am Chem Soc 107(13):3902–3909
10. Blackburn EH (2001) Switching and signaling at the telomere. Cell 106(6):661–673
11. Guyen B et al. (2004) Synthesis and evaluation of analogues of 10H-indolo[3,2-b]quinoline as G-quadruplex stabilising ligands and potential inhibitors of the enzyme telomerase. Org Biomol Chem 2(7):981–988
12. Hanahan D, Weinberg RA (2000) The hallmarks of cancer. Cell 100(1):57–70
13. Hahn WC et al. (1999) Inhibition of telomerase limits the growth of human cancer cells. Nat Med 5(10):1164–1170
14. Herbert B et al. (1999) Inhibition of human telomerase in immortal human cells leads to progressive telomere shortening and cell death. Proc Natl Acad Sci USA 96(25):14276–14281
15. Chen Z, Koeneman KS, Corey DR (2003) Consequences of telomerase inhibition and combination treatments for the proliferation of cancer cells. Cancer Res 63(18):5917–5925
16. Neidle S, Parkinson GN (2003) The structure of telomeric DNA. Curr Opin Struct Biol 13(3):275–283
17. Neidle S, Parkinson GN (2002) Telomere maintenance as a target for anticancer drug discovery. Nat Rev Drug Discov 1(5):383–393
18. Rezler EM, Bearss DJ, Hurley LH (2002) Telomeres and telomerases as drug targets. Curr Opin Pharmacol 2(4):415–423
19. Rezler EM, Bearss DJ, Hurley LH (2003) Telomere inhibition and telomere disruption as processes for drug targeting. Annu Rev Pharmacol Toxicol 43:359–379
20. Mergny JL et al. (2002) Natural and pharmacological regulation of telomerase. Nucleic Acids Res 30(4):839–865
21. Wright WE et al. (1997) Normal human chromosomes have long G-rich telomeric overhangs at one end. Genes Dev 11(21):2801–2809
22. Sun D et al. (1997) Inhibition of human telomerase by a G-quadruplex-interactive compound. J Med Chem 40(14):2113–2116
23. Perry PJ et al. (1998) 1,4- and 2,6-disubstituted amidoanthracene-9,10-dione derivatives as inhibitors of human telomerase. J Med Chem 41(17):3253–3260
24. Perry PJ et al. (1998) Human telomerase inhibition by regioisomeric disubstituted amidoanthracene-9,10-diones. J Med Chem 41(24):4873–4884
25. Shi DF et al. (2001) Quadruplex-interactive agents as telomerase inhibitors: synthesis of porphyrins and structure–activity relationship for the inhibition of telomerase. J Med Chem 44(26):4509–4523
26. Heald RA et al. (2002) Antitumor polycyclic acridines. 8.(1) Synthesis and telomerase-inhibitory activity of methylated pentacyclic acridinium salts. J Med Chem 45(3):590–597

27. Gavathiotis E et al. (2003) Drug recognition and stabilisation of the parallel-stranded DNA quadruplex d(TTAGGGT)4 containing the human telomeric repeat. J Mol Biol 334(1):25–36
28. Heald RA, Stevens MF (2003) Antitumour polycyclic acridines. Palladium(0) mediated syntheses of quino[4,3,2-kl]acridines bearing peripheral substituents as potential telomere maintenance inhibitors. Org Biomol Chem 1(19):3377–3389
29. Hutchinson I et al. (2004) Synthesis and properties of bioactive 2- and 3-amino-8-methyl-8H-quino[4,3,2-kl]acridine and 8,13-dimethyl-8H-quino[4,3,2-kl]acridinium salts. Org Biomol Chem 2(2):220–228
30. Mergny JL et al. (2001) Telomerase inhibitors based on quadruplex ligands selected by a fluorescence assay. Proc Natl Acad Sci USA 98(6):3062–3067
31. Riou JF et al. (2002) Cell senescence and telomere shortening induced by a new series of specific G-quadruplex DNA ligands. Proc Natl Acad Sci USA 99(5):2672–2677
32. Shin-ya K et al. (2001) Telomestatin, a novel telomerase inhibitor from *Streptomyces anulatus*. J Am Chem Soc 123(6):1262–1263
33. Kim MY et al. (2002) Telomestatin, a potent telomerase inhibitor that interacts quite specifically with the human telomeric intramolecular g-quadruplex. J Am Chem Soc 124(10):2098–2099
34. Harrison RJ et al. (1999) Human telomerase inhibition by substituted acridine derivatives. Bioorg Med Chem Lett 9(17):2463–2468
35. Read M et al. (2001) Structure-based design of selective and potent G quadruplex-mediated telomerase inhibitors. Proc Natl Acad Sci USA 98(9):4844–4849
36. Parkinson GN, Lee MP, Neidle S (2002) Crystal structure of parallel quadruplexes from human telomeric DNA. Nature 417(6891):876–880
37. Case DA et al. (2005) The Amber biomolecular simulation programs. J Comput Chem 26(16):1668–1688
38. Ponder JW, Case DA (2003) Force fields for protein simulations. Adv Protein Chem 66:27–85
39. Haider SM, Parkinson GN, Neidle S (2003) Structure of a G-quadruplex–ligand complex. J Mol Biol 326(1):117–125
40. Fedoroff OY et al. (1998) NMR-based model of a telomerase-inhibiting compound bound to G-quadruplex DNA. Biochemistry 37(36):12367–12374
41. Shiino M, Watanabe Y, Umezawa K (2001) Synthesis of N-substituted *N*-nitrosohydroxylamines as inhibitors of mushroom tyrosinase. Bioorg Med Chem 9(5):1233–1240
42. Lee HS (2002) Tyrosinase inhibitors of *Pulsatilla cernua* root-derived materials. J Agric Food Chem 50(6):1400–1403
43. Khan KM et al. (2005) A facile and improved synthesis of sildenafil (Viagra) analogs through solid support microwave irradiation possessing tyrosinase inhibitory potential, their conformational analysis and molecular dynamics simulation studies. Mol Divers 9(1–3):15–26
44. Oussaid A, Thach LN, Loupy A (1997) Tetrahedron Lett 38:2451
45. McEachern MJ, Krauskopf A, Blackburn EH (2000) Telomeres and their control. Annu Rev Genet 34:331–358
46. Ren J et al. (2001) Molecular recognition of a RNA : DNA hybrid structure. J Am Chem Soc 123(27):6742–6743
47. Raevsky OA (1999) Russ Chem Rev 68:437
48. Raevsky OA, Skvortsov VS (2002) 3D hydrogen bond thermodynamics (HYBOT) potentials in molecular modelling. J Comput Aided Mol Des 16(1):1–10

49. Sultankhodzhaev MN et al. (2005) Tyrosinase inhibition studies of diterpenoid alkaloids and their derivatives: structure–activity relationships. Nat Prod Res 19(5):517–522
50. Casanola-Martin GM et al. (2006) New tyrosinase inhibitors selected by atomic linear indices-based classification models. Bioorg Med Chem Lett 16(2):324–330
51. Tan CY, Wainman D, Weaver DF (2003) N-, alpha-, and beta-substituted 3-aminopropionic acids: design, syntheses and antiseizure activities. Bioorg Med Chem 11(1):113–121
52. Bazil CW, Pedley TA (1998) Advances in the medical treatment of epilepsy. Annu Rev Med 49:135–162
53. Perucca E (1996) Established antiepileptic drugs. Baillieres Clin Neurol 5(4):693–722
54. Perucca E (1996) The new generation of antiepileptic drugs: advantages and disadvantages. Br J Clin Pharmacol 42(5):531–543
55. Eadie MJ (1984) Anticonvulsant drugs. An update. Drugs 27(4):328–363
56. Leppik IE (1994) Antiepileptic drugs in development: prospects for the near future. Epilepsia 35(Suppl 4):S29–40
57. Wagner ML (1994) Felbamate: a new antiepileptic drug. Am J Hosp Pharm 51(13):1657–1666
58. Davies-Jones GAB (1988) Anticonvulsants: side effects of drugs, 11th edn. Elsevier, New York
59. Riddall DR, Leach MJ, Davison AN (1976) Neurotransmitter uptake into slices of rat cerebral cortex in vitro: effect of slice size. J Neurochem 27(4):835–839
60. Martin DL, Shain W (1979) High affinity transport of taurine and beta-alanine and low affinity transport of gamma-aminobutyric acid by a single transport system in cultured glioma cells. J Biol Chem 254(15):7076–7084
61. Hosli E, Hosli L (1980) Cellular localization of the uptake of [3H]taurine and [3H]beta-alanine in cultures of the rat central nervous system. Neuroscience 5(1):145–152
62. Sandberg M, Jacobson I (1981) Beta-alanine, a possible neurotransmitter in the visual system? J Neurochem 37(5):1353–1356
63. Paluchowska MH et al. (2002) Active conformation of some arylpiperazine postsynaptic 5-HT(1A) receptor antagonists. Eur J Med Chem 37(4):273–283
64. Glennon RA et al. (1988) Arylpiperazine derivatives as high-affinity 5-HT1A serotonin ligands. J Med Chem 31(10):1968–1971
65. Glennon RA et al. (1988) NAN-190: an arylpiperazine analog that antagonizes the stimulus effects of the 5-HT1A agonist 8-hydroxy-2-(di-*n*-propylamino)tetralin (8-OH-DPAT). Eur J Pharmacol 154(3):339–341
66. van Steen BJ et al. (1995) A series of N4-imidoethyl derivatives of 1-(2,3-dihydro-1,4-benzodioxin-5-yl)piperazine as 5-HT1A receptor ligands: synthesis and structure–affinity relationships. J Med Chem 38(21):4303–4308
67. Mokrosz JL et al. (1995) Structure–activity relationship studies of CNS agents, XIX: Quantitative analysis of the alkyl chain effects on the 5-HT1A and 5-HT2 receptor affinities of 4-alkyl-1-arylpiperazines and their analogs. Arch Pharm (Weinheim) 328(2):143–148
68. Paluchowska MH et al. (1996) Structure–activity relationship studies of CNS agents, part 31: Analogs of MP 3022 with a different number of nitrogen atoms in the heteroaromatic fragment—new 5-HT1A receptor ligands. Arch Pharm (Weinheim) 329(10):451–456

69. Caliendo G et al. (1999) Synthesis and biological activity of pseudopeptide inhibitors of Ras farnesyl transferase containing unconventional amino acids. Farmaco 54(11–12):785–790
70. Lopez-Rodriguez ML et al. (1996) Synthesis and structure–activity relationships of a new model of arylpiperazines. 1. 2-[[4-(*o*-Methoxyphenyl)piperazin-1-yl]methyl]-1, 3-dioxoperhydroimidazo[1,5-alpha]pyridine: a selective 5-HT1A receptor agonist. J Med Chem 39(22):4439–4450
71. Ismaiel AM et al. (1997) 2-(1-Naphthyloxy)ethylamines with enhanced affinity for human 5-HT1D beta (h5-HT1B) serotonin receptors. J Med Chem 40(26):4415–4419
72. Blaauboer BJ (2003) Biokinetic and toxicodynamic modelling and its role in toxicological research and risk assessment. Altern Lab Anim 31(3):277–281
73. Blaauboer BJ (2003) The integration of data on physicochemical properties, in vitro-derived toxicity data and physiologically based kinetic and dynamic as modelling a tool in hazard and risk assessment. A commentary. Toxicol Lett 138(1–2):161–171
74. Clark RD et al. (2004) Modelling in vitro hepatotoxicity using molecular interaction fields and SIMCA. J Mol Graph Model 22(6):487–497
75. Font M et al. (2006) Structural characteristics of novel symmetrical diaryl derivatives with nitrogenated functions. Requirements for cytotoxic activity. Bioorg Med Chem 14(6):1942–1948
76. Hewitt M et al. (2007) Structure-based modelling in reproductive toxicology: (Q)SARs for the placental barrier. SAR QSAR Environ Res 18(1–2):57–76
77. Mekenyan O et al. (2003) In silico modelling of hazard endpoints: current problems and perspectives. SAR QSAR Environ Res 14(5–6):361–371
78. Mekenyan OG et al. (2004) A systematic approach to simulating metabolism in computational toxicology. I. The TIMES heuristic modelling framework. Curr Pharm Des 10(11):1273–1293
79. Toropov AA, Benfenati E (2007) Optimisation of correlation weights of SMILES invariants for modelling oral quail toxicity. Eur J Med Chem 42(5):606–613
80. Toropov AA, Benfenati E (2006) QSAR models of quail dietary toxicity based on the graph of atomic orbitals. Bioorg Med Chem Lett 16(7):1941–1943
81. Budzisz E et al. (2003) Cytotoxic effects, alkylating properties and molecular modelling of coumarin derivatives and their phosphonic analogues. Eur J Med Chem 38(6):597–603
82. Croni MT et al. (2000) Structure–toxicity relationships for aliphatic compounds encompassing a variety of mechanisms of toxic action to *Vibrio fischeri*. SAR QSAR Environ Res 11(3–4):301–312
83. Fouchecourt MO, Beliveau M, Krishnan K (2001) Quantitative structure–pharmacokinetic relationship modelling. Sci Total Environ 274(1–3):125–135
84. Gia O et al. (1997) Some new methyl-8-methoxypsoralens: synthesis, photobinding to DNA, photobiological properties and molecular modelling. Farmaco 52(6–7):389–397
85. Onwurah IN (2002) Quantitative modelling of crude oil toxicity using the approach of cybernetics and structured mechanisms of microbial processes. Environ Monit Assess 76(2):157–166
86. Bello-Ramírez AM, Nava-Ocampo AA (2004) A QSAR analysis of toxicity of Aconitum alkaloids. Fund Clin Pharmacol 18:699–704
87. N'Goka V et al. (1991) GABA-uptake inhibitors: construction of a general pharmacophore model and successful prediction of a new representative. J Med Chem 34(8):2547–2557
88. Leeson PD, Iversen LL (1994) The glycine site on the NMDA receptor: structure–activity relationships and therapeutic potential. J Med Chem 37(24):4053–4067

Top Heterocycl Chem (2007) 10: 99–122
DOI 10.1007/7081_2007_068

Published online: 20 June 2007

Microbial Transformation of Nitrogenous Compounds

Mahmud Tareq Hassan Khan[1,2] (✉) · Arjumand Ather[3]

[1]PhD School of Molecular and Structural Biology, and Department of Pharmacology, Institute of Medical Biology, University of Tromsø, 9037 Tromsø, Norway
mahmud.khan@fagmed.uit.no

[2]Pharmacology Research Lab., Faculty of Pharmaceutical Sciences, University of Science and Technology, Chittagong, Bangladesh

[3]Norwegian Structural Biology Center (NorStruct), Institute of Chemistry, University of Tromsø, 9037 Tromsø, Norway

Abstract Alkaloids are very much important molecules, not only for chemical reasons but also because of their diverse biological activities. Up to now several reviews have been published explaining the use of biotransformation or microbial transformation techniques to modify alkaloids, which added several advantages over the classical chemical transformation systems. This chapter is a critical update of the microbial transformations reported in the last couple of years, targeting novel biocatalysts from microbes.

Keywords Alkaloid · Microbial transformation · Biotransformation · Cinchona · Nicotine · Morphine

Abbreviations

GC	Gas chromatography
HPLC	High-performance liquid chromatography
LC-MS	Liquid chromatography coupled with mass spectrometry

LC-MS/MS	Liquid chromatography coupled with tandem mass spectrometry
LC-NMR	Liquid chromatography coupled with nuclear magnetic resonance
TLC	Thin-layer chromatography

1 Introduction

Alkaloids continue to provide mankind with a plethora of medicines, poisons, and tonics. As many precious drugs result from such natural compounds, there is much interest in their transformation to provide new compounds or intermediates for the synthesis of new or improved drugs [1].

Alkaloids are a group of complex nitrogen-containing compounds derived from a variety of sources, including microbes, marine organisms, and plants, via complex biosynthetic pathways. They are used in the treatment of a wide range of clinical problems, for example as anticancer agents, antihypertensives, antimalarials, and analgesics, and for the treatment of parkinsonism and central nervous system disorders [1].

As alkaloids contain a nitrogen atom, they react mostly alkaline and are able to form soluble salts in aqueous environments. In plants, however, they can occur in the free state, as a salt, or as an *N*-oxide, and accumulate in the plant vacuole as a reservoir or often coupled to phenolic acids like chlorogenic acid or caffeic acid [2].

In the mid-1950s, microbial transformation technology was introduced into the field of synthetic chemistry as a new methodology. There was a sudden interest in research on the problems of producing steroid hormones by microbial transformation [3]. The biotransformation of alkaloids by microbes and plants has been reviewed by several authors in recent years [1, 4–7]. The selective use of individual enzymatic transformations utilizing microorganisms in chemical production pathways, in particular by biotransformations of steroids in 1950, expanded the field of biotechnological production of pharmaceuticals [8]. The increasing knowledge of the regulation of the biosynthesis of primary, and especially secondary, metabolites, the growing experience in the use of microorganisms as biocatalysts and a source of valuable enzymes, and the development of new economical technical approaches raised the number and volume of drugs prepared by microbial biosynthesis and biotransformation [8]. The modern method of genetic engineering supported by chemical DNA synthesis enabled the preparation of important proteohormones and physiologically active peptides in microorganisms. Finally, the development of monoclonal antibodies, although at present still formed in mammalian cells, will lead to new ways of therapy in the future [8–10].

Figure 1 shows the molecular structures of some alkaloids that have been studied in recent years for their transformations through microorganisms.

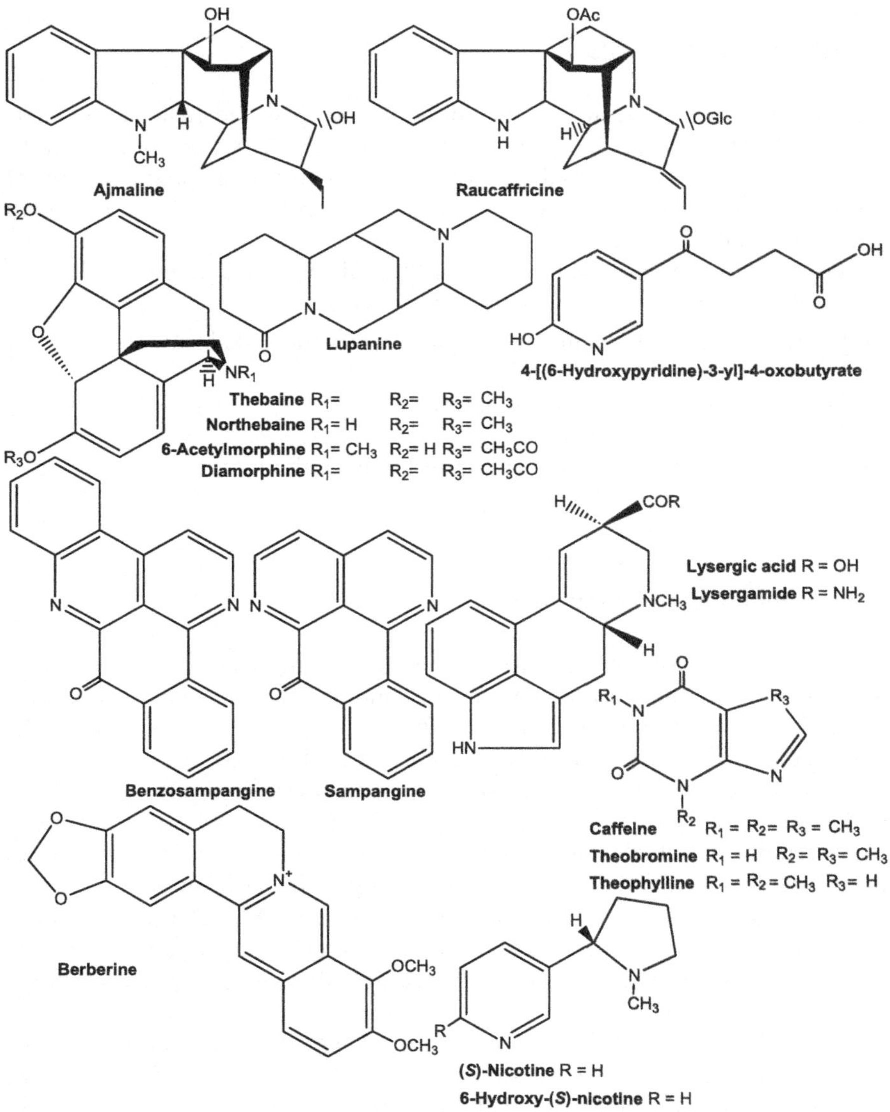

Fig. 1 Molecular structures of some alkaloids that have been used and reported in microbial transformations in the last few years

This review presents a critical analysis of recent updates of the research work that has been reported in last couple of years in the fields of microbial transformation or biotransformation using microbes to modify alkaloids.

2
Classification of Alkaloids

According to Julsing et al. [2], alkaloids can be classified in terms of their biological activities, their chemical structures, or their biosynthetic pathways [2]. Over 12 000 alkaloids are known so far from plants, and several of these are being used medicinally in the global pharmaceutical market worth 4 billion US dollars [2].

Based on the amino acid of origin in their biosynthetic pathway, alkaloids can be divided into the following five major classes [2]:

- From ornithine—tropane, pyrrolidine, pyrrolizide alkaloids, etc.
- From tyrosine—benzylisoquinoline
- From tryptophane—indolequinoline
- From pyridine—pyridine
- From lysine—quinolizidine and piperidine alkaloids

3
Why Microbial (Bio)transformation?

There have been lots of endeavors aimed at the discovery of novel biological catalysts for the transformation of alkaloids. The microbial (bio)transformation systems have plus points above the plant systems in that the manufacture of biomass can be accomplished rapidly and microbial genetic systems are commonly well understood. Transformations utilizing microbial organisms could imitate mammalian catabolism, and so might allow the assembly of valuable intermediates or metabolites in huge amounts to permit recognition and exploit them in drug preclinical trials as well as toxicity studies. The individual expression of the enzymes involved in alkaloid biosynthesis in heterologous hosts, such as bacteria, allows the detailed assessment of catalytic mechanisms that may be unknown in classical synthetic organic chemistry [1].

The microorganisms do not necessarily always form the same metabolites as humans in a similar manner, but they can be useful for studies of drug interactions, disposition, etc. [11]. Although there is no correlation between the mammalian and microbial isozymes, the mechanism involved in microorganisms is still unknown, but may be similar to that involved in animals. Additionally, the fungal metabolism of the compounds is often affected by the concentration, nutritional factors, inducers, and environmental factors [12].

4
Development of Microbial Models

The techniques and different approaches used in microbial transformations have been mentioned and reviewed by several authors [1, 4, 11, 13–17]. The compounds can be metabolized by using either different pure enzymes or simply cultivated microbial whole cells [18–22]. A more general scheme is shown in Fig. 2 for the development of a standard system.

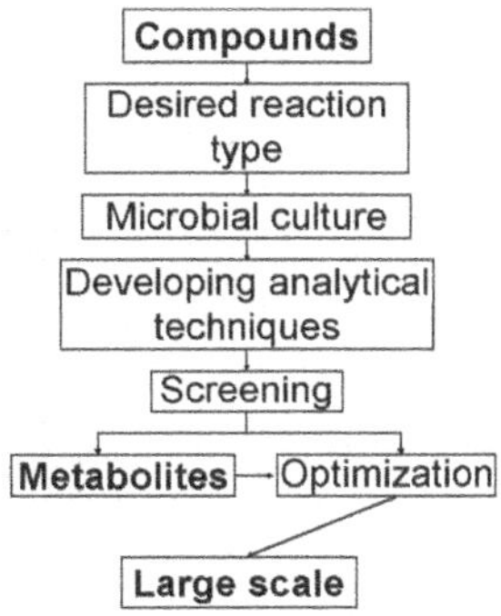

Fig. 2 Scheme for the development of microbial transformation of interesting chemical entities (modified from [11])

The development of appropriate analytical techniques is essential for success and to proceed toward screening, such as TLC, GC, HPLC, LC-MS, and LC-NMR [11, 23].

5
Some Interesting Examples

Table 1 shows examples of some of the microorganisms involved in different biotransformation processes [11]. The following sections cite some more recent examples of microbial transformation of biologically important compounds, especially alkaloids and nitrogen-containing compounds.

5.1
Cinchona Alkaloids

Cinchona alkaloids have been used as drugs for the treatment of several diseases. Quinine is very popular as an antimalarial drug against the erythrocyte stage of the parasite [34]. Recently, Shibuya et al. (2003) reported the microbial transformation of four *Cinchona* alkaloids (quinine, quinidine, cinchonidine, and cinchonine) by endophytic fungi isolated from *Cin-*

Table 1 Some examples of the microorganisms involved in different biotransformation processes (modified from [11])

Compound	Structure of the compound	Microorganisms utilized	Type of reaction	Refs.
Papavarine	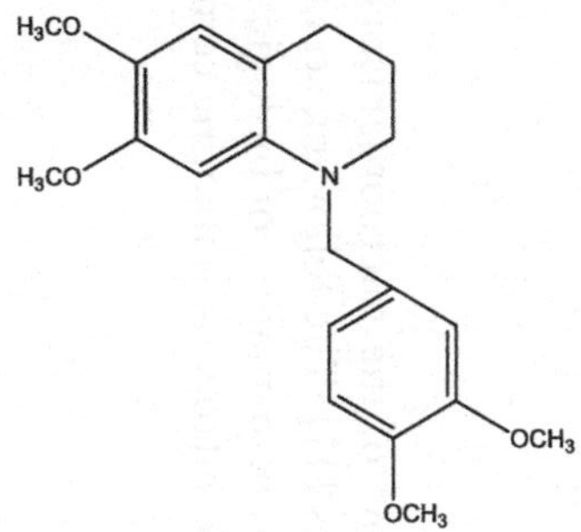	*Aspergillus alliaceus* NRRL 315 *Cunninghamella echinulata* ATCC 9244 *Cunninghamella echinulata* NRRL 3655 *Cunninghamella bainieri* ATCC 9244	O dealkylation	[24]
Acronycine	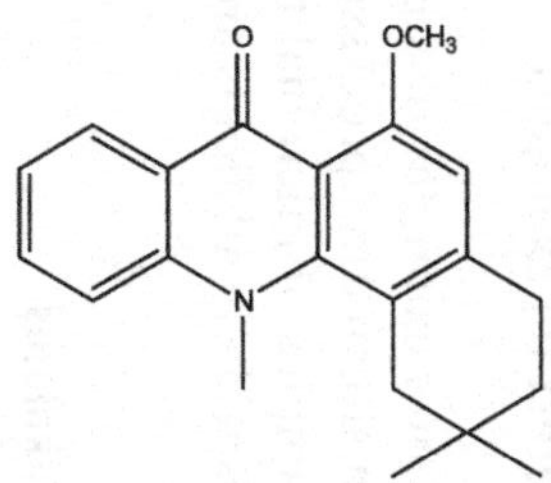	*Cunninghamella echinulata* NRRL 3655 *Cunninghamella bainieri* ATCC 9244	Aromatic hydroxylation	[25]
		Streptomyces spectabilis NRRL 2494	O dealkylation	

Table 1 (continued)

Compound	Structure of the compound	Microorganisms utilized	Type of reaction	Refs.
Yohimbin	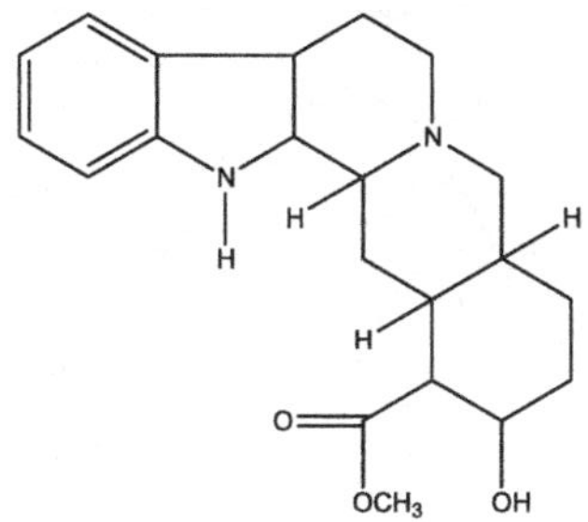	*Streptomyces platensis* NRRL 2364	Aromatic hydroxylation	[26]
Thebaine	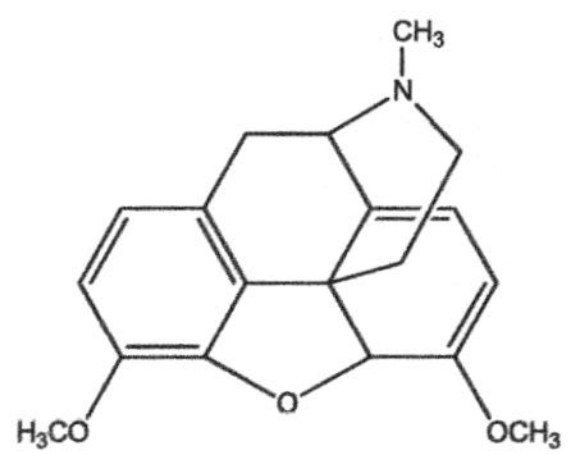	*Tremetes cinnabarina*	Aromatic hydroxylation O dealkylation Oxidation	[27]

Table 1 (continued)

Compound	Structure of the compound	Microorganisms utilized	Type of reaction	Refs.
Tetrandrine	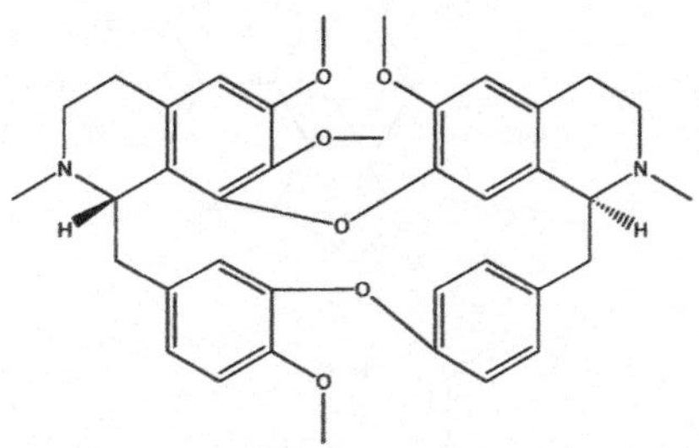	*Cunninghamella echinulata* NRRL 3655 *Cunninghamella bainieri* ATCC 9244	N dealkylation	[28]
Colchicine	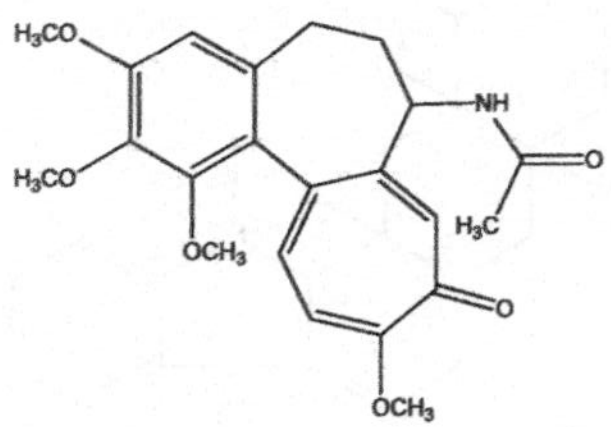	*Streptomyces griseus* ATCC 13968	O dealkylation	[29]
Nicotine	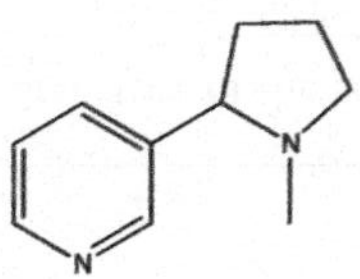	*Microsporum gypseum* ATCC 11395	N dealkylation	[30]

Table 1 (continued)

Compound	Structure of the compound	Microorganisms utilized	Type of reaction	Refs.
Sampangine	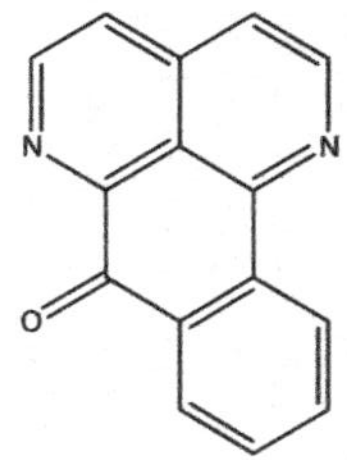	*Cunninghamella elegans* ATCC 9245	Glucuronidation	[31]
Artemisinin	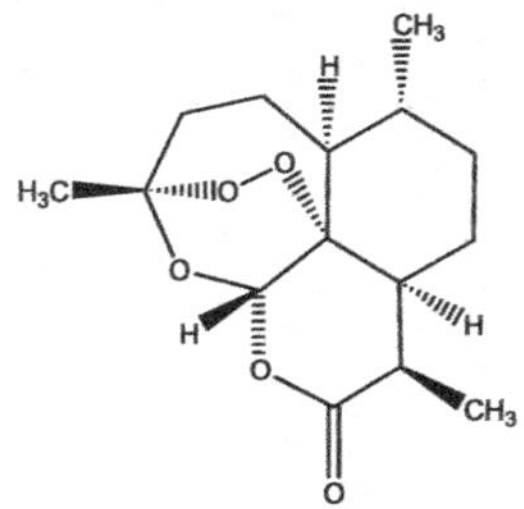	*Nocardia coralline* ATCC 19070	Reduction	[32]
Veratramine	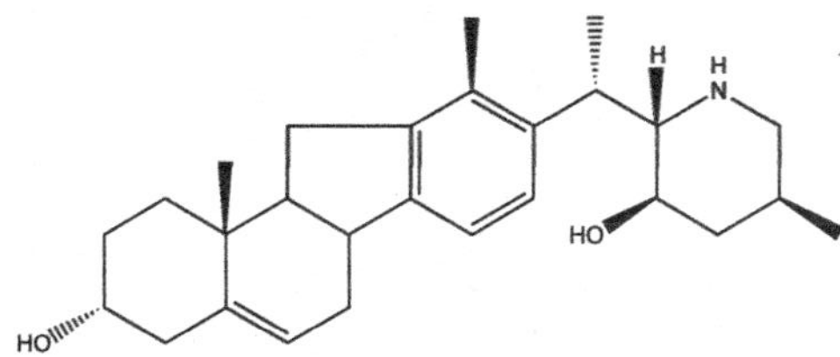	*Nocardia* Sp.	Oxidation	[33]

chona pubescens [34]. The endophytic filamentous fungus *Xylaria* sp. has been found to transform the *Cinchona* alkaloids into their 1-*N*-oxide derivatives [34]. Their structures are shown in Fig. 3.

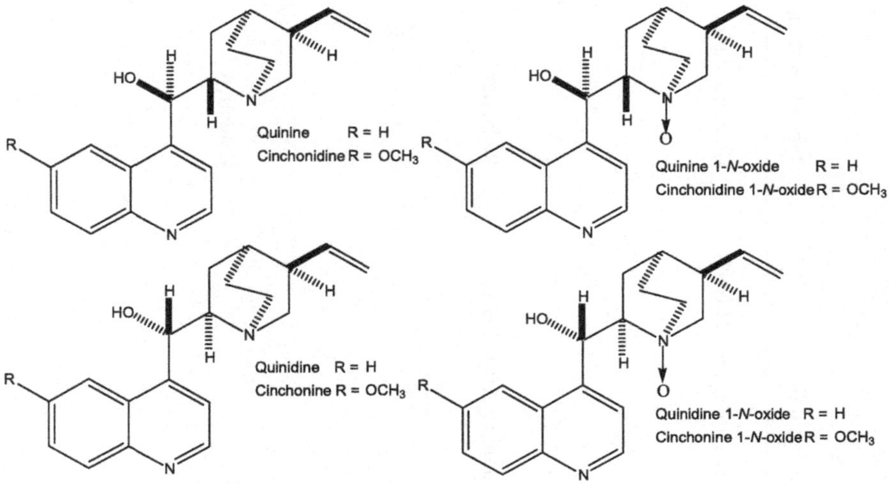

Fig. 3 Structures of *Cinchona* alkaloids (quinine, quinidine, cinchonidine, and cinchonine) transformed into their corresponding 1-*N*-oxide derivatives [34]

5.2 *Veratrum* Alkaloids

Veratrum alkaloids are a group of potent hypotensive agents that act by reflex suppression of the cardiovascular system [35, 36]. Rubijervine is one of the most common *Veratrum* alkaloids possessing the 22,26-epiminocholestane (solanidane) skeleton [37]. The preparative-scale fermentation of rubijervine with *Cunninghamella echinulata* ATCC 9244 has been reported together with the resulting new metabolites 7α-hydroxyrubijervine and solanid-5-ene-3β,12α-diol-1-one [38]. The structures of rubijervine and its metabolites are shown in Fig. 4. The microbe *C. echinulata* ATCC 9244 was able to metabolize rings A and B of rubijervine but failed to metabolize rings C, D, or its N-containing side chain, a finding which is analogous to the results of previous fermentation studies of steroidal alkaloids [38].

5.3 Nicotine

In a recent publication, Wang and coworkers (2005) reported the "green" synthetic process of 6-hydroxy-3-succinoyl-pyridine from *S*-nicotine in to-

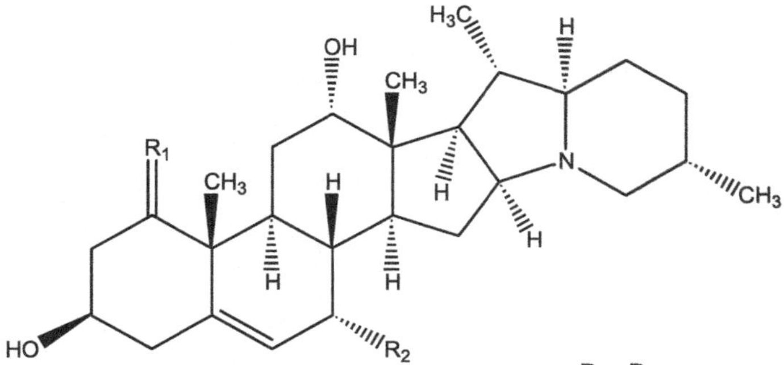

	R_1	R_2
Rubijervine	H_2	H
7alpha-hydroxyrubijervine	H_2	OH
Solanid-5-ene-3beta,12alpha-diol-1-one	O	H

Fig. 4 Structures of rubijervine and its metabolites [38]

bacco waste utilizing whole cells of a *Pseudomonas* sp. [39]. The synthetic pathway is shown in Scheme 1. The reported biotransformation method of nicotine to 6-hydroxy-3-succinoyl-pyridine utilizing *Pseudomonas* sp. made it possible to convert tobacco wastes with high nicotine content into valuable compounds [39].

5.4 Palmatine Analogs

L-Tetrahydropalmatine (L-THP), a naturally occurring neuroactive alkaloid from *Stephania ainiaca* Diels, has been widely used as an analgesic in China for many years [40]. It is not only an analgesic, but also a hypnotic [40]. Modern pharmaceutical practices have made it available worldwide as a single pure chemical of high therapeutic potency [40]. In recent years, toxic effects, including depression of neurological, respiratory, and cardiac function in pediatric poisonings, as well as acute or chronic hepatitis after regular use in adults, have been reported by several authors [41–45].

Very recently Li et al. (2006) reported comparative studies of the similarities and differences between the microbial and mammalian metabolisms of L-THP, the microbial transformation by *Penicillium janthinellum*, and metabolism in rats [46]. The biotransformation of L-THP by *P. janthinellum* AS 3.510 resulted in the formation of three metabolites. Their structures (shown in Fig. 5) were identified as L-corydalmine, L-corypalmine, and 9-*O*-desmethyl-L-THP, by comprehensive NMR and MS analysis [46].

Six metabolites have been detected from an in vivo study in rats, three of which (L-corydalmine, L-corypalmine, and 9-*O*-desmethyl-L-THP) were identified as the same compounds as those obtained from microbial metabolism by LC-MS/MS analysis and comparison with reference standards obtained

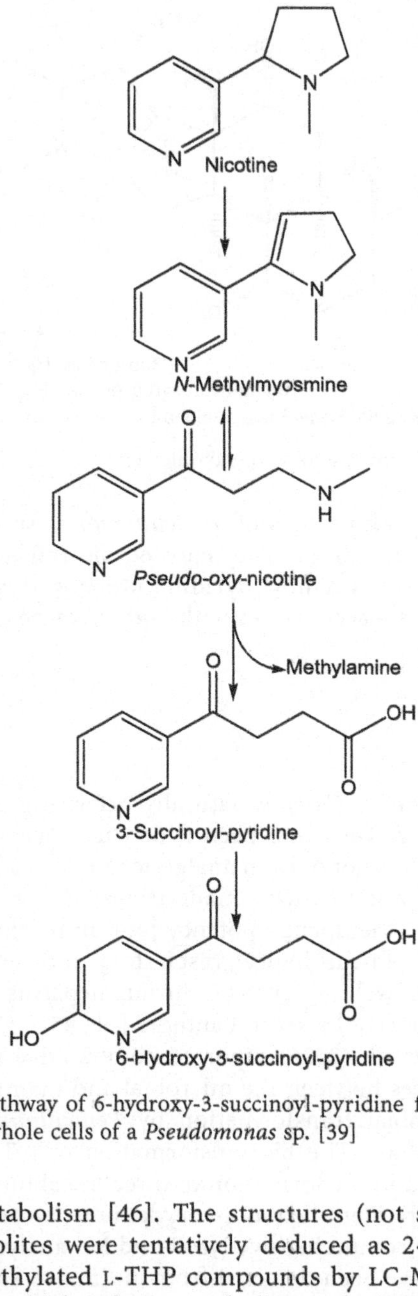

Scheme 1 Synthetic pathway of 6-hydroxy-3-succinoyl-pyridine from (*S*)-nicotine in tobacco waste utilizing whole cells of a *Pseudomonas* sp. [39]

from microbial metabolism [46]. The structures (not shown) of the additional three metabolites were tentatively deduced as 2-*O*-desmethyl-L-THP and two di-*O*-demethylated L-THP compounds by LC-MS/MS analysis. The time courses of microbial and rat metabolisms of L-THP were also reported by the authors [46].

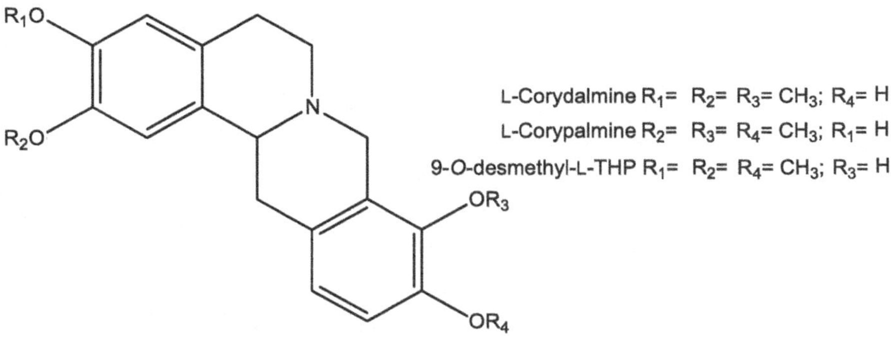

Fig. 5 Structural features of the transformed products of L-THP by *P. janthinellum* [46]

5.5 Benzylisoquinoline Alkaloids

Canonica et al. reported the microbial transformation of the benzylisoquinoline alkaloid laudanosine (**1**, Scheme 2), a minor opium alkaloid with the benzylisoquinoline skeleton, by a strain of *Pseudomonas putida*, which gives a metabolite in which O demethylation of one methoxyl group of ring C, and addition of one ketonic oxygen at the C-9 position and one phenolic oxygen at ring C, have occurred. Additionally, *O*-methylcoripalline was formed in the same transformation [47].

The chemical and biological importance of benzylisoquinoline alkaloids has stimulated many studies on their regiospecific oxygenation and the oxidative cyclization reaction [48, 49]. Morphine is the most important member of the group of benzylisoquinoline alkaloids and is a natural product with high medicinal significance [2]. Other benzylisoquinoline alkaloids are also pharmaceutically important. Like morphine, codeine is used as an analgesic. Berberine and sanguinarine are used as antimicrobials and others as muscle relaxants, such as parpaverine and (+)-tubocurarine [2]. The morphine alkaloids are the major alkaloid components of the opium poppy, *Papaver somniferum*, and these compounds still provide some of the most potent analgesic compounds in clinical use [50]. Many synthetic derivatives of morphine have been reported, and subtle changes in functionality can significantly alter the pharmacological activity of these compounds [50]. The microbial transformations of morphine alkaloids have been investigated as a means of producing opiates and pivotal intermediates that are difficult to produce by chemical synthesis [50].

The oxidation of morphine by *Pseudomonas putida* M10 gave rise to a large number of transformation products including hydromorphone (dihydromorphinone), 14β-hydroxymorphine, 14β-hydroxymorphinone, and dihydromorphine. Similarly, in incubations with oxymorphone (14β-hydroxy-

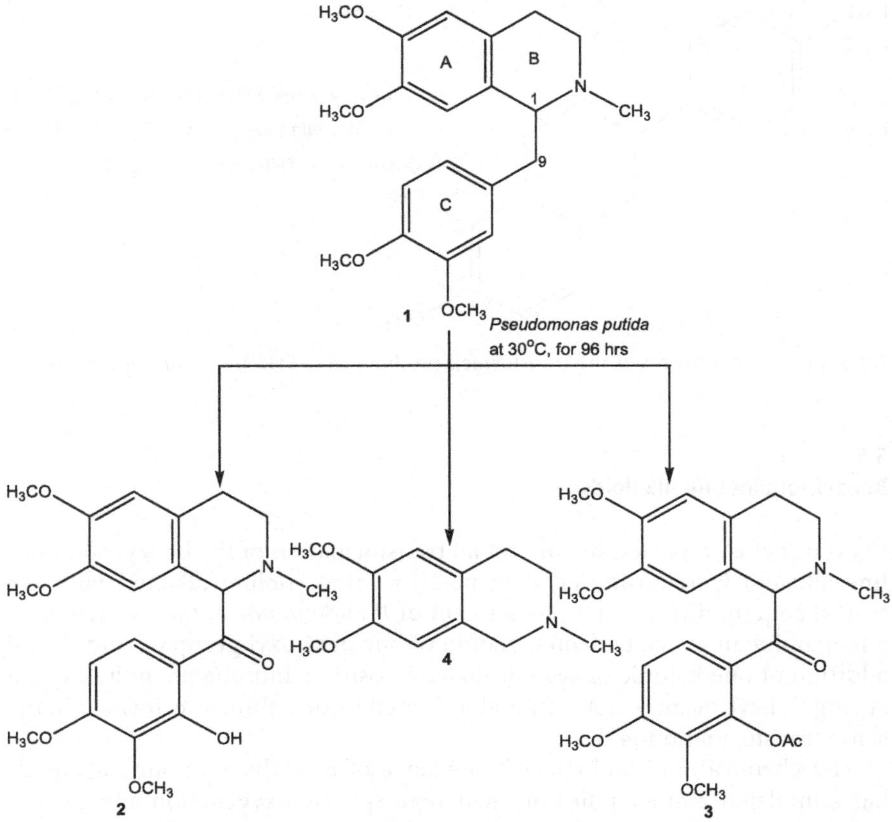

Scheme 2 Structures of the benzylisoquinoline alkaloid laudanosine and its transformed products (2, 3, and 4) by utilizing *Pseudomonas putida* incubated at 30 °C for 96 h [47]

dihydromorphinone) as substrate, the major transformation product was identified as oxymorphol (14β-hydroxydihydromorphine) [51, 52]. The transformation steps are shown in Scheme 3. The identities of all these biological products were confirmed by the usual MS and NMR techniques [52]. This was the first report describing structural evidence for the biological synthesis of 14β-hydroxymorphine and 14β-hydroxymorphinone [52]. These products have applications as intermediates in the synthesis of semisynthetic opiate drugs [51, 52].

The initial steps in the metabolism of morphine and codeine by *Pseudomonas putida* M10 involve oxidation of the C-6 hydroxy group and subsequent reduction of the 7,8-olefinic bond, forming hydromorphone (dihydromorphinone) and hydrocodone (dihydrocodeinone), respectively (Scheme 4) [52]. These products have important industrial applications: hydromorphone is an analgesic some seven times more potent than morphine [53],

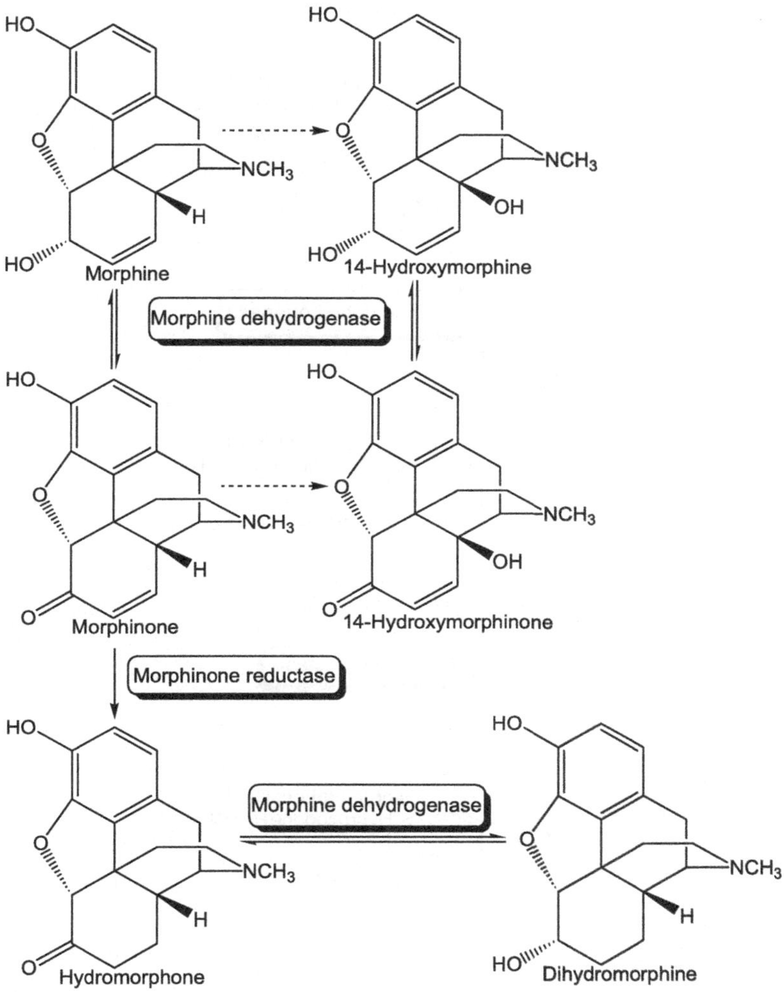

Scheme 3 Transformation steps involved in the oxidation of morphine by incubation with *Pseudomonas putida* M10, which gave hydromorphone (dihydromorphinone), 14β-hydroxymorphine, 14β-hydroxymorphinone, and dihydromorphine [52]

while hydrocodone is widely used as an antitussive [54]. Existing methods for the synthesis of hydromorphone are unsatisfactory because of the difficulty in specifically oxidizing the C-6 hydroxy group. The synthesis of the above-mentioned opiates seems to be valuable utilizing the biocatalytic approaches [52].

The oxidation of morphine by whole-cell suspensions and cell extracts of *Cylindrocarpon didymum* gave rise to the formation of 2,2′-bimorphine (for structure, see Fig. 6). The identity of 2,2′-bimorphine was confirmed by clas-

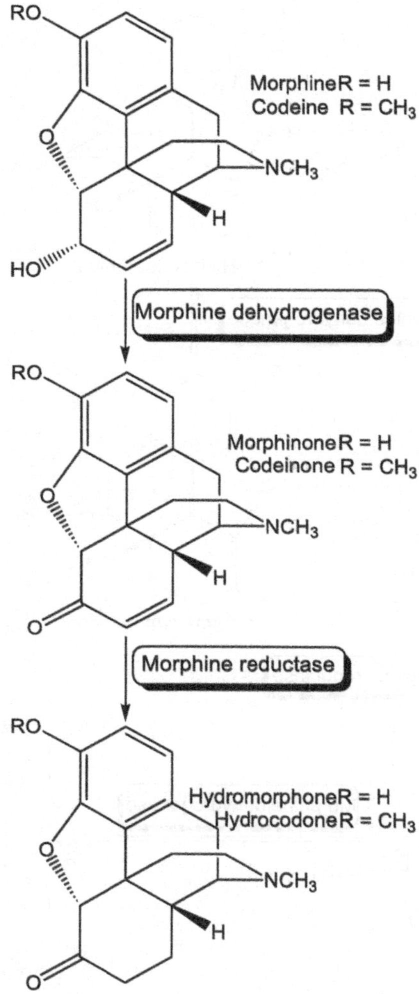

Scheme 4 The initial steps in the metabolism of morphine and codeine by *Pseudomonas putida* M10 involved in the oxidation process

sical spectroscopic methods [55]. The *C. didymum* also displayed activity on the morphine analogs hydromorphone, 6-acetylmorphine, and dihydromorphine, but not codeine or diamorphine, suggesting that a phenolic group at C-3 is essential for activity [55].

The same 2,2′-bimorphine has been further transformed into 10-α-*S*-monohydroxy-2,2′-bimorphine and 10,10′-α,α'-*S*,*S*′-dihydroxy-2,2′-bimorphine by the same group of researchers, again utilizing *C. didymum* whole-cell suspensions [50]. These compounds were confirmed using classical spectroscopic

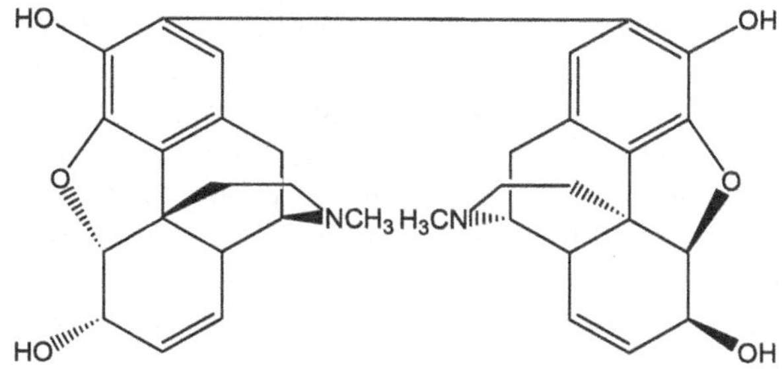

Fig. 6 Molecular structure of 2,2′-bimorphine

methods. Scheme 5 shows the steps involved in the transformation of the 2,2′-bimorphine.

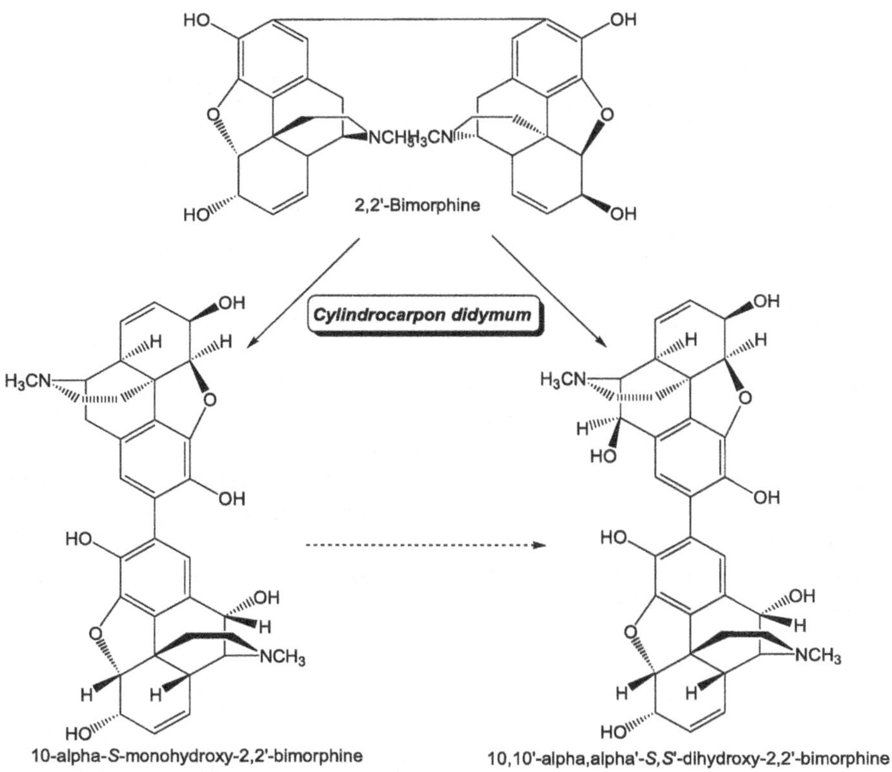

Scheme 5 Steps involved in the transformation of 2,2′-bimorphine to 10-α-*S*-monohydroxy-2,2′-bimorphine and 10,10′-α,α'-*S*,*S*′-dihydroxy-2,2′-bimorphine in the presence of *Cylindrocarpon didymum* [50]

Preparative-scale fermentation of papaveraldine, the known benzylisoquinoline alkaloid, with *Mucor ramannianus* 1839 (sih) has resulted in a stereoselective reduction of the ketone group and the isolation of *S*-papaverinol and *S*-papaverinol *N*-oxide [56]. The structure elucidations of both metabolites were reported to be based primarily on 1D and 2D NMR analyses and chemical transformations [56]. The absolute configuration of *S*-papaverinol has been determined using Horeau's method of asymmetric esterification [56]. The structures of the compounds are shown in Fig. 7.

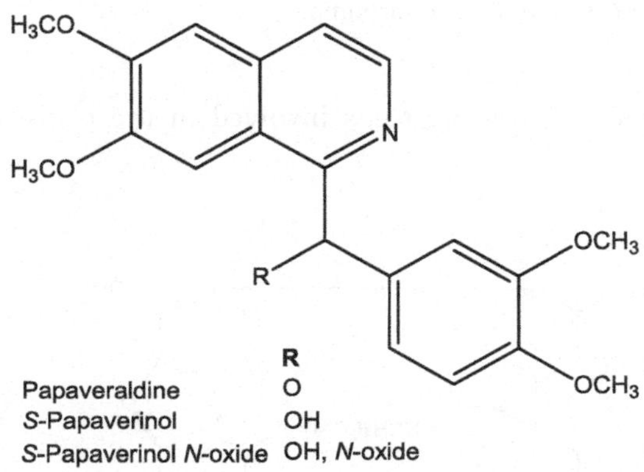

Fig. 7 Structures of papaveraldine, *S*-papaverinol, and *S*-papaverinol *N*-oxide [56]

5.6 Gluco-indole Alkaloids

Takayama et al. reported three monoterpenoid gluco-indole alkaloids, 3β-isodihydrocadambine, cadambine, and 3α-dihydrocadambine, which were isolated from *Nauclea cadamba* ROXB growing in Thailand. The stereochemistry at C-19 in 3β-isodihydrocadambine was elucidated to be *R* by spectroscopic analysis. Treatment of 3α-dihydrocadambine with β-glucosidase in aqueous ammonium acetate solution gave an indolopyridine alkaloid, 16-carbomethoxynaufoline, and an unusually rearranged compound [57]. The reaction steps are shown in Scheme 6.

5.7 Sampangines

Microbial transformation studies of the synthetic antifungal alkaloid benzosampangine have revealed that the compound is metabolized by a number of microorganisms. Using a standard two-stage fermentation technique, *Absidia glauca* (ATCC 22752), *Cunninghamella blakesleeana* (ATCC 8688a), *Cun-*

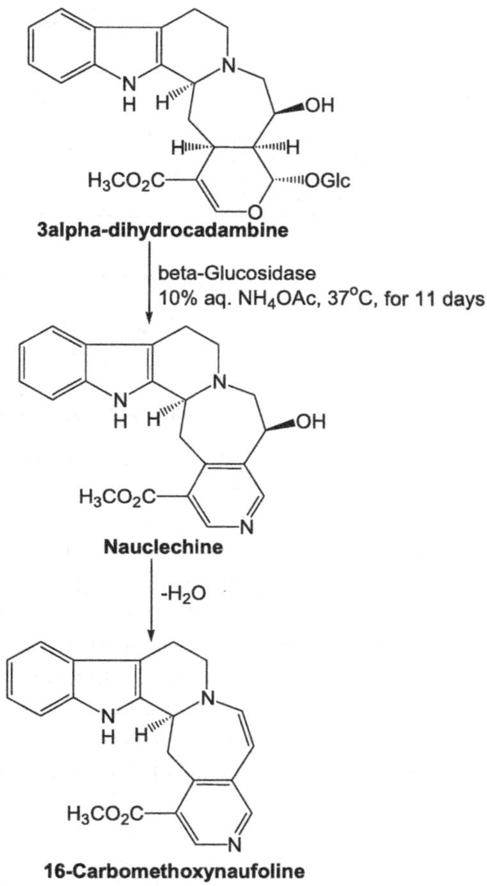

Scheme 6 Biotransformation of 3α-dihydrocadambine to 16-carbomethoxynaufoline with β-glucosidase in aq. NH_4OAc [57]

ninghamella sp. (NRRL 5695), *Fusarium solani* f. sp. *cucurbitae* (CSIH #C-5), and *Rhizopogon* sp. (ATCC 36060) each produced a β-glucopyranose conjugate of benzosampangine [58]. The identity of β-glucopyranose benzosampangine has been established on the basis of spectroscopic data [58]. The reaction is shown in Scheme 7.

The same research group previously (in 1999) reported microbial transformation studies of the antifungal alkaloid sampangine, which showed that the compounds can also be metabolized by a large number of microorganisms [59]. They utilized a standard two-stage fermentation technique; *Beauvaria bassiana* (ATCC 7159), *Doratomyces microsporus* (ATCC 16225), and *Filobasidiella neoformans* (ATCC 10226) produced the 4′-*O*-methyl-β-glucopyranose conjugate, while *Absidia glauca* (ATCC 22752), *C. elegans*

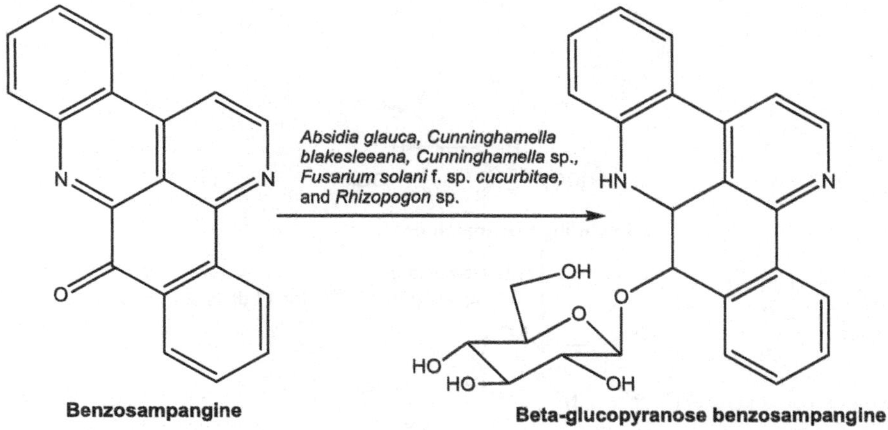

Scheme 7 Conversion of benzosampangine into the β-glucopyranose conjugate of benzosampangine utilizing different microorganisms [58]

(ATCC 9245), *Cunninghamella* sp. (NRRL 5695), and *Rhizopus arrhizus* (ATCC 11145) produced the β-glucopyranose conjugate [59]. The structures of the related starting and transformed molecules are shown in Fig. 8.

The metabolites 4′-*O*-methyl-β-glucopyranose and β-glucopyranose conjugates have been characterized on the basis of spectral data [59]. Both of them exhibited significant in vitro activity against *Cryptococcus neoformans* but did not show activity against *Candida albicans*. The β-glucopyranose conjugate was inactive in vivo in a mouse model of cryptococcosis [59].

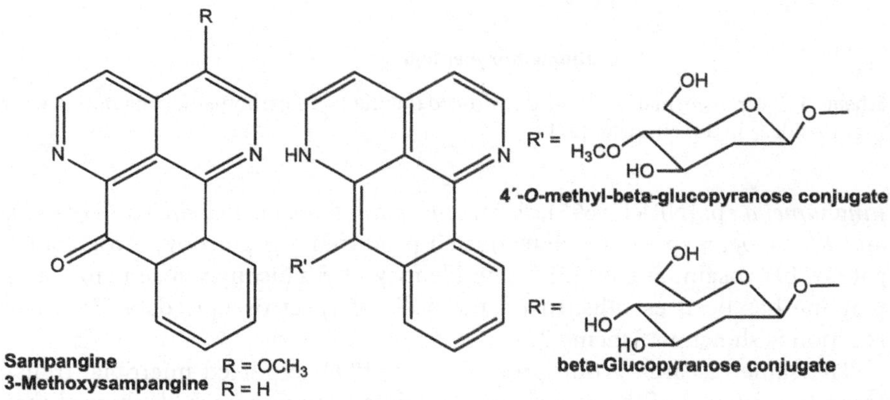

Fig. 8 Structures of sampangine, 3-methoxysampangine, and the transformed 4′-*O*-methyl-β-glucopyranose and β-glucopyranose conjugates [59]

5.8 Azacarbazoles

Peczynska-Czoch et al. reported the microbial transformation of cytotoxic 5,11-dimethyl-5*H*-indolo[2,3-*b*]quinoline, which is known as a compound displaying antitumor activity and affecting the activity of calf thymus DNA topoisomerase II, utilizing the *Rhizopus arrhizus* strain which yielded a 9-hydroxy derivative [60]. The metabolite obtained displayed a stronger cytotoxity against KB cells than the parent compound ($ID_{50} = 0.001$ μmol/mL), and also stimulated the formation of calf thymus topoisomerase II mediated pSP65 DNA cleavage in vitro at 3 μM concentration [60]. The structure (shown in Scheme 8) of the transformed compound has been confirmed by ^{1}H NMR. The authors also claimed, as the compound is analogous to 9-hydroxyellipticine, which is an antitumor alkaloid, that this indolo[2,3-*b*]quinoline derivative can be regarded as a novel potential antitumor agent [60].

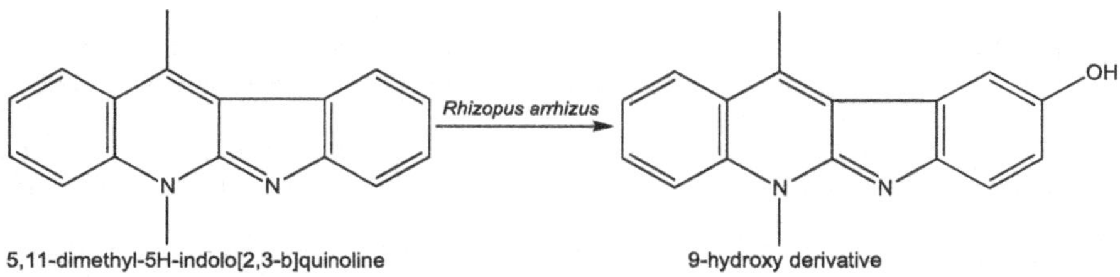

Scheme 8 Microbial transformation of the 5,11-dimethyl-5*H*-indolo[2,3-*b*]quinoline to its 9-hydroxy derivative in the presence of *Rhizopus arrhizus* [60]

6 Conclusion

Microbial biotransformations of pharmacologically important molecules are becoming a complementary tool in the study of drug metabolism in mammals, for the reason that most families of human liver cytochrome P450 have been expressed in microorganisms as individual enzymes. To obtain large amounts of metabolites for pharmacological and toxicological studies, microbial metabolism is clearly useful [16, 61–66].

Chemical synthetic methods of some biochemically essential compounds sometimes involve difficult, lengthy, and expensive steps, as well as low yield [51]. Utilizing biocatalytic approaches, either pure enzymes or living cells of microorganisms could be obtained in these cases through simple and short steps [52]. Recombinant DNA technology can provide biological routes for the synthesis of known and novel semisynthetic drug molecules [51].

References

1. Rathbone DA, Bruce NC (2002) Microbial transformation of alkaloids. Curr Opin Microbiol 5(3):274–281
2. Julsing MK et al. (2006) Combinatorial biosynthesis of medicinal plant secondary metabolites. Biomol Eng 23(6):265–279
3. Naito A (2000) Tracing the past half century of my microbial transformation studies. Yakugaku Zasshi 120(10):839–848
4. Rathbone DA, Lister DL, Bruce NC (2001) Biotransformation of alkaloids. In: Cordell GA (ed) The alkaloids: chemistry and biology, vol 57. Academic, San Diego, pp 1–74
5. Rathbone DA, Lister DL, Bruce NC (2003) Biotransformation of alkaloids. In: Cordell GA (ed) The alkaloids, vol 58. Elsevier, Amsterdam
6. Caponigro F, French RC, Kaye SB (1997) Protein kinase C: a worthwhile target for anticancer drugs? Anticancer Drugs 8(1):26–33
7. Rosazza JP (1978) Antitumor antibiotic bioactivation, biotransformation and derivatization by microbial systems. Recent Results Cancer Res 63:58–68
8. Kieslich K (1986) Production of drugs by microbial biosynthesis and biotransformation. Possibilities, limits and future developments (1st communication). Arzneimittelforschung 36(4):774–778
9. Kieslich K (1986) Production of drugs by microbial biosynthesis and biotransformation. Possibilities, limits and future developments (3rd communication). Arzneimittelforschung 36(6):1006–1009
10. Kieslich K (1986) Production of drugs by microbial biosynthesis and biotransformation. Possibilities, limits and future developments (2nd communication). Arzneimittelforschung 36(5):888–892
11. Venisetty RK, Ciddi V (2003) Application of microbial biotransformation for the new drug discovery using natural drugs as substrate. Curr Pharmaceut Biotechnol 4:153–167
12. Foster GR et al. (1991) Can J Microbiol 37(5):791–795
13. Rosazza JP et al. (1975) Microbial models of mammalian metabolism. O-dealkylation of 10,11-dimethoxyaporphine. J Med Chem 18(8):791–794
14. Smith RV, Rosazza JP (1975) Microbial models of mammalian metabolism. J Pharm Sci 64(11):1737–1759
15. Smith RV et al. (1975) Gas–liquid and thin-layer chromatographic determinations of xylenols in microbial extracts. J Chromatogr 106(1):235–237
16. Smith RV, Rosazza JP (1974) Microbial models of mammalian metabolism. Aromatic hydroxylation. Arch Biochem Biophys 161(2):551–558
17. Smith RV, Rosazza JP (1983) Microbial models of mammalian metabolism. J Nat Prod 46(1):79–91
18. Izuka H, Naito A (1981) Microbial conversion of steroids and alkaloids. University of Tokyo Press/Springer, Berlin
19. Faber K (1995) Biotransformations in organic chemistry, 2nd edn. Springer, Berlin
20. Fonken GS, Johnson RA (1972) Chemical oxidations with microorganisms. In: Belew JS (ed) Oxidation in organic chemistry, vol 2. Marcel Dekker, New York, pp 185–212
21. Kieslich KC (1976) Microbial transformation of non-steroid cyclic compounds. Wiley Interscience, New York
22. De Raddt A et al. (1992) Microbial reagents in organic synthesis—microbial and enzymatic transformation of nitriles. In: Servi S (ed) NATO ASI Series C, vol 381. Kluwer, Dordrecht, pp 209–253

23. Cannell RJ et al. (1997) Xenobiotica 27(1):147–157
24. Rosazza JP, Kammer M, Youel L (1977) Microbial models of mammalian metabolism: O-demethylations of papaverine. Xenobiotica 7(3):133–143
25. Ronald EB, David EW, Rosazza JP (1974) J Med Chem 17(6):599–602
26. Hartman RE et al. (1964) Microbial hydroxylation of indole alkaloids. Appl Microbiol 12:138–140
27. Groger D, Schumander HP (1969) Experientia 25(1):95–96
28. Davis PJ, Rosazza JP (1976) J Org Chem 41:2548–2551
29. Zeitler HJ, Niemer H (1969) Hoppe-Seyler's Z Physiol Chem 350(2):366–372
30. Sinderlar RD, Rosazza JP, Barfknecht CF (1979) Appl Environ Microbiol 38(5):208–211
31. Khaled YO et al. (1999) J Nat Prod 62(7):988–992
32. Soolee IR et al. (1989) J Nat Prod 52(2):337–341
33. El Sayed KA (1998) J Nat Prod 61(1):149–151
34. Shibuya H et al. (2003) Transformation of Cinchona alkaloids into 1-N-oxide derivatives by endophytic *Xylaria* sp. isolated from *Cinchona pubescens*. Chem Pharm Bull (Tokyo) 51(1):71–74
35. Kupchan SM, Zimmerman JH, Afonso A (1961) Lloydia 24:1–26
36. Honerjager P (1982) Rev Physiol Biochem Pharmacol 92:1–74
37. El Sayed KA et al. (1996) Int J Pharmacogn 34:161–173
38. El Sayed KA, Dunbar DC (2002) Microbial transformation of rubijervine. Chem Pharm Bull (Tokyo) 50(11):1427–1429
39. Wang SN et al. (2005) "Green" route to 6-hydroxy-3-succinoyl-pyridine from (*S*)-nicotine of tobacco waste by whole cells of a Pseudomonas sp. Environ Sci Technol 39(17):6877–6880
40. Desmet P, Elferink F, Verpoorte R (1989) Left-turning tetrahydropalmatine in Chinese tablets. Ned Tijdschr Geneeskd 133:308
41. Picciotto A et al. (1998) Chronic hepatitis induced by jin bu huan. J Hepatol 28(1):165–167
42. Woolf GM et al. (1994) Acute hepatitis associated with the Chinese herbal product jin bu huan. Ann Int Med 121(10):729–735
43. Divinsky M (2002) Case report: jin bu huan—not so benign herbal medicine. Can Fam Physician 48:1640–1642
44. McRae CA et al. (2002) Hepatitis associated with Chinese herbs. Eur J Gastroenterol Hepatol 14(5):559–562
45. Stickel F, Egerer G, Seitz HK (2000) Hepatotoxicity of botanicals. Public Health Nutr 3(2):113–124
46. Li L et al. (2006) Liquid chromatography–tandem mass spectrometry for the identification of L-tetrahydropalmatine metabolites in *Penicillium janthinellum* and rats. Biomed Chromatogr 20(1):95–100
47. Canonica L et al. (1983) The microbial oxygenation of the benzylisoquinoline alkaloid laudanosine. Cell Mol Life Sci 39(11):1273–1275
48. Shamma M (1972) The isoquinoline alkaloids. Academic Press, New York
49. Dhingra P (1978) In: Trahanovsky WS (ed) Oxidations in organic chemistry. Academic Press, New York
50. Stabler PJ, Holt PJ, Bruce NC (2001) Transformation of 2,2′-bimorphine to the novel compounds 10-alpha-S-monohydroxy-2,2′-bimorphine and 10,10′-alpha,alpha′-S,S′-dihydroxy-2,2′-bimorphine by *Cylindrocarpon didymum*. Appl Environ Microbiol 67(8):3716–3719
51. French CE et al. (1995) Biological production of semisynthetic opiates using genetically engineered bacteria. Biotechnology (NY) 13(7):674–676

52. Long MT et al. (1995) Transformations of morphine alkaloids by *Pseudomonas putida* M10. Appl Environ Microbiol 61(10):3645–3649
53. Melmon KL, Morrelli HF (1972) Clinical pharmacology: basic principles in therapeutics. Macmillan, New York
54. Moffat AC et al. (1986) Clarke's isolation and identification of drugs. The Pharmaceutical Press, London
55. Stabler PJ, Bruce NC (1998) Oxidation of morphine to 2,2′-bimorphine by cylindrocarpon didymum. Appl Environ Microbiol 64(10):4106–4108
56. El Sayed KA (2000) Microbial transformation of papaveraldine. Phytochemistry 53(6):675–678
57. Takayama H et al. (2003) Gluco-indole alkaloids from *Nauclea cadamba* in Thailand and transformation of 3alpha-dihydrocadambine into the indolopyridine alkaloid, 16-carbomethoxynaufoline. Chem Pharm Bull (Tokyo) 51(2):232–233
58. Orabi KY, Clark AM, Hufford CD (2000) Microbial transformation of benzosampangine. J Nat Prod 63(3):396–398
59. Orabi KY et al. (1999) Microbial transformation of sampangine. J Nat Prod 62(7):988–992
60. Peczyfiska-Czoch W et al. (1996) Microbial transformation of azacarbazoles X: regioselective hydroxylation of 5,11-dimethyl-5*H*-indolo[2,3-*b*]quinoline, a novel DNA topoisomerase II inhibitor, by *Rhizopus arrhizus*. Biotechnol Lett 18(2):123–128
61. Azerad R (1999) Microbial models for drug metabolism. Adv Biochem Eng Biotechnol 63:169–218
62. Lacroix I, Biton J, Azerad R (1999) Microbial models of drug metabolism: microbial transformations of trimegestone (RU 27987), a 3-keto-delta(4,9(10))-19-norsteroid drug. Bioorg Med Chem 7(11):2329–2341
63. Maurs M et al. (1999) Microbial hydroxylation of natural drimenic lactones. Phytochemistry 52(2):291–296
64. Isabelle L, Jacques B, Robert A (1999) Microbial models of drug metabolism: microbial transformations of trimegestone (RU 27987), a 3-keto-4,9(10)-19-norsteroid drug. Bioorg Med Chem 7:2329–2341
65. Moussa C et al. (1997) Microbial models of mammalian metabolism. Fungal metabolism of phenolic and nonphenolic *p*-cymene-related drugs and prodrugs. II. Metabolites of nonphenolic derivatives. Drug Metab Dispos 25(3):311–316
66. Moussa C et al. (1997) Microbial models of mammalian metabolism. Fungal metabolism of phenolic and nonphenolic *p*-cymene-related drugs and prodrugs. I. Metabolites of thymoxamine. Drug Metab Dispos 25(3):301–310

Top Heterocycl Chem (2007) 10: 123–153
DOI 10.1007/7081_2007_055

Published online: 9 June 2007

Synthesis of Triazole and Coumarin Compounds and Their Physiological Activity

Naceur Hamdi[1,2] (✉) · Pierre H. Dixneuf[3]

[1]Borj Cedria Higher Institute of Sciences and Technology of Environment, Ecopark of Borj Cedria Touristic road of Soliman, 2050 Hamman Lif, Tunisia
hamdi_naceur@yahoo.fr

[2]*Present address:*
maison no J99, DIAR BENMAHMUD, Elakba, 2011 Denden, Tunisia

[3]Institut de Chimie de Rennes, UMR 6509 CNRS, Université de Rennes, campus de Beaulieu, 35042 Rennes, France

Abstract Seventy three coumarinc derivatives were synthesized. The structures of these obtained products were proved on the basis of spectral and analytical data. A comparative pharmacological study showed that these compounds have different anticoagulant and antiinflammatory activities. The most prospective compounds are with enhance activity.

Keywords Anti-inflammatory activity · Benzofuranyl ether · Coumarinyl ether · Triazoles

1 Introduction

The biological importance of coumarins as an anticoagulant, aflatoxins, coumesterol as an estrogen, and phytoalexin has led to a considerable amount of synthetic work in the field of coumarins with 3 : 4-carbocyclic and 3 : 4-heterocyclic fused ring systems. These systems sometimes serve as useful synthetic intermediates as, for example, in the synthesis of analogues of the naturally occurring citromycetin, tetrahydrocannabinol, and 6-ketorotenoids. Although much data is available on the synthetic and pharmacological properties of these systems, there seems to be no organized review on this subject.

It would therefore be useful to bring the data together in a review and make it available for ready and easy reference.

2 Synthetic Ring Systems

A series of novel 1-substituted-4-phenyl-1,2,3-triazolo(4,3-a)quinazolin-5(4H)-ones **1** were synthesized by the cyclization of 2-hydrazino-3-phenyl-quinazolin-4(3H) **2** with various one carbon donors. The starting material 2-hydrazino-3-phenylquinazolin-4(3H)-one **2**, was synthesized from aniline **7** by a novel innovative route. When tested for their in vivo H_1-antihistaminic activity on conscious guinea pigs all the test compounds protected the animals from histamine-induced bronchospasm significantly, whereas the compound 1-methyl-4-phenyl-1,2,3-triazolo(4,3-a)quinazolin-5(4H)-one **1b** (percentage protection 70.7%) was found to be equipotent with the reference standard chlorpheniramine maleate (percentage protection 71%). These compounds show negligible sedation (5%) when compared to the reference standard (26%). Hence they could serve as prototype molecules for future development [1, 4, 5].

The key intermediate 2-thioxo-3-phenylquinazolin-4(3H)-one was prepared by adding carbon disulfide and sodium hydroxide solution simultaneously to a vigorously stirred solution of aniline **7** in dimethylsulfoxide over 30 min; stirring was then continued for an additional 30 min. Dimethylsulfate was added to the reaction mixture whilst stirring at 5–10 °C after which it was stirred for another 2 h and then poured into ice water to obtain a solid dithiocarbamic acid methylester **6**. The compound **6** and methylanthranilate **5** when refluxed in ethanol for 18 h yielded the desired 2-thioxo-3-substituted quinazolin-4(3H)-one **4**. The product obtained was cyclic and not an open chain thiourea **5a**. It was confirmed by its R_f value, high melting point, and its solubility in sodium hydroxide solution. The IR spectra of these compounds show intense peaks at 3220 cm^{-1} for amino (NH), 1660 cm^{-1} for carbonyl (CO), and 1200 cm^{-1} for thioxo (CS) stretching.

The NMR spectrum of **4** showed signals at 7–9 (m,9H,ArH) and 10.5 (s,1H,NH). Data from the elemental analyses have been found to be in conformity with the assigned structures. Furthermore, the molecular ion recorder in the mass spectrum is also in agreement with the molecular weight of the compound.

The 2-methylthio-3-substituted quinazolin-4(3H)-one **3** was obtained by dissolving **4** in 2% alcoholic sodium hydroxide solution and methylating with dimethylsulfate whilst stirring at room temperature (yield 88%, m.p 124–126 C). The IR spectrum of **3** showed the disappearance of the amino (NH) and thioxo (CS) stretching signals of the starting materials. It showed a peak for carbonyl (CO) stretching at 1680 cm^{-1}. The NMR spectrum of compound **3** showed signals at 2.5 (s,3H,SCH_3) and 7.0–8.6 (m,9H,ArH). Data from the elemental analyses and molecular ion recorded in the mass spectrum further confirmed the assigned structure.

Nucleophilic displacement of the methylthio group of **3** with hydrazine hydrate was carried out using ethanol as the solvent to afford 2-hydrazino-3-substituted quinazolin4(3H)-one **2** (yield 81%, m.p 158–160 C).

The long duration of the reaction (22 h) might be due to the presence of a bulky aromatic ring at position 3, which might have reduced the reactivity of the quinazoline ring system at the C_2 position. The formation of **6** was confirmed by the presence of NH and NH_2 peaks around 3334–3280 cm^{-1} in the IR spectrum. It also showed a peak for carbonyl (CO) at 1680 cm^{-1}. The NMR spectrum of the compound **6** showed signals at 5.0 (s,2H,$NHNH_2$), 7.0–8.1 (M,9H,ArH), and 8.7 (s,1H,$NHNH_2$). Data from the elemental analyses have been found to be in conformity with the assigned structure. Furthermore, the molecular ion recorded in the mass spectrum is also in agreement with the molecular weight of the compound.

The title compounds **1a–e** were obtained in fair to good yields through the cyclization of **2** with a variety of one carbon donors such as formic acid, acid acetic, propionic acid, buturic acid, and chloroacetyl chloride at reflux (Scheme 1).

The formation of cyclic product is indicated by the disappearance of peaks due to the NH and NH_2 of the starting material at 3400–3200 cm^{-1} in the IR

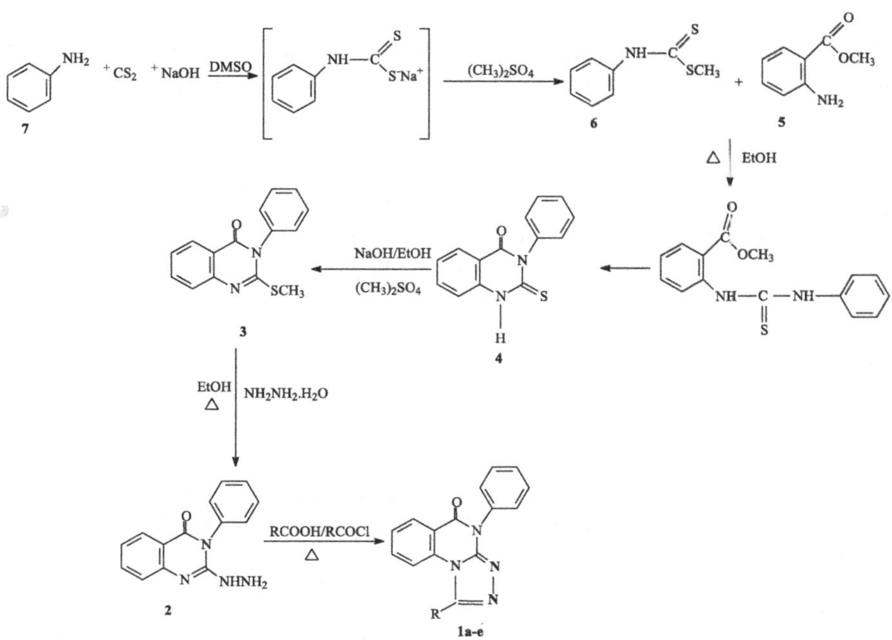

Scheme 1 Synthesis protocol of the compounds **1a–e**

spectrum of all the compounds **1a–e**. The NMR spectrum of **1a–e** showed the absence of NH and NH_2 signals. A multiplet at 7.0–8.0 integrating for aromatic protons was observed. The molecular ion recorded in the mass spectrum is in agreement with the compounds. Elemental (C, H, N) analysis indicated that the calculated and observed values were within the acceptable limits (±0.4%). Physical data of the title compounds are represented in Table 1.

The in vivo antihistaminic activity results indicate that all test compounds protected the animals from histamine-induced bronchospasm significantly. Structural activity relationship (SAR) studies indicated that different alkyl substituents on the first position of the triazoloquinazoline ring exerted varied biological activity.

Compound **1a** with no substitution, showed good activity with increased lipophilicity activity. Further increases in lipophilicity led to a decrease in activity. Replacement of a proton of the methyl group by a lipophobic group (chloro) resulted in a further decrease in activity. The order of activity of substituents at the first position was methyl, ethyl, unsubstituted, propyl, and chloromethyl. Compounds with a small substituent at C_1 seem to provide optimum activity. As the test compounds could not be converted to water soluble form, in vitro evaluation for antihistaminic activity could not be performed.

Table 1 Physical and pharmacological data of compounds **1a–e**

Compd no	R	Yield, %	Mp,C (re-cryst solv.[a])	Molecular formula[b]	Mol. wt.[c]	% Protection[d]	% CNS depression[d]
1a	H	89	262–265(C–E)	$C_{15}H_{10}N_4O$	262	69.9	4.2
1b	CH_3	87	276–279(C–E)	$C_{16}H_{12}N_4O$	276	70.7	4.8
1c	CH_2CH_3	80	190–193(E)	$C_{17}H_{14}N_4O$	290	70.0	5.3
1d	$(CH_2)_2CH_3$	76	190–194(C–E)	$C_{18}H_{16}N_4O$	304	69.2	7.3
1e	CH_2Cl	78	258–260(C–E)	$C_{16}H_{11}N_4OCl$	310	68.7	3.7
Chlorphenir-amine maleate						71.0	26
Cetiri-zine						78.9	8.5

[a] Abbreviations for the solvents used are as follows: C = Chloroform, E = Ethanol
[b] Elemental (C,H,N) analysis indicated that the calculated and observed values were within the acceptable limits (±0.4%)
[c] Molecular weight determination by mass spectral analysis
[d] Values are the means from six separate experiments. SE was less than 10% of the mean. Dose of test compounds, chlorpheniramine maleate and cetirizine are 10 mg/kg for antihistaminic activity, and 5 mg/kg for sedative-hypnotic activity

As sedation is one of the major side effects associated with antihistamines, the test compounds were also evaluated for their sedative potentials. This was determined by measuring the reduction in locomotor activity using an actophotometer [6, 7]. The test compounds and the reference standards (chlorpheniramine maleate and cetirizine) were administrated orally at a dose of 5 mg/kg in 1% CMC.

The (7-hydroxy-2-oxo-2H-chromen-4-yl)acetic acid starting material was originally prepared by condensing resorcinol with acetonedicarboxylic acid in the presence of concentrated sulfuric acid, a procedure later simplified by Dey and Row [8]. Applying the hydrazinolysis of (7-hydroxy-2-oxo-2H-chromen-4-yl)acetic acid ethyl ester **8**, with 100% hydrazine hydrate in methanol at room temperature, (7-hydroxy-2-oxo-2H-chromen-4-yl)acetic acid hydrazide **9** was prepared in good yields. The carbohydrazide **9** was then condensed with different aromatic aldehydes in ethanol/acetic acid (24:1) to give the corresponding Shiffs bases, i.e., (7-hydroxy-2-oxo-2H-chromen-4-yl)acetic acid arylidenehydrazides **10a–l**, in very good yields (Scheme 2).

The IR spectra of carbohydrazide **9** showed absorption bands at 3317 cm^{-1} (OH,Hydrazide NH_2), 3269 cm^{-1} (aromatic CH), 1711 cm^{-1} (CO stretching), and 1621–1640 cm^{-1} ($CO-NH-NH_2$ groups). The ^1H NMR spectra exhibited a singlet due to the $CONHNH_2$, NH proton at 9.32 ppm. Methylene protons resonated as a singlet at 4.23 ppm. The structures of the products **10a–l** were inferred from their analytical and spectral data. Thus, their IR spectra showed characteristic absorption bands at 3400–3240 cm^{-1} (NH,OH), 1710–1700 cm^{-1} (lactone CO), and NHCO at 1650–1600 cm^{-1}.

The ^1H NMR spectra did not only show the absence of the NH_2 protons at 3.34, but also the presence of the NCH proton at 8.16 ppm.

On the other hand, refluxing **9** in formic acid for 5 h afforded the N-formyl derivative **11** in high yield. Acetylation of **9** by refluxing in acetic acid, afforded acetic acid *N*′-(2-(7-hydroxy-2-oxo-2H-chromen-4-yl)-acetyl)-hydrazide **12** in good yield. Compound **13** was also obtained by refluxing **9** with 3-(2-bromoacetyl)-4-hydroxy-2H-chromen-2-one in ethanol. Reaction of compound **9** with phenyl isothiocyanate in ethanol at room temperature gave 4-phenyl-1-(7-hydroxy-2-oxo-2H-chromen-4-acetyl-)thiosemicarbazide **14**.

Condensation of **9** with ethyl acetoacetate without a solvent gave ethyl 3-(2-(2-(7-hydroxy-2-oxo-2H-chromen-4-yl)-acetyl)hydrazono) butanoate **15** in 48% yield (Scheme 3).

The structures of compounds **11–15** were established by their analytical data and their IR and ^1H NMR spectra. The IR absorptions due to the NH, OH, and CO functions appeared at 3450–3000 and 1728–1610 cm^{-1}, respectively. The ^1H NMR spectra of compounds **11–15** exhibited singlets in the 8.03–10, 58 ppm region corresponding to the NH and OH protons.

When **11** was refluxed with equimolar amounts of 4-trifluoro methylaniline or 2,3,4-trifluoro aniline in acetonitrile with a few drops of acetic acid, the compounds (7-hydroxy-2-oxo-2H-chromen-4-yl) acetic acid-

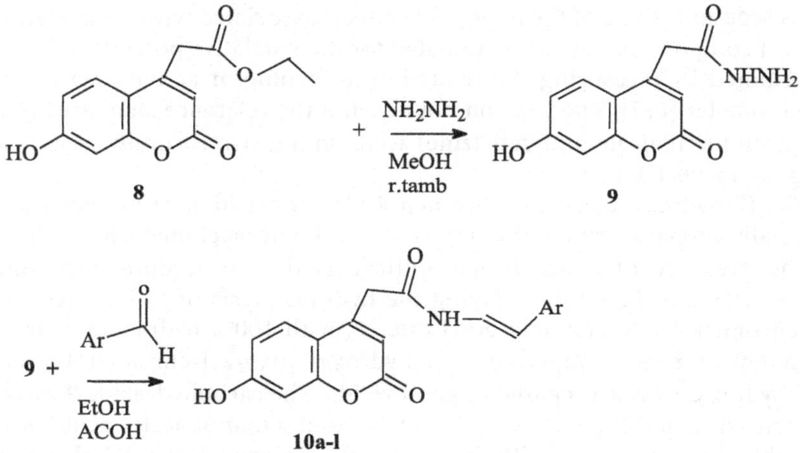

Entry	Ar	Entry	Ar
a	phenyl	g	2,5-dihydroxyphenyl
b	2-hydroxyphenyl	h	3-phenoxyphenyl
c	2-chlorophenyl	i	3-methoxy-4-hydroxphenyl
d	3-chlorophenyl	j	styryl
e	2,3-dihydroxyphenyl	k	4-N,N-dimethylaminophenyl
f	2,4-dihydroxyphenyl	l	2-hydroxy-5-nitrophenyl

Scheme 2 Synthesis protocol of the compounds **10a–l**

N′-[(4trifluoromethylphenylimino)methyl]hydrazide **16** and (7-hydroxy-2-oxo-2H-chromen-4-yl)-aceticacide-*N′*-[(2,3,4-trifluorophenylimino)methyl] hydrazide **17** were obtained in good yields (Scheme 4).

Cyclization of thiosemicarbazide **14** with chloroacetylchloride in chloroform afforded thiazolidinone derivative **22** (Scheme 5).

The structures of compounds **18–22** were confirmed by their analytical data and their IR and ^{1}H NMR spectra.

Compounds **10a–l** and **11–22** described here were examined for their antimicrobial activity. Good results were obtained in compounds **13, 14, 18,** and **19**; all these compounds were found to possess high antimicrobial activity against *Staphylococcus pneumoniae* and were slightly less active against *Pseudomonas aeruginosa*, *Bacillus subtilis*, *Bacillus cereus*, and *Salmonella panama*. The other compounds showed either moderate or no activity against these organisms [9].

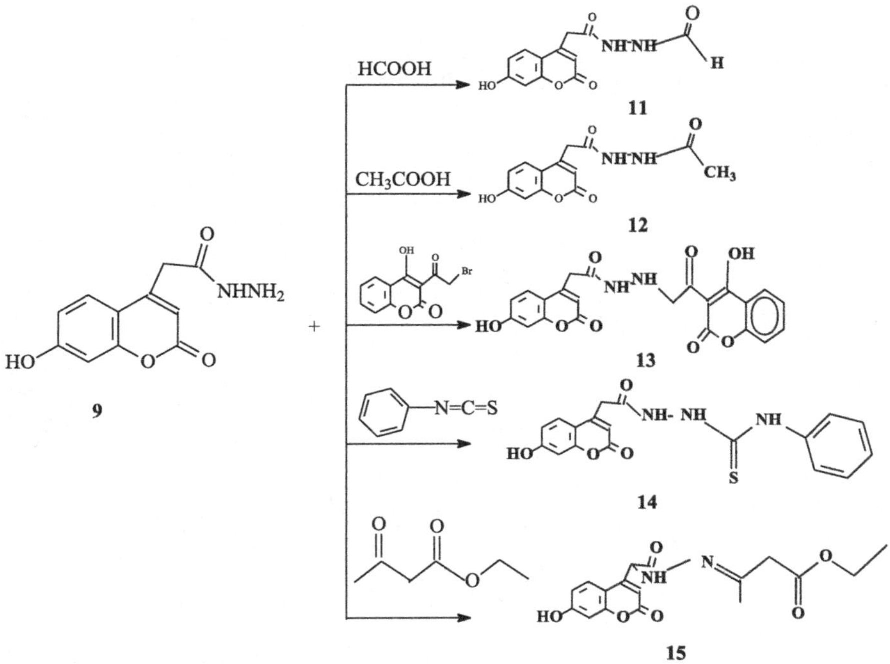

Scheme 3 Synthesis protocol of the compounds 11–15

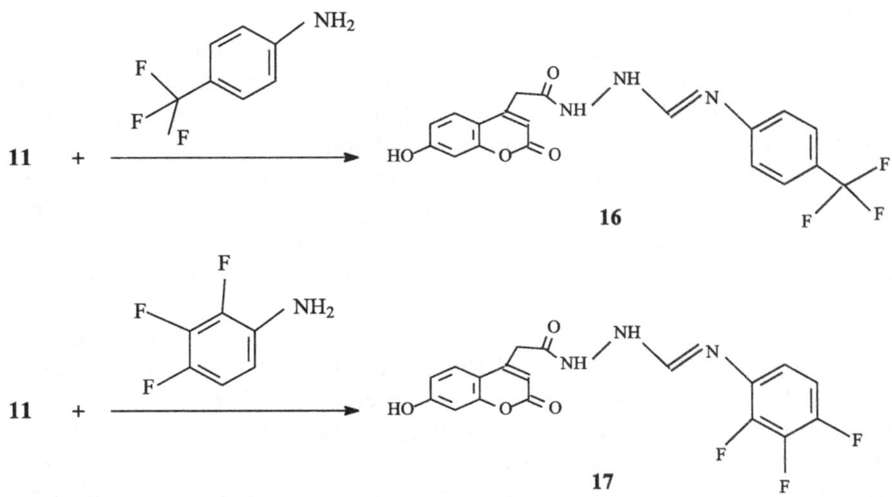

Scheme 4 Synthesis protocol of the compounds 16–17

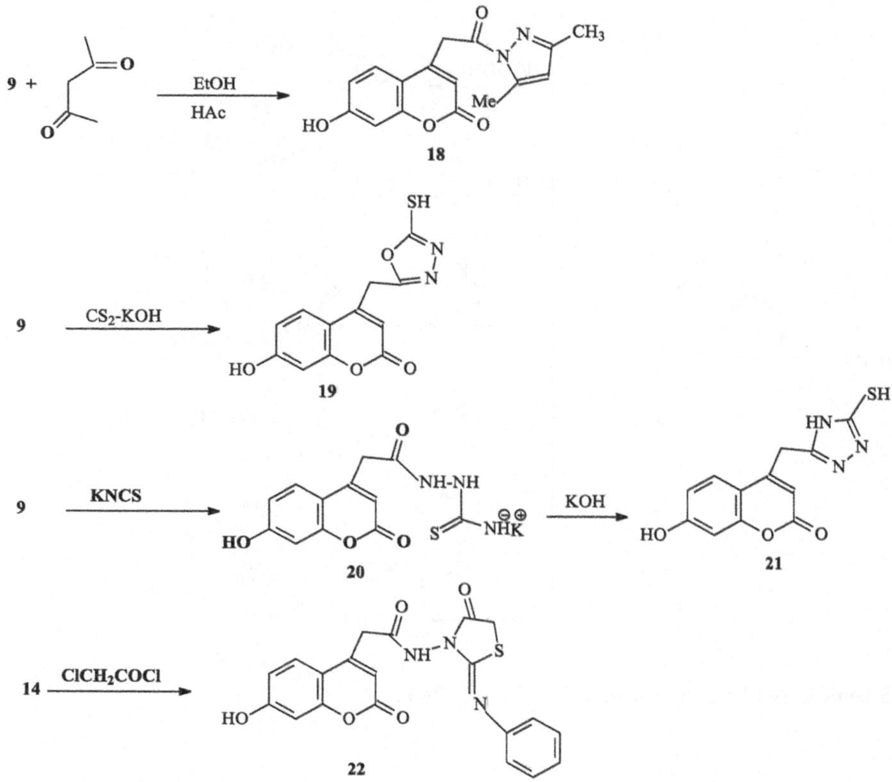

Scheme 5 Synthesis protocol of the compounds **18–22**

It is known that certain coumarins are transformed into 5-(2-hydroxyphenyl)-3H-pyrazol-3-ones by reaction with hydrazines [10]. Upon application of this method, the expected pyrazolones **25** were synthesized starting from 3-nitrocoumarins [11] as known in (Scheme 6).

Treatment of 4-methoxy-3-nitrocoumarin **23** with hydrazine hydrate and methylhydrazine in ethanol at room temperature for 3 h gave 1,2-dihydro-5-(2-hydroxyphenyl)-4-nitro-3H-pyrazol-3-one **24a** and 1,2-dihydro-5-(2-hydroxyphenyl)-2-methyl-4-nitro-3H-pyrazol-3-one **24b** in 70 and 51% yields, respectively.

Compound **24a** was also obtained in 17% yield by heating of 4-hydroxy-3-nitrocoumarin **27** [12] with hydrazine hydrate in ethanol. A similar reaction of **27** with methylhydraine in boiling ethanol did not afford **24b** because of the decomposition of the starting coumarin. However, when **27** was treated with methylhydrazine at room temperature for 24 h without solvent, the ring-opened methylhydrazine adduct **28** (36% yield) and 1,2-dihydro-5-(2-hydroxyphenyl)-1-methyl-4-nitro-3H-pyrazol-3-one **26** (11%yield) were

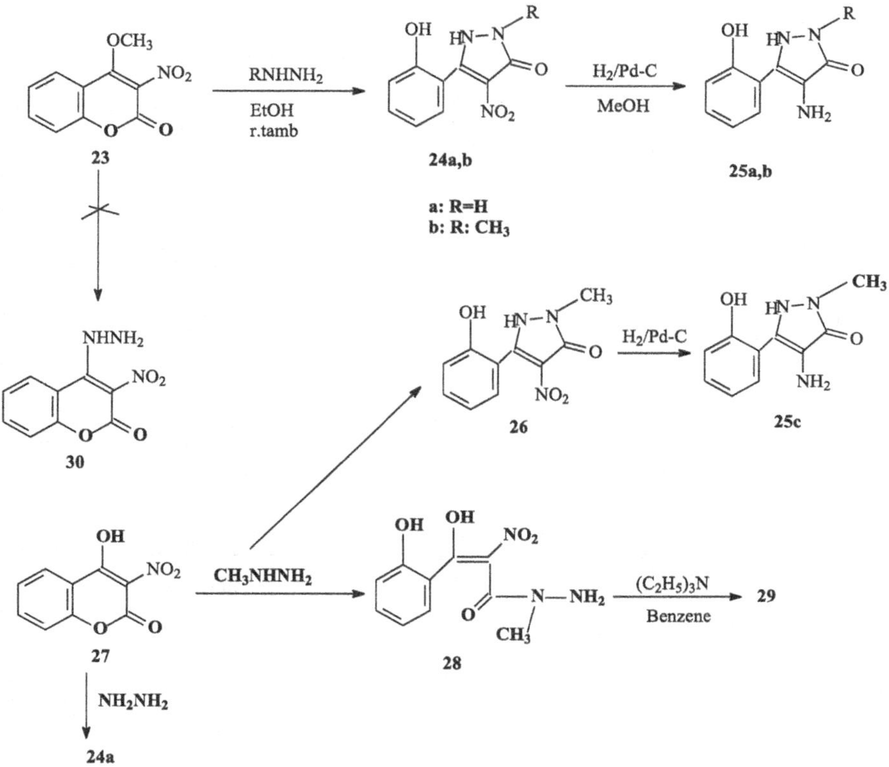

Scheme 6 Synthesis protocol of the compounds **24–29**

obtained. Compound **28** was cyclized to **29** when heated in benzene in the presence of triethylamine. The formation of **29** by the reaction of **23** with methylhydrazine resulted from selective attack of the secondary amino group of methylhydrazine at the 2-position of **23**. This selectivity may be due to the methoxyvinylcarbonyl moiety in **23**. It is worth noting that the reaction of **23** with hydrazines did not give 4-hydrazino-3-nitrocoumarins **30** which could arise by simple condensation of the hydrazines at the 4-position of **23**. This also suggests initial attack of the hydrazines at the 2-position of **23** in the pyrazolone formation.

Compounds **24a,b** were reduced to the corresponding amino pyrazolones **25a–c** by catalytic hydrogenation over 5% Pd – C in methanol. These amino pyrazolones were isolated as hydrochlorides because the free bases were decomposed in air with red coloring. The structural assignments of **24a,b** and **25a–c** were based on elemental analysis and spectral data.

The position of the methyl group of **25b,c** was deduced from high-resolution mass analysis as shown in Figs. 1 and 2. The mass fragmentation pattern of **25b** was quite different from that of **25c**, thus assuring the differ-

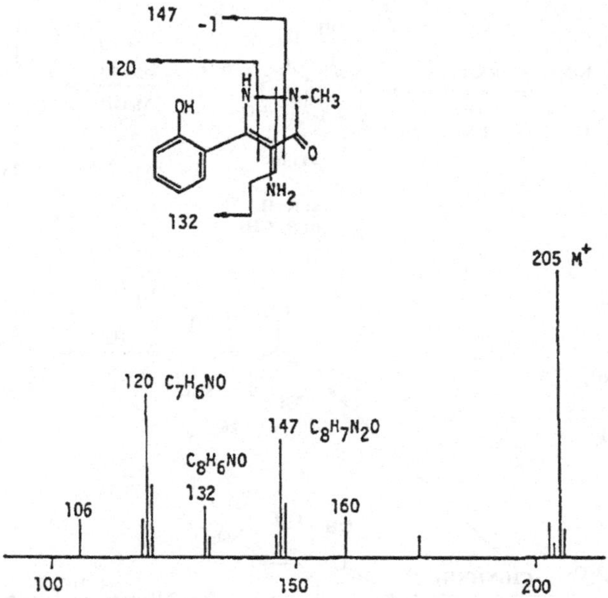

Fig. 1 MS of **25b**

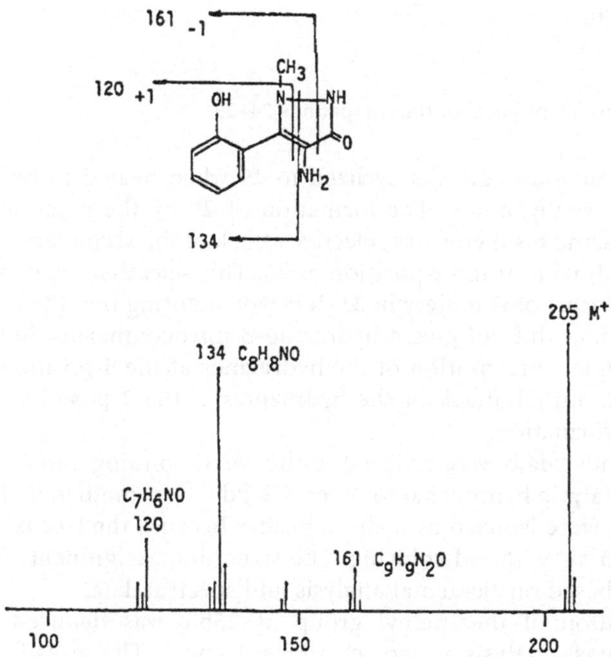

Fig. 2 MS of **25c**

ent position of the methyl group of these two isomers. The pyrimidinones **33** were also synthesized from **23** as shown in (Scheme 7).

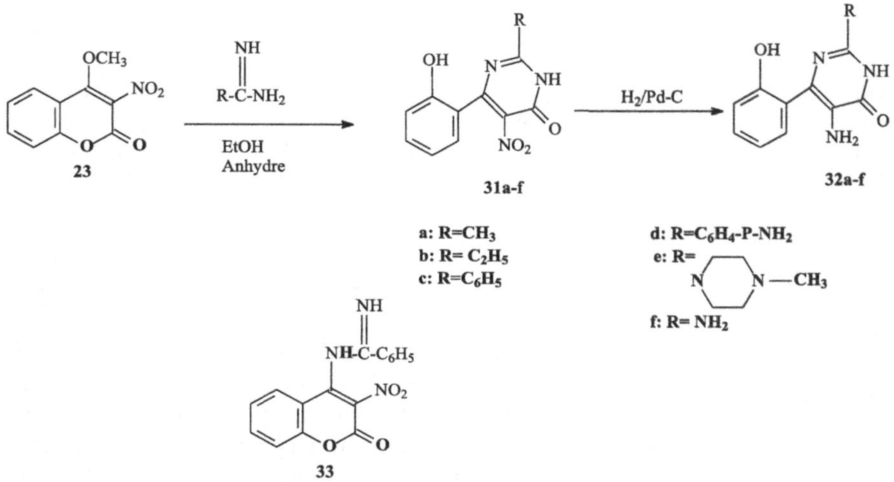

Scheme 7 Synthesis protocol of the compounds **31–33**

Pene et al. [13] reported that the coumarin **23** reacted with guanidine to give 2-amino-6-(2-hydroxy-phenyl)-5-nitropyrimidin-4(3H)-one **31**. However, this method was not suitable for the pyrimidinone formation when amidines were employed in place of guanidine. We therefore used **23** as a starting material, which readily gave the expected pyrimidinones by treatment with amidines. Heating of **23** with the appropriate amidine and guanidine (Scheme 7) in anhydrous ethanol for 1 h gave 2-substituted-6-(2-hydroxy-phenyl)-5-nitropyrimidin-4(3H)-ones **32a–f** (Table 2). Catalytic hydrogenation of these nitropyrimidines over 5% Pd – C in ethanol provided the corresponding aminopyrimidinones **33a–f** (Table 3). The structures of

Table 2 6-(2-Hydroxyphenyl)-5-nitropyrimidin-4(3H)-ones **32a–c**

Compound	Yield (%)	Mp (C)	Formula	MS (m/z) (M+)	IR (KBr) cm-1
32a	70	239–241	$C_{11}H_9N_3O_4$	247	1660,1530
32b	72	196–199	$C_{12}H_{11}N_3O_4$	261	1660,1535
32c	75	248–249	$C_{16}H_{11}N_3O_4$	309	1675,1535
32d	33	285–288	$C_{16}H_{12}N_4O_4$	324	3380,1650
32e	83	260–261	$C_{15}H_{17}N_5O_4$	331	1670,1530
32f	81	268–270	$C_{10}H_8N_4O_4$	248	3370,1690

Table 3 5-Amino-6-(2-hydroxyphenyl)pyrimidin-4(3H)-ones **33a–f**

Compound	Yield (%)	Mp (C)	Formula	MS (m/z) (M+)	IR (KBr) cm-1
33a	49	252–253	$C_{11}H_{11}N_3O_2$	217	3420,3350
33b	54	222–224	$C_{12}H_{13}N_3O_2$	231	3450,3360
33c	38	237–238	$C_{16}H_{13}N_3O_2$	279	3370,3250
33d	56	273–275	$C_{16}H_{14}N_4O_2$	294	3350,3450
33e	90	232–235	$C_{15}H_{19}N_5O_2$	301	3350,1630
33f	26	243–245	$C_{10}H_{10}N_4O_2$	218	3350,3200

32a–f and **33a–f** were determined on the basis of elemental analysis and spectral data.

The compounds **25a–c** and **33a–f** were tested for analgesic activity [14] by oral administration in mice in terms of the inhibition of the writhing syndrome induced by acetic acid. The ED_{50} which represents the dose producing 50% inhibition of the writhing induced by acetic acid was determined. The ED_{50} values of the compounds tested are summarized in Table 4. The pyrazolones **25a–c** showed more potent analgesic activity than **1** and aminopyrine.

In particular, when compared with aminopyrine the activity of **25b** was more than four-times greater. Replacement of the pyrazole ring in **1** by the pyrazolone nucleus resulted in an increase of the activity.

Table 4 Analgesic activity of compounds **25a–c** and **33a–c**

Compound	ED_{50}, mg/kg, p.o (95% CL)	Compound	ED_{50}, mg/kg, p.o (95% CL)
25a*	15.8 (11.0–23.7)	**33d**	14.4 (11.2–18.4)
25b*	9.8 (7.5–12.1)	**33e**	31.6 (29.1–34.4)
25c*	12.0 (8.8–16.1)	**33f**	35.0 (29.7–41.1)
33a	16.5 (13.4–20.8)	**1***	20.2 (18.4–22.2)
33b	34.5 (28.7–41.8)	aminopyrine	44.8 (38.7–51.8)
33c	39.5 (33.5–46.7)		

* Administrated in the form of hydrochloride

* Structure of **1**:

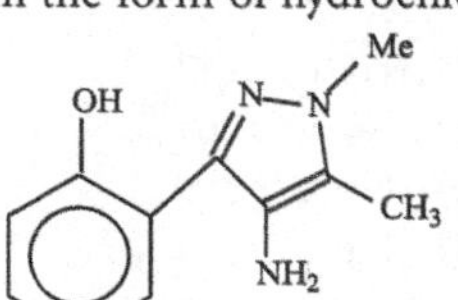

1

The pyrimidinones **33a–f** also exhibited analgesic activity which was comparable or superior to that of aminopyrine. Among these, **33a–d** showed prominent activity which was superior to that of **1** and about three times as potent as that of aminopyrine. From the data shown in (Table 4), it's difficult to estimate the contribution of the substituent R in the compounds **25** and **33** to the activity.

However, 4-amino-1,2-dihydro-5-(2-hydroxyphenyl)-3H-pyrazol-3-one and 5-amino-6-6-(2-hydroxyphenyl)pyrimidin-4(3H)-one appear to be interesting structures for developing new analgesic agents.

4-(Bromomethyl) coumarins **34** were synthesized by the Pechmann cyclization of phenols with 4-bromoethylacetoacetate [15]. They were reacted with vanillin **35** to give ethers **36**, which underwent Knovenagel condensation with ethyl cyanoacetate resulting in the formation of unsaturated cyanoesters **37**. Orthovanillin (**35A**) reacted with compounds **34** under similar conditions resulting in the formation of ethers **39**, which contain an active methylene group and the ortho carbonyl group. Ethers **36**, when refluxed in alcoholic potassium carbonate underwent an intermolecular aldol reaction leading to the formation of 4-(7′)-methoxy-2′-benzo(b)furanyl coumarins **38**, which is in accordance with earlier observations [16]. The reactions are outlined in (Scheme 8) and the compounds prepared are presented in Table 5 with their physical data.

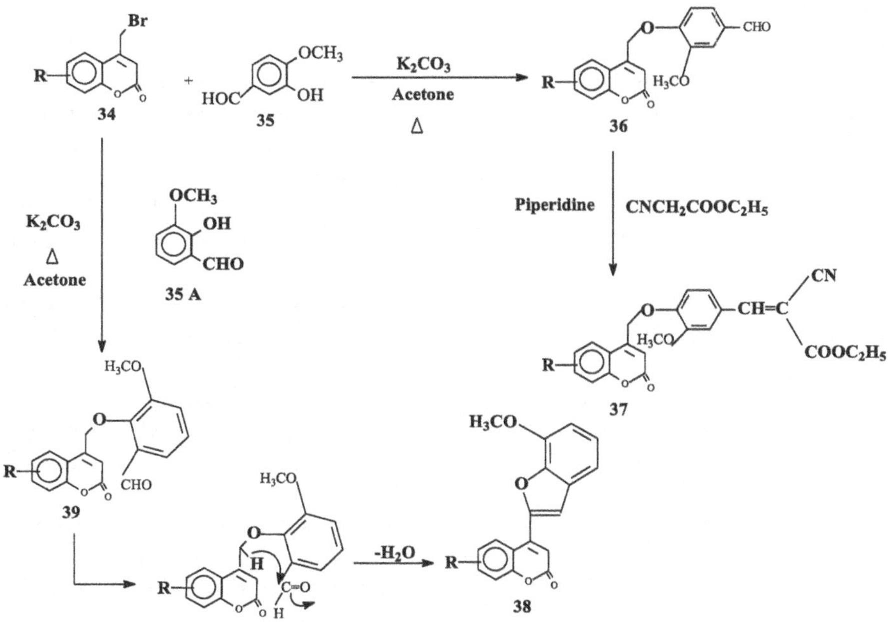

Scheme 8 Synthesis protocol of the compounds **34–38**

Table 5 Physical data of compounds **36–39**

Compound	R	Molecular formula	Yield (%)	M.P.(C)	Solvent for crystallization
36a	6-CH3	$C_{19}H_{16}O_5$	78	168–169	Chloroform
36b	7-CH3	$C_{19}H_{16}O_5$	80	198–199	Ethanol
36c	5,6-Benzo	$C_{22}H_{16}O_5$	80	224–225	DMF
36d	7,8-Benzo	$C_{22}H_{16}O_5$	75	232–233	DMF
36e	6-OCH3	$C_{19}H_{16}O_6$	79	204–205	Ethanol
36f	6-Cl	$C_{18}H_{13}O_5Cl$	60	212–213	Ethanol
37a	6-CH_3	$C_{24}H_{21}NO_6$	80	172–173	Ethanol
37b	7-CH3	$C_{24}H_{21}NO_6$	82	178–179	Ethanol
39a	6-CH3	$C_{19}H_{16}O_5$	72	160–161	Ethanol
39b	7-CH3	$C_{19}H_{16}O_5$	72	178–179	Ethanol
39c	5,6,Benzo	$C_{22}H_{16}O_5$	74	226–227	DMF
39d	7,8-Benzo	$C_{22}H_{16}O_5$	74	182–183	Dioxane
39e	6-OCH3	$C_{19}H_{16}O_6$	76	204–205	Ethanol
39f	6-Cl	$C_{18}H_{13}O_5Cl$	75	203–204	Ethanol
38a	6-CH3	$C_{19}H_{14}O_4$	78	164–165	Ethanol
38b	7-CH3	$C_{19}H_{14}O_4$	74	179–180	Ethanol
38c	5,6-Benzo	$C_{22}H_{14}O_4$	82	220–221	Ethanol
38d	7,8-Benzo	$C_{22}H_{14}O_4$	76	186–187	Dioxane
38e	6-OCH3	$C_{19}H_{14}O_5$	82	202–203	DMF
38f	6-Cl	$C_{18}H_{11}O_4Cl$	72	196–197	DMF

The vanillin ethers **36** and **39** exhibited the IR band of the lactone carbonyl group at 1710–1720 cm^{-1} whereas the aldehydic carbonyl stretching was observed in the range of 1680–1690 cm^{-1}. In the NMR spectra all the protons resonated at expected fields. The aldehydic proton appeared downfield around 9–10 aromatic protons in the range of 7–8 and the C_3 – H of coumarin around 6.5. The methylene, methoxy, and methyl protons resonated around 5, 3.8, and 2.2, δ respectively.

The cyano esters **37** exhibited a broad band around 1728 cm^{-1} due to carbonyl groups whereas the CN stretching band was observed around 2200 cm^{-1}.

Compound **37a** showed the absence of an aldehydic proton and the singlet around 8.15 ppm was assigned to the ethylenic proton located β with respect to the electron-withdrawing cyano and ester groups. The benzofuranyl coumarins **38** exhibited the carbonyl-stretching band around 1690 cm^{-1} in the IR spectra (Table 6). PMR data for 13 compounds are given in Table 2. The EI mass spectrum of **36a** showed a molecular ion peak at m/z 324 (41%).

All the compounds showed a good safety profile until the highest dose [17]. There was no sedation, convulsions and tremors upon inspection, no ulceration and no hemorrhagic spots were observed.

Table 6 NMR spectral data of compounds **36–39**

Compound	R	^{1}H-NMR (δ, ppm)
36a	6-CH_3	2.45(3H,s,CH3), 3.96(3H,s,OCH_3), 5.45(2H,s,OCH_2), 6.66(1H,s,C_3-H), 7.2–7.9(6H,m,Ar-H), (9.2(1H,s,CHO))
36b	7-CH3	2.47(3H,s,CH3), 4.0(3H,s,OCH_3), 5.36(2H,s,OCH_2), 6.63(1H,s,C_3-H), 7.0–7.5(6H,m,Ar-H), (9.86(1H,s,CHO))
36c	5,6-Benzo	3.96(3H,s,OCH_3), 5.3(2H,s,OCH_2), 6.5(1H,s,C_3-H), 6.8–7.9(8H,m,Ar-H), (10.0(1H,s,CHO))
36d	7,8-Benzo	3.98(3H,s,OCH_3), 5.48(2H,s,OCH_2), 6.8(1H,s,C_3-H), 6.81-9.91(8H,m,Ar-H), (9.89(1H,s,CHO))
36e	6-OCH3	3.9(3H,s,OCH_3), 4.0(3H,s,OCH3), 5.4(2H,s,OCH_2), 6.3(1H,s,C_3-H), 6.5–7.5(6H,m,Ar-H), (9.9(1H,s,CHO))
37a	6-CH_3	1.41(3H,t,CH_3), 4.40(2H,q,CH2), 2.5(3H,CH_3), 3.99(2H,s,OCH_3), 5.37(2H,s,OCH_2), 5.25(1H,s,C-H), 6.64(1H,s,C_3-H), 6.8–8.0(6H,m,Ar-H), (8.15(1H,s,CHO))
39a	6-CH_3	2.5(3H,s,CH_3), 4.0(3H,s,OCH_3), 5.45(2H,s,OCH_2), 6.63(1H,s,C_3-H), 6.5–7.5(6H,m,Ar-H), (10.0(1H,s,CHO))
39b	7-CH_3	2.6(3H,s,CH_3), 4.0(3H,s,OCH_3), 5.2(2H,s,OCH_2), 6.61(1H,s,C_3-H), 7.5–8.2(6H,m,Ar-H), (10.0(1H,s,CHO))
39c	5,6-Benzo	3.94(3H,s,OCH_3), 5.36(2H,s,OCH_2), 6.47(1H,s,C_3-H), 6.9–7.9(8H,m,Ar-H), (9.9(1H,s,CHO))
38a	6-CH_3	2.29(3H,s,CH_3), 4.0(3H,s,OCH_3), 6.62(1H,s,C_3-H), 6.5–7.9(7H,m,Ar-H)
38d	7,8-Benzo	4.0(3H,s,OCH_3), 6.66(1H,s,C_3-H), 6.9–8.1(9H,m,Ar-H)
38e	6-OCH_3	3.9(3H,s,OCH_3), 4.0(3H,s,OCH_3), 6.46(1H,s,C_3-H), 6.9–8.1(9H,m,Ar-H)
38f	6-Cl	4.0(3H,s,OCH_3), 6.63(1H,s,C_3-H), 7.0-7.5(6H,m,Ar-H)

The mean reaction time was around 5.5–6 s, which did not differ significantly from the control group. Hence it was inferred that these compounds did not show good analgesic activity.

The mean ulcer index was measured for the compounds **36c**, **37a**, and **38a**. The mean ulcer index for the phenylbutazone treated group is 40, which is highly significant. The value for control and the compounds were 3.33, 1.67, 6.56, and 10.00, respectively. These observations indicate that said compounds do not cause any ulceration.

The present study reports the anti-inflammatory activity of vanillin ethers, benzofuranyl coumarins, and unsaturated cyanoesters of 4-aryloxymethyl coumarins **36**. When compared with the control all the compounds showed reduction in edema volume.

Amongst the ethers **36** with p-formyl group the 5,6-benzo-substituted compound **36c** was found to exhibit maximum activity. The ethers with the ortho formyl group **39a** showed a quick onset of action initially, then the activity showed a decreasing trend at the end of 5 h. However, the corresponding benzofuranyl coumarin **38a** exhibited uniform activity comparable with the standard. The unsaturated cyanoester **37a** obtained from **36a** was found to be the most active in the series along with **36c** and **38a** as the other active compounds. The results indicating edema volume and percentage inhibition of inflammation at various time intervals have been summarized in Table 7.

It can be seen that though the aldehydic function is important for anti-inflammatory activity as observed in compound **36**, its conversion to the unsaturated cyanoesters **37** shows an increase in the activity. Similarly, the ethers **39** with the ortho aldehydic group upon conversion to benzofurans 6 show a slight increase in activity. None of the compounds tested showed any analgesic activity.

The 1H-1,2,4-triazole compounds possess important pharmacological activities such as antifungal and antiviral activities [18–20]. In the present study, the reactive intermediates **45a–c**, prepared in situ from the dichlorides **44a–c**, were reacted via the cycloaddition reaction with ethyl cyanoacetate **40** to give, after spontaneous rearrangement, the triazole hydrazides **41a–c**. These compounds were used as starting materials for the synthesis of the

Table 7 Comparison of edema volume at different time intervals

Group	Compound	Edema volume at different time intervals** ± S.E. and (% inhibition) 1 h	2 h	3 h
1	Control	0.76±0.15	1.03±0.14	1.08±0.16
2	Phenylbutazone	0.22 ± 0.07^{a} (71)	0.22 ± 0.07^{a} (78)	0.29 ± 0.05^{a} (75)
3	**3a**	0.18 ± 0.04^{a} (76)	0.34 ± 0.03^{a} (67)	0.29 ± 0.05^{a} (75)
4	**3b**	0.14±0.03 a(81)	0.32 ± 0.04^{a} (69)	0.31 ± 0.04^{a} (73)
5	**3c**	0.24 ± 0.04^{a} (68)	$0.18\pm0.4^{a,c}$ (82)	0.20 ± 0.07^{a} (83)
6	**3d**	0.30 ± 0.05^{a} (62)	$0.40\pm0.06^{a,b}$ (61)	$0.50\pm0.06^{a,c,d,e}$ (57)
7	**4a**	$0.09\pm0.05^{a,b}$ (88)	$0.09\pm0.07^{a,c,d,f}$ (90)	$0.16\pm0.05^{a,h}$ (86)
8	**5a**	0.13 ± 0.02^{a} (83)	$0.30\pm0.05^{a,c}$ (72)	$0.34\pm0.03^{a,f}$ (70)
9	**6a**	$0.12\pm0.03^{a,b}$ (84)	$0.19\pm0.02^{a,c}$ (81)	$0.24\pm0.02^{a,h}$ (79)
	F Value*	51.86	81.69	84.49
		$P < 0.01$	$P < 0.001$	$P < 0.001$

* The superscript a,b,c,d,e,f,g,h indicates significance difference from 1,2,3,4,5,6,7,8 groups, respectively. The data is analyzed by one-way ANOVA(F-test) followed by Newmann Keul's Studentized range test

** Index for anti-inflammatory activity. Model: acute inflammation, method: Carragennan induced edema, test animal: albino rats, number of animals per group: 6, route of administration: oral, standard: phenylbutazone (100 $mg\,kg^{-1}$), test compounds: 100 $mg\,kg^{-1}$

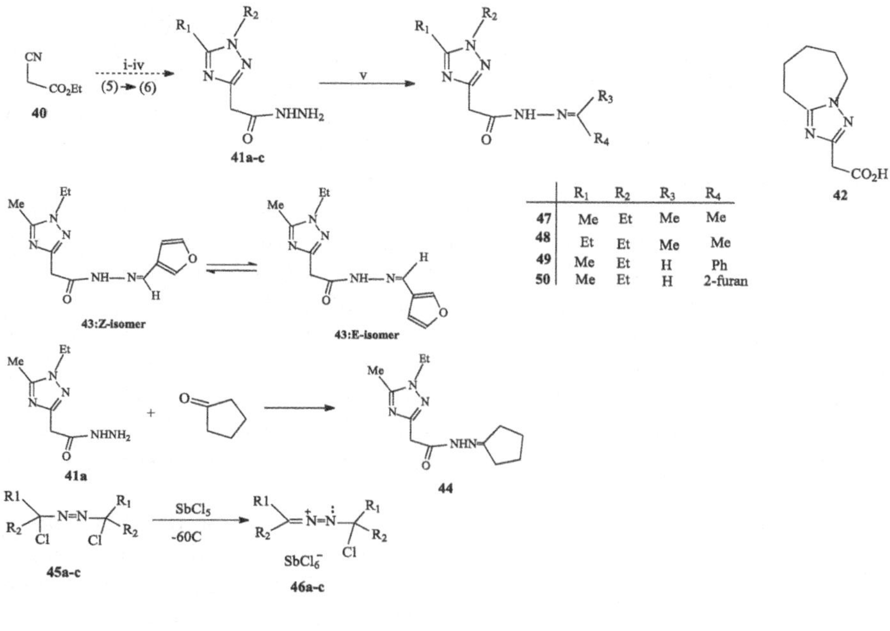

Scheme 9 Synthesis protocol of the compounds **41–50**. Conditions and reagents: (i) **45a-c**, (ii) $SbCl_5$, CH_2Cl_2, -60 °C to 23 °C, 7 h, CH_2Cl_2, (iii) aq. $NaHCO_3$, NH_3, MeCN, 0 °C, 2 h, (iv) NH_2NH_2, (v) RCHO or RCOR, E*t*OH

alkyidene derivatives **47–50**, by heating with the appropriate aldehydes or ketones in EtOH for 4–5 h in 70–83% (Scheme 9).

The structures of the new compounds were established by their ^{1}H, ^{13}C NMR, and mass spectra. Interestingly, the ^{1}H NMR spectra of **47** and **48** showed one isomer only, meanwhile compounds **49** and **50** were characterized by the presence of Z- and E-isomers because of the different groups around the unsaturated center.

Compound **50** was selected for further study by ^{1}H NMR measurements at different temperatures to establish the effect of the free rotation around the double bond between 30 and 130 °C. At 30 °C, the two Z/E isomers have been clearly confirmed, when the NH signal appeared as two half-singlets at δ 11.52/11.36, representing each a 1/2H, and the = CH furan as two half-singlets at δ 8.11/7.90, as well as the CH_2-1′ at δ 3.93/3.50 as two half singlets. At higher temperature (ca. 120 °C), there are changes in the ^{1}HNMR spectrum, which shows one isomer, proved from the signals of NH, CH furan, and CH_2-1′, oriented as three singlets at δ 10.93, 8.06, and 4.00, respectively (Scheme 1). The same observation was recorded by the ^{13}C NMR spectra at 30 and 120 °C (see Sect. 4). The two isomers in the ^{1}H NMR spectrum of **44** is explained in terms of the endo and exo forms of the cyclopentane ring.

Next, the hydrazides **41a–e** were used in synthesis of different 1,2,3-triazole-3-thiones and 1,3,4-oxadiazole-2-thiones. Thus, treatment of **41e** with phenyl isothiocyanate in boiling temperature for 4 h gave the thiosemicarbazide solid, which separated and boiled directly with 5% NaOH for 3 h to provide, after neutralization with dilute HCl, **55** (67%). Similarly, treatment of **41d** with 4-methoxyphenylisocyanate afforded **51** (62%).

By applying the reported procedure [21] the 1,3,4-oxadiazole derivatives **53**, **57**, and **58** were prepared in 83%, 77%, and 79% yields, respectively, from the boiling of a solution of **45c**, **45a**, and **45b** in EtOH containing KOH and CS_2 for 8 h, followed by neutralization with dilute HCl.

Treatment of **41b** with bromocyanide (BrCN) at 70 °C for 3 h afforded the amino compound **53** (76%). The structures of all the newly synthesized compounds were confirmed by the ^1H, ^{13}C NMR, and mass spectra. Compounds **55** and **52** were identified by the homo- and heteronuclear spectra, as well as from comparison to the 1,2,4-triazolo-azipene derivatives prepared previously [22] (Scheme 10).

Compounds **44**, **51**, and **52–56** were evaluated for their antitumor activity according to NCI in vitro protocols [23]. They were assayed in vitro against three cancer types: breast, lung, and central nervous system (CNS) cancers,

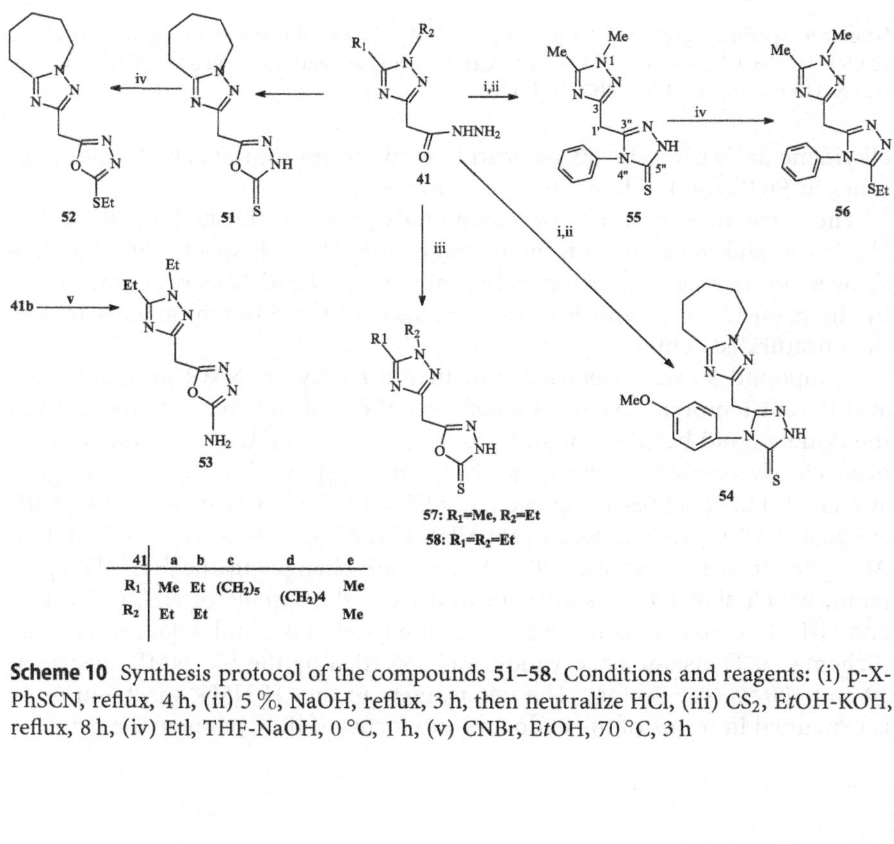

Scheme 10 Synthesis protocol of the compounds **51–58**. Conditions and reagents: (i) p-X-PhSCN, reflux, 4 h, (ii) 5 %, NaOH, reflux, 3 h, then neutralize HCl, (iii) CS_2, *Et*OH-KOH, reflux, 8 h, (iv) Etl, THF-NaOH, 0 °C, 1 h, (v) CNBr, *Et*OH, 70 °C, 3 h

while **56** was screened against a panel consisting of 60 human tumor cell lines, derived from nine cancer types (leukemia, non-small cell lung, colon, CNS, melanoma, ovarian, renal, prostate, and breast cancers, at five, 10-fold dilutions from a maximum of 10^{-4} M.)

Only compound **56** showed marked activity against colon (HCC-2998), and melanoma (UACC-257) cancers, with low percentage growth of log10 concentration =– 2.0 at 10^{-5} and – 83 at 10^{-4} M, respectively, since the negative value indicates cells killed at the mentioned concentrations. In addition, the same compound exhibited remarkable activity against individual cell lines e.g., melanoma (LOX IMVI), ovarian (OVCAR-3), prostate (PC-3) and breast (NCI/ADR-RES) cancers with log 10 values of – 69, 62, 64, and – 68, respectively.

Also [1,2,3]-triazoles have found wide use in pharmaceuticals, agrochemicals, dyes, photographic materials, and corrosion inhibition [24]. There are numerous examples in the literature including anti-HIV activity [25], antimicrobial activity against Gram-positive bacteria [26], and selective adrenergic receptor agonism by means of triazole compounds [27]. Several methods have been described for the synthesis of [1,2,3]-triazoles. Among them, the most important and useful one is the cycloaddition of azides with alkynes [28]. However, this reaction usually needs elevated temperatures and also forms a mixture of 1,4 and 1,5 regioisomers when unsymmetrical alkynes are employed. Recently, studies on 1,4 versus 1,5 regioselectivity were reported. Sharpless used Cu(I) salt as a catalyst to promote the reaction of azide with terminal alkynes in order to generate 1,4-substituted products with high regioselectivity [29]. Meldal [30] also regioselectively synthesized 1,4-substituted [1,2,3]-triazoles by 1,3-dipolar reaction of azides with polymer-supported terminal alkynes. The initial regioselective 1,3-dipolar addition of diazopropane to imidates **55** constitutes a novel route for the synthesis of

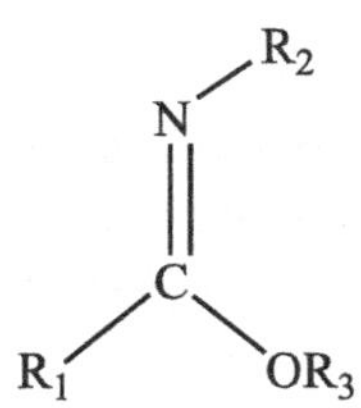

60a-c

	60a	60b	60c
R_1	Ph	Ph	Ph
R_2	H	COmNH$_2$C$_6$H$_4$	CO(CH$_2$)$_2$-CH$_3$
R_3	Me	Me	Me

Scheme 11 Synthesis protocol of the compounds **60**

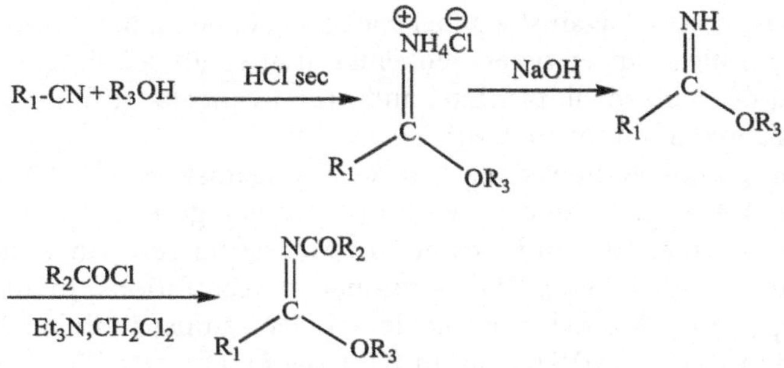

60a-c

Scheme 12 Synthesis protocol of the compounds **60**

[1,2,3]-triazoles. We now report the synthesis of new [1,2,3]-triazoles by regioselective 1,3-dipolar cycloaddition of the versatile 2-diazopropane **59** to imidates **60**. To the best of our knowledge, this reaction has never been reported before (Scheme 11).

Imidates **60** were prepared in two steps by first reacting nitrile derivatives with various alcohols. The condensation of the obtained iminoester with appropriate acetyl chloride resulted in the formation of the title compounds **60a–c** (Scheme 12). The structures of the products **60** were elucidated by means of spectroscopic analysis.

3 Cycloaddition Reaction of 2-Diazopropane with Imidates 60a–c

3.1 Synthesis of [1,2,3]-Triazoles

The 2-diazopropane **59** reacts at 0 °C in dichloromethane with the imidate **60a** to give exclusively the adduct **61a** after 10 h of reaction. This compound results from the regioselective 1,3-dipolar cycloaddition of the 2-diazopropane to the imidate C – N bond (Scheme 13).

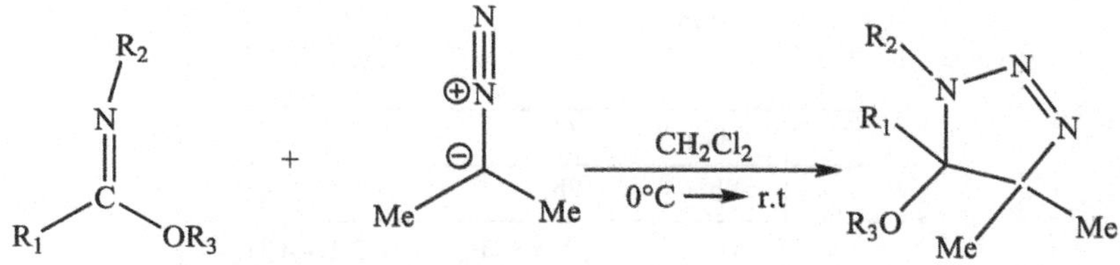

60a-c **59** **61a-c**

Scheme 13 Synthesis protocol of the compounds **61**

The structure of compound **61a** was determined by ^1H and ^{13}C NMR spectroscopy as well as 2-D NMR experiments. The ^1H NMR spectrum shows two singlets at 1.92 and 1.97 ppm for the methyl protons and at 3.81 ppm for the methoxylic protons. ^{13}C NMR showed a signal at 50.2 ppm corresponding to the methoxy groups and the aromatic carbon resonances appeared between 126.4 and 143.2 ppm.

The methyl protons correlate with C_5 and with the carbon C_4 consistent with the neighboring C_4–C_5 connection. The NOESY spectrum shows a nOe correlation between the methoxylic protons and the methyl protons.

Under similar conditions, reaction of imidate **60b** with 2-diazopropane performed at 0 °C in dichloromethane, was completed in less than 10 h and gave mainly product **61b**. As before, the structure of **61b** was determined via a detailed mono- and bidimensional NMR study.

We also investigated the reaction of imidate **60c** with 2-diazopropane to get the corresponding regioisomer **61c**. Data from the elemental analysis indicated that the calculated and observed values are within the acceptable limits (±0.4%) and have been found to be in conformity with assigned structure. Furthermore, the ^{13}C NMR spectrum in $CDCl_3$ is also in agreement with this structure and shows the absence of signal corresponding to the iminic carbone of the starting imidate **60c** at 160 ppm [31].

These results show for the first time, the reactivity of the double bond C = N with the 2 diazopropane that constitutes an efficient route for the preparation of new heterocyclic systems. In all cases, the reaction is periselective: only the double bond C = N is affected; diazo carbon attacks the quaternary carbon of the imidate **60** and not the double bond C = O (substrates **60b** and **60c**). Indeed, diazopropane reacts with ketones with inverse regioselectivity (with regards to imidates **60**) to yield oxadiazolines [32, 33] (Scheme 14).

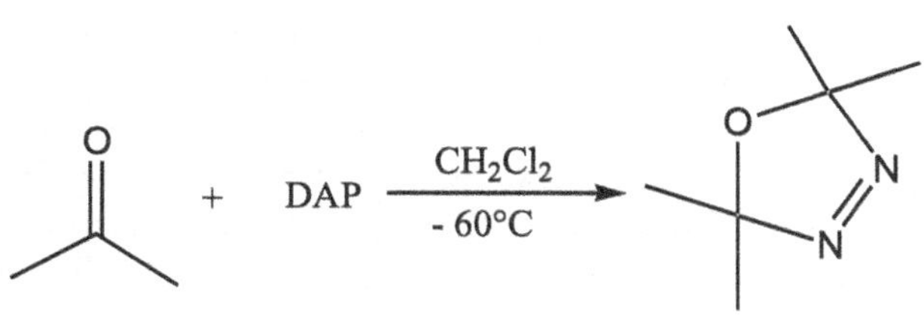

Scheme 14 Synthesis protocol of oxadiazoline

In conclusion, we have been successful in developing a new method for the synthesis of [1,2,3]-triazoles by regioselective 1,3-dipolar cycloaddition of 2-diazopropane with imidates **60** in good yields.

Three-membered strained ring systems constitute an attractive class of molecules as synthetic intermediates [34–36]. Among them the rigid, unsaturated cyclopropenes are the key to selective useful transformations [37, 38] such as regioselective additions of organometallics [39–41] or cycloadditions [42, 43]. Three-membered cyclic derivatives also have potential biological activity since their structural feature is included in valuable natural products such as phorbol and chrysanthemic acid [44, 45]. An approach for the synthesis of cyclopropene derivatives is offered via the formation of cyclopropenyl lithium, generated from 2,2-dibromo-1-chlorocyclopropane [46]. The metal-catalyzed addition of carbene sources to alkynes constitutes the most direct access to cyclopropene derivatives [47], most of them involving rhodium [48] (Hamdi et al., unpublished results) and copper [49] catalysts. However, the metal-catalyzed addition of diazoalkane is not always satisfactory with bulky and functional alkynes. The carbene easily inserts into heteroatom-hydrogen bonds [50] and the insertion of the triple bond into the metal-carbene bond is in competition with cyclopropenation [51]. The initial regioselective 1,3-dipolar addition of diazoalkane to functional alkynes followed by photochemical elimination of dinitrogen [52, 53], constitutes an alternative for the preparation of cyclopropenes with an α-functional group on the condition that carbene insertion into the heteroatom-hydrogen bond can be prevented. Moreover, when the functional group of alkynes controls the regioselectivity of diazoalkane 1,3-cycloaddition, this approach allows simple access to functional 3H-pyrazoles with potential biological properties [54].

We now report the synthesis of new antibacterial 3H-pyrazoles by regioselective 1,3-dipolar cycloaddition of the versatile 2-diazopropane to non-protected disubstituted propargyl alcohols and that the unsubstituted propargyl alcohol allows the double addition of 2-diazopropane and gives a 3H-pyrazole with formal insertion of the dimethylcarbene into a carbon-carbon bond. We also show that the photolysis of the 3H-pyrazoles leads to new alcohols containing the cyclopropenyl unit.

Whereas the $Rh_2(OAc)_4$-catalyzed addition of diazoalkanes to propargyl alcohols readily gives the insertion of the carbene into the O – H bond, with only a small amount of cyclopropenation of the resulting propargylic ether [54] the 2-diazopropane **59** reacts at 0 °C with 1,1-diphenyl-2-propyn-1-ol **62a** in dichloromethane and exclusively gives, after 10 h of reaction, only the adduct **63a** isolated in 75% yield and corresponding to the regioselective 1,3-dipolar cycloaddition of the 2-diazopropane to the alkyne C – C bond (Scheme 15).

The structure of this compound **63a** was determined by ^1H and ^{13}C NMR spectroscopy. The ^1H NMR spectrum shows singlets at 1.45 ppm for the methyl protons, and at 4.17 and 6.30 ppm for the hydroxylic and ethylenic protons. The addition regioselectivity in the formation of **63a** was established by ^1H-^{13}C HMBC 2D-NMR, which shows the C_5-C_4-C_3-Me linkages. The ethylenic proton correlates only with three carbon atoms C_6, C_5, and C_3. The methyl protons correlate with C_3 and with the ethylenic carbon C_4, consistent with the neighboring C_3-C_4 connection. The NOESY spectrum shows

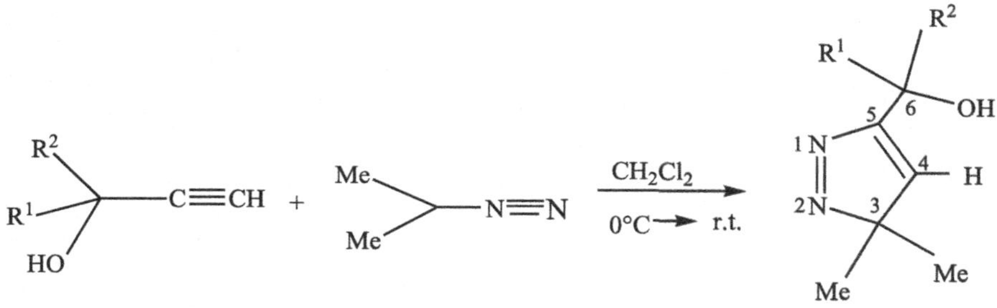

62a-b 59 63a-b

a : R^1, R^2 = Ph

b : R^1= Ph , R^2= Me

Scheme 15 Synthesis of 3H-pyrazoles **63**

a nOe cross peak between the ethylenic and aromatic protons. Whereas the addition of diazoalkane to alkynes is not usually regioselective, the cycloaddition of diazopropane to **62a** leads only to derivative **63a**, with the linkage of the nitrogen atom with the unsaturated carbon connected to the functional group, and corresponds to that observed for the 1,3-dipolar cycloaddition reaction of simple diazoalkanes with α,β-unsaturated ketones [55].

Analogously, the 1,3-dipolar cycloaddition reaction of 2-diazopropane with propargyl alcohol **62b**, performed at 0 °C in dichloromethane, was completed in less then 10 h and led to a monoadduct **63b** with the same regioselective addition mode of **59** to the triple bond. The HMBC spectrum showed correlations between the ethylenic proton and the carbons C_3 and C_5 and between the methyl protons and the carbons C_3 and C_4.

In contrast the unsubstituted propargylic alcohol **62c** reacted with an excess of 2-diazopropane and was completely transformed after 10 h at 0 °C to surprisingly give the tetrasubstituted 3H-pyrazole **64** isolated in 73% yield (Scheme 16). The ^{1}H NMR spectra of **64** showed the presence of two exocyclic equivalent and two intracyclic different methyl groups, and singlets at δ 4.01 ppm for the OH group, 4.86 ppm for the methylenic and 6.70 ppm for the ethylenic protons. The formation of **64** which includes the incorporation of two CMe_2 groups arising from the diazoalkane can be explained via the formation of the expected cycloadduct intermediate (**I**), followed by a second cycloaddition of diazoalkane to the remaining double C=C bond to give the intermediate (**II**). The latter is not stable at room temperature, looses dinitrogen and undergoes a rearrangement of the carbon skeleton leading to **64**.

It is noteworthy that the addition of an excess of the 2-diazopropane to the alkynes **62a,b** did not give the corresponding bisadduct of diazoalkane. It

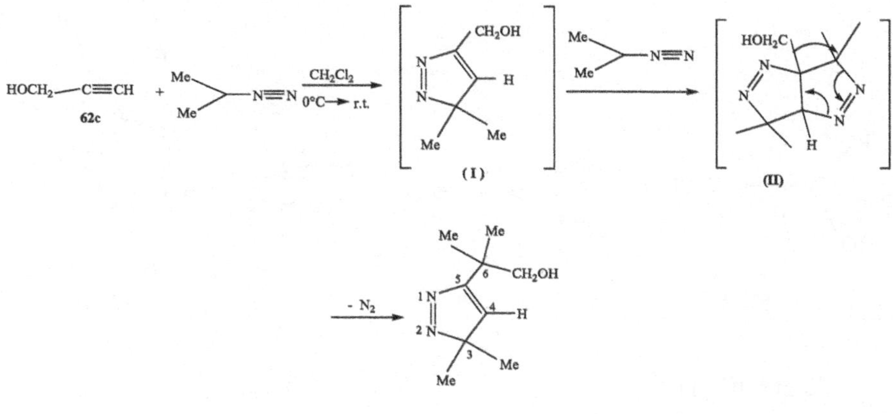

64

Scheme 16 Synthesis of the tetrasubstituted 3H-pyrazole **64**

is likely that the bulkiness of the CR^1R^2OH group, closed to the C = C bond in the intermediate (I), prevents the second addition of diazopropane, that is allowed by the smaller propargyl alcohol CH_2OH group.

The antibacterial activity of the obtained 3H-pyrazoles **63a,b** and **64** has been studied. They have been tested opposite a pathogenic bacterial stump and have shown antibacterial activity against the original *Staphyloccocusaureus.* The 3H-pyrazole **63b** offers the strongest antibacterial activity [56].

3.2
Photochemical Transformation of the Obtained 3H-Pyrazoles into Cyclopropene Derivatives

The photochemical study of 3H-pyrazoles was carried out in the search for a route to cyclopropenyl tertiary alcohols. Irradiation of **63a** in dry dichloromethane at 300 nm and at room temperature for 0.5 h led to the exclusive formation of the gem-dimethylcyclopropene **65** (Scheme 17). The formation of cyclopropene **65** arises from the loss of N_2 and cyclization of the vinylcarbene intermediate (III).

The structure of **65** was determined via a detailed mono- and bidimensional NMR study. In ^{13}C NMR a signal at 20.3 ppm corresponds to the methyl groups and the carbon nucleus $= C_2 - H$ carbon appeared at 142.3 ppm. The gem-dimethylcyclopropene structure of **65** was consistent with an analysis of the ^1H-^{13}C HMBC spectrum.

The analogous photochemical reaction (300 nm) of 3H-pyrazole **64** in dichloromethane at room temperature led to cyclopropene derivative **66**, possessing a α-hydroxy group, isolated in 70% yield (Scheme 18).

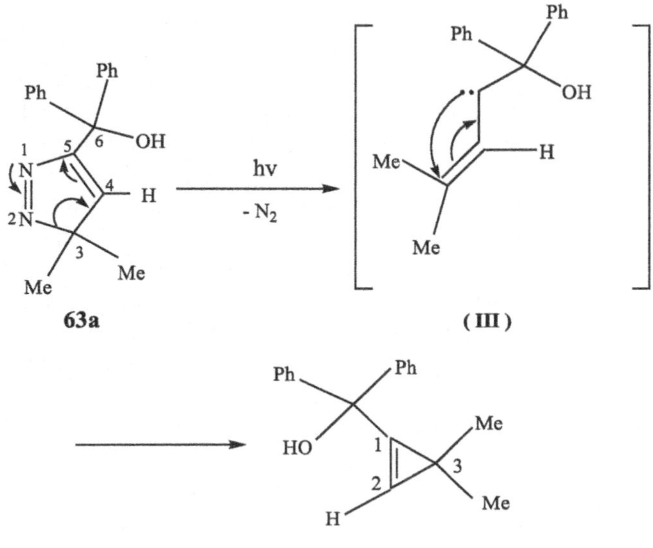

63a (III)

65 (75 %)

Scheme 17 Preparation of α-hydroxycyclopropene **65**

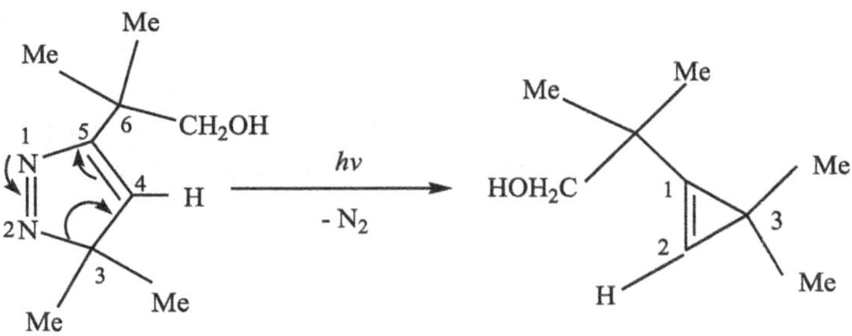

64 **66** (70 %)

Scheme 18 Preparation of α-hydroxycyclopropene **66**

The above two consecutive transformations provide straightforward access from propargyl alcohols to cyclopropene derivatives with an α- or β-hydroxy group. This simple method is complementary to the access to 3-hydroxymethylcyclopropenes, via $Rh_2(OAc)_4$ catalyzed addition of diazoacetate to alkynes followed by reduction of the ester group, a route that is restricted to the access of primary cyclopropenyl alcohols [57], and is an alternative to the use of 2,2-dibromo-1-chlorocyclopropane via cyclopropenyl lithium.

This study demonstrates that the addition of the 2-diazopropane with the triple bond of propargyl alcohols is regioselective, and affords new antibacterial 3H-pyrazoles. The photochemical reaction of these 3H-pyrazoles selectively leads to α- and β-hydroxy cyclopropenes. The overall transformation constitutes a simple straightforward route to substituted cyclopropenyl alcohols without initial protection of the propargyl alcohol hydroxyl group.

Several coumarin derivatives have revealed pronounced medicinal value as antibacterial and antifungal agents [57, 58]. Others have displayed antitubercular activity [59] and some have insecticidal properties [60]. This prompted us to investigate the preparation of a new series of compounds containing coumarin moieties with different side chains or fused rings.

In this work, we have studied the reaction of 4-hydroxycoumarin (**67**) and 3-(dimethylamino-methylene)chromane-2,4-dione (**68**) with aromatic binucleophilic compounds (Scheme 19).

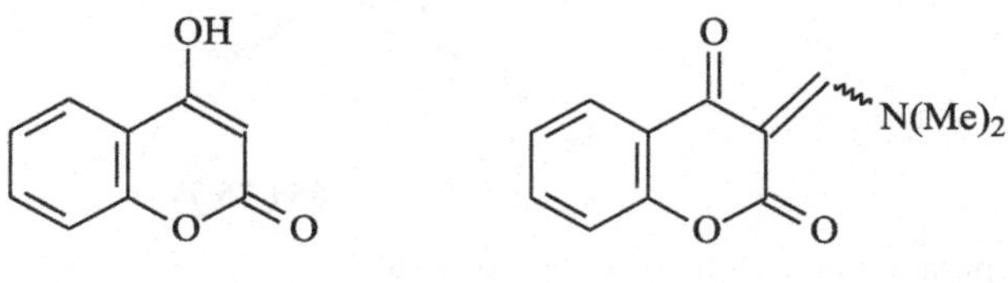

Scheme 19 Preparation of the starting compounds **67** and **68**

4-Hydroxycoumarin (**67**) was found to react with *o*-phenylenediamine on refluxing in toluene to give product **69** (Scheme 20). According to the elemental analysis, IR and NMR spectroscopy data, the structure of 4-(2-

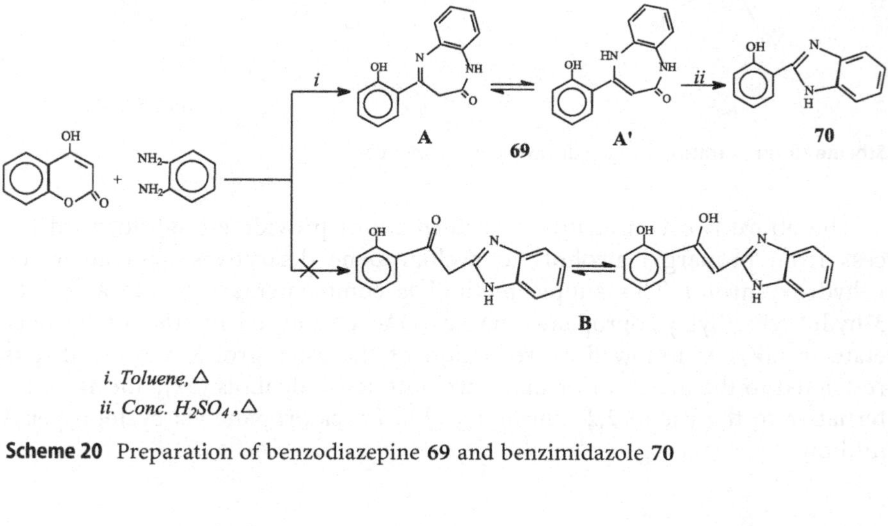

i. Toluene, Δ
ii. Conc. H_2SO_4*,* Δ

Scheme 20 Preparation of benzodiazepine **69** and benzimidazole **70**

hydroxyphenyl)-2,3-dihydro-1H-1,5-benzodiazepin-2-one can be ascribed to compound **69**.

A mass spectrometric study was carried out to establish the structure of compound **69**. Its mass spectrum contains the molecular ion peak m/z 252 (16.98%) and a base peak (100%) at m/z 210, corresponding to 2-(2-hydroxyphenyl)benzimidazole (**70**). A tendency towards decreasing the heterocycle size is characteristic of the mass spectrometric behavior of 1,5-benzodiazepin-2-ones [61] and consequently the mass spectra of these compounds contains intense peaks of the corresponding benzimidazoles. It is also known that the mass spectrometric fragmentation of 1,5-benzodiazepines is similar to their thermal or acid decomposition. In fact, refluxing compound **69** in concentrated sulfuric acid yields benzimidazole **70** as the main product.

According to the ^{1}H-NMR spectra in DMSO – d_6, benzodiazepine **69** exists as a 4 : 1 mixture of tautomers **A** and **A′**. Benzodiazepin-2-one **69** is formed due to the substitution of the hydroxyl group of coumarin **67** by one of the amino groups of *o*-phenylenediamine and the C – O bond cleavage in the pyrone ring upon reaction with the second amino group.

In a similar manner, 4-hydroxycoumarin reacts with equimolar amounts of 4-aminothiophenol and 4-aminophenol to give the corresponding coumarin derivatives **71a,b** (Scheme 21).

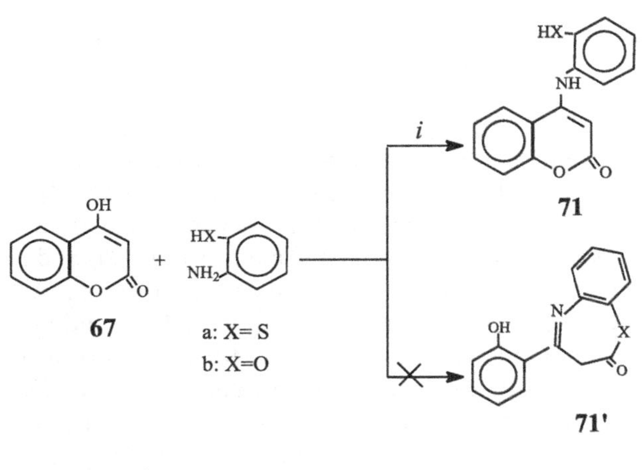

i. Toluene,△

Scheme 21 Preparation of benzodiazepine 71

The ^{1}H-NMR spectra of compound **71a** in DMSO – d_6 showed the presence of a signal at 12.5 ppm corresponding to the exchangeable NH proton, the ethylenic proton as a singlet at δ 5.6 ppm, and the aromatic protons appear between 7.27 and 7.80 ppm. The elemental and spectral analysis was in agreement with the structures of these compounds.

We also investigated reaction of 4-hydroxycoumarin with an excess of *N*,*N*-dimethylformamide dimethyl acetal (DMFDMA) which afforded the corresponding 3-(dimethylaminomethylene)-chromane-2,4-dione derivative **72**. The structure was again confirmed by IR, NMR, and MS analyses.

The possibility that diamines might further react intermolecularly or intramolecularly [62] at either of the two carbonyl groups, prompted us to examine the interaction of aromatic diamines with compound **72**. The reactions of equimolar amounts of **72** and 4-phenylenediamine and 4-aminothiophenol were carried out by refluxing in toluene for 4 h (Scheme 22). Thin-layer chromatography showed the formation of a single product in both cases. The ^{1}H-NMR spectra of **73a** showed the presence of two doublets of the same intensity at δ 10–12, corresponding to two exchangeable protons, the ethylenic protons of the Z and E isomers [62] as two doublets at δ 7–8, and signals corresponding to the expected Ar protons.

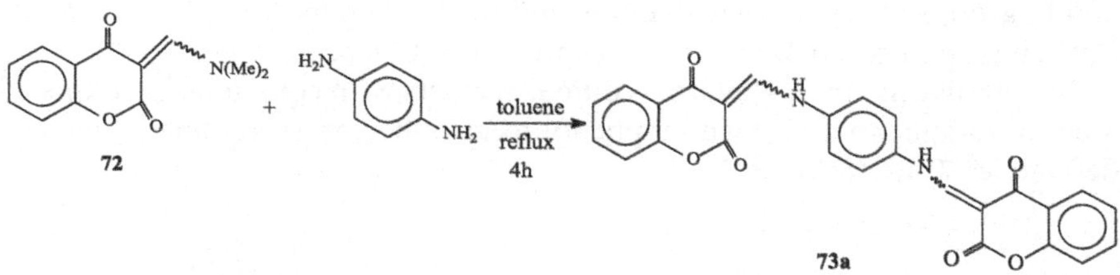

Scheme 22 Preparation of compound **73**

These observations indicated that an intermolecular double condensation to give a bis *N*-(methylene-4-oxocoumarinyl)-1,4 aromatic diamine had occurred. Data from the elemental analysis indicated that the calculated and observed values were within the acceptable limits (±0.4%) and in conformity with the assigned structure. In the addition of molar equivalents of 1,4-aromatic binucleophilic compounds to compound **72** we did not observe any heterocyclic compounds resulting from the further intermolecular nucleophilic attack on the single condensation product. Since the condensation of 3-(dimethylaminomethylene)-chromane-2,4-dione with aromatic binucleophilic compounds is the only route to the new coumarinic compounds, this represents a useful synthetic method.

Some representative examples of the new compounds were tested against a pathogenic bacterial strain and have shown activity against *Staphylococcus aureus* ATCC 25923. The benzodiazepin-2-one **69** offers the strongest antibacterial activity. The results are summarized in Table 8.

Coumarin and their derivatives have been found to exhibit different biological and pharmacological activities [63–66]. Owing to the widespread applications, synthetic and biological activity evaluation of coumarins and their derivatives has been a subject of intense investigations [67–69].

Table 8 Antibacterial screening of products **69** and **71**

Compound	Concentration (mg/disk)	Inhibition zone (mm)
69	1	50
	2	58
	4	58
71a	1	30
	2	33
	4	33
71b	1	39
	2	41
	4	40

4
Conclusion

In this review, general and efficient approaches to a diverse series of triazoles and coumarin derivatives have been developed and discussed. The obtained products have been characterized with the help of spectroscopic techniques and were screened for their antiviral and antitumor activity. The methods could provide valuable routes to various coumarin derivatives and enrich the organic and medicinal chemistry of coumarins. The synthesized triazoles and coumarin derivatives showed moderate to good antiviral and antitumor activities.

We have also been successful in developing a new method for the synthesis of [1,2,3]-triazoles by regioselective 1,3-dipolar cycloaddition of 2-diazopropane with imidates in good yields. These compounds may find wide use in medicinal chemistry.

It can be seen that though the aldehydic function is important for anti-inflammatory activity as observed in compound **36**, its conversion to the unsaturated cyanoesters **37** shows an increase in activity. Similarly, the ethers **39** with an ortho aldehydic group upon conversion to benzofurans **38** show a slight increase in activity.

References

1. Alagarsamy V, Giridhar R, Yadav MR (2005) Bioorg Med Chem Lett 15:1877–1880
2. Wade JJ (1986) US Patent 4 528 288
3. Wade JJ (1986) Chem Abstr 104:5889
4. West WR, July WR (1982) Eur Patent 34 529
5. West WR, July WR (1982) Chem Abstr 96:20114

6. Dews BB (1953) J Pharmacol 8:46
7. Kuhn WL, Van Mannen EF (1961) J Pharmacol Exp Ther 134:60
8. Dey BB, Row KK (1924) J Ind Chem Soc 1:107–123
9. Takagi K, Tanaka M, Morita H, Ogura K (1987) Eur J Med Chem 22:239–242
10. Mustapha A, Hismat OH, Wassef ME, El-Ebrashi NMA, Nawar AA (1966) Liebigs Ann Chem 692:166
11. Tabakovic K, Tabakovic I, Trkovnik M, Trinajstic N (1983) Liebigs Ann Chem 1901
12. Huebner CF, Link KP (1945) J Am Chem Soc 67:99
13. Elderfiled RC (ed) (1957) Heterocyclic Compounds, Vol. 5. Wiley, New York, pp 49–50
14. Takagi K, Hubert-Habart M (1980) Bull Soc Chim Fr II 445
15. Shankarnarayan Y, Dey BB (1934) J Ind Chem Soc 11:687–693
16. Khan IA, Kulkarni MV (1999) Ind J Chem 34B:491–494
17. Ghate M, Manohar D, Kulkarni V, Shobha R, Kattimani SY (2003) Eur J Med Chem 38:297–302
18. Tsukuda T, Shiratori Y, Watanabe M, Ontsuka H, Hattori K, Shirai M, Shimma N (1998) Bioorg Med Chem Lett 8:1819–1824
19. Witkoaski JT, Robins RK, Sidwell RW, Simon LN (1972) J Med Chem 15:1150–1154
20. Heubach G, Sachse B (1975) Chem Abstr 92:181200h
21. Reid JR, Heindl ND (1976) J Heterocycl Chem 13:925–926
22. Al-Soud YA, Al-Masoudi NA (2003) Heteroatom Chem 14:298–303
23. Al-Soud YA, Al-Dweri MN, Al-Masoudi NA (2004) IL Farmaco 59:775–783
24. Katsuki T (2002) Adv Synth Catal 344:131–147
25. Reissig HU, Zimmer R (2003) Chem Rev 103:1151–1196
26. Lebel H, Marcoux JF, Molinaro C, Charrette AB (2003) Chem Rev 103:977–1050
27. Baird MS, Schmidt T (1996) In: de Meijere A (ed) Carbocyclic Three-Membered Ring Compounds. Georg Thieme Verlag, Stuttgart, p 114
28. Baird MS (1997) Cyclopropenes Transformations; Addition Reactions. Houben-Weyl, Thieme, Stuttgart, E17D/2
29. Liao L, Fox JM (2002) J Am Chem Soc 124:14322–14323
30. Nakamura M, Hirai A, Nakamura E (2000) J Am Chem Soc 122:978–979
31. Lehmkuhl H, Mehler K (1978) Justus Liebigs Ann Chem 1841–1853
32. Nakamura E, Machii D, Imubushi T (1989) J Am Chem Soc 111:6849–6852
33. Yamago S, Ejiri S, Nakamura E (1994) Chem Lett 1889–1892
34. Rigby JH, Kierkus PC (1989) J Am Chem Soc 111:4125–4126
35. Zhu YF, Yamazaki T, Tsang JW, Lok S, Goodmann M (1992) J Org Chem 57:1074–1081
36. Nuske H, Brase S, de Meijere A (2000) Synlett 1467–1469
37. Doyle MP, McKervey MA, Ye T (1998) Modern Catalytic Methods for Organic Synthesis with Diazo Compounds. Wiley, New York
38. Doyle MP, Ene D, Peterson CS, Lynch V (1999) Angew Chem Int Ed 38:700–702
39. Davies HML, Lee HG (2004) Org Lett 6:1233–1236
40. Diaz-Requejo MM, Mairena MA, Belderrain TR, Nicasio MC, Trofimenko S, Perez PJ (2001) Chem Commun 1804–1805
41. Noels AF, Demonceau A, Petiniot N, Hubert AJ, Teyssie P (1982) Tetrahedron Lett 38:2733–2739
42. Hoye TR, Dinsonore GJ (1991) Tetrahedron Lett 32:3755–3758
43. Day AC, Raymond P, Southam RM, Whiting MC (1966) J Chem Soc C 467
44. Closs GL, Böll WA, Heya H, Dev V (1968) J Am Chem Soc 90:175
45. Franck-Neumann M (1968) Angew Chem Int Ed Engl 7:65
46. Bastide J, Lematre J (1969) CR Acad Sci 269:358
47. Vidal M, Vincens M, Arnaud P (1972) Bull Soc Chim Fr 657–665

48. Hamdi N, Khemiss A (2002) J Soc Alger Chim 12:45–52
49. Richey HG Jr, Bension RM (1980) J Org Chem 45:5036–5042
50. Kelkar RM, Joshi UK, Paradkar MV (1986) Synthesis 214
51. Krisha-Rao KSR, Subba-Rao NV (1964) Curr Sci 33:614
52. Moppet RB (1964) J Med Chem 7:446
53. Redighiero G, Antonello C (1985) Bull Chim Farm 97:592
54. El-Naggar AM, Ahmed FSM, Abd El-Salam AM, Radi MA, Latif MSA (1981) J Heterocylic Chem 18:1208
55. Merchant CR, Gupita AS, Shah PJ, Shirali SS (1979) Chem Ind 351
56. Mali RS, Yeola SN, Kulkarn BK (1983) Ind J Chem 22B:352
57. Terent'ev PB, Stnakyavichus AP (1987) Mass-spektrometriya biologicheski aktivnykh azotistykh asnovanii [Mass Spectrometry of Biologically Active Nitrogen Bases]. Mokslas, Vilnyus (in Russian)
58. Barchet VR, Merz KW (1964) Tetrahedron Lett 2239
59. El Abbassi M, Essassi EM, Fifani J (1987) Tetrahedron Lett 1389
60. Steinfurhrer T, Hantschmann A, Pietsch M, Weibenfels M (1992) Liebigs Ann Chem 23
61. Wolfbeis OS, Ziegler E (1976) Z Naturforsch 31b:1519
62. Bauer SW, Kirby WM, Sherris JC, Thurck M (1966) Am J Pathol 45:408
63. Egan D, Kennedy RO, Moran E, Cox D, Prosser E (1999) Drug Metab Rev 22:503–529
64. Bhalla M, Nathani PK, Kumar A, Bhalla TN, Shander K (1992) J Ind Chem 31(B):183–186
65. Takeda S, Aburada M (1981) J Pharmacobiodyn 4:724–729
66. Yang CH, Chiang C, Liu K, Perg S, Wang R (1980) Yao Huseh Tung Pao 15:48–54
67. Tyagi A, Dixit VP, Joshi BC (1980) Naturwissenschaften 67:104–109
68. Deana AJ (1983) Med Chem 26:580–585
69. Tyagi YK (2000) Studies on the synthesis of azides and heterocyclic triazoles and investigation on biotransformation of polyphenols. Thesis. CCS University, Meerut

Top Heterocycl Chem (2007) 10: 155–209
DOI 10.1007/7081_2007_071

Published online: 4 July 2007

Protoberberine Alkaloids: Physicochemical and Nucleic Acid Binding Properties

Motilal Maiti (✉) · Gopinatha Suresh Kumar

Biophysical Chemistry Laboratory, Indian Institute of Chemical Biology,
700 032 Kolkata, India
mmaiti@iicb.res.in

Abstract Protoberberine alkaloids and related compounds represent an important class of molecules and have attracted recent attention for their various pharmacological activities. This chapter deals with the physicochemical properties of several isoquinoline alkaloids (berberine, palmatine and coralyne) and many of their derivatives under various environmental conditions. The interaction of these compounds with polymorphic DNA structures (B-form, Z-form, H^L-form, protonated form, triple helical form and quadruplex form) and polymorphic RNA structures (A-form, protonated form, triple helical form and quadruplex form) reported by several research groups, employing various analytical techniques such as spectrophotometry, spectrofluorimetry, circular dichroism, NMR spectroscopy, viscometry as well as molecular modelling and thermodynamic analysis to elucidate their mode and mechanism of action for structure–activity relationships, are also presented.

Keywords Berberine · Palmatine · Coralyne · Polymorphic nucleic acid structures · Alkaloid–nucleic acid interactions

1 Nucleic Acid Polymorphism: An Introduction

Since the discovery of the double helical structure of deoxyribonucleic acid (DNA) by Watson and Crick in 1953 [1], there has been considerable belief that the canonical right-handed B-DNA may adopt a wide range of different conformations depending on the nucleotide sequences and environmental conditions. This speculation turned out to be a reality [2–10]. In living systems, the conformational flexibility of DNA resides primarily in the polymorphs of the DNA double helix (including right-handed and left-handed double helical DNA) and occurs under various environmental conditions [4]. The main family of DNA forms identified, based on circular dichroic and crystallographic data, are the right-handed A- and B-forms [2, 3] and the left-handed Z-form [6–8]. Beside these, several variants of right-handed DNA (like C-, D-, V-form) and left-handed DNA (like Z_1- and Z_2-forms etc.) have also been discovered [2, 3, 5, 9, 11]. In addition to the well-known global changes in DNA secondary structures, the formation and properties of certain non-canonical structures like loops, bulges, hairpins, cruciforms [2, 3] and higher order structures like triple helical form, quadruplexes, i-motif etc. [12–14] are also known to exist. By contrast, double-stranded helical ribonucleic acid (RNA) structure is confined to two very similar polymorphs of the A-form (A and A′) and a wide range of single-stranded non-helical RNA folds that introduces the essential structural variability. RNA can undergo various conformational changes, like to the left-handed Z-form, the protonated form and to higher structures like the triple helical form and the quadruplex structure, under various environmental conditions [15–18].

1.1
B-Form DNA

The native conformation of the double helical right-handed structure of DNA in solution is the so-called B-form that contains the richest source of information within a living organism. It is composed of the four nucleotides, deoxyadenosine (dA), deoxyguanosine (dG), deoxycytidine (dC) and deoxythymidine (dT), and consists of two polynucleotide strands wound about a common axis with a right-handed twist, originally proposed by Watson and Crick [1]. B-DNA is conformationally complex due to the large number of torsional angles that define the orientation of the glycosal bond, sugar-ring bonds and polynucleotide backbone bonds. The best molecular model for this, obtained by supplementing 0.3 nm resolution X-ray diffraction intensities with more precise stereochemical data, consists of a right-handed antiparallel bihelix with stranded base pairs located in planes roughly perpendicular to the helix axis with Watson–Crick base pairing. The turn of the helix is 3.4 nm long and consists of ten base pairs that occupy the core of the structure with the circumferential sugar phosphate backbone defining a major and minor groove (Fig. 1). The furanose rings have C2′ *endo* pucker in the stereochemically optimal structure [19]. In common with the other DNA forms, the base assumes an *anti* conformation about the glycosyl linkage and the configuration about the exocyclic C(4′)–C(5′) bond in gauche–gauche [3].

Most DNA, natural and synthetic, can adopt the B-form as defined by its characteristic X-ray pattern [3]. There are few a species of synthetic DNA, namely poly(dA)·poly(dT), poly(dI)·poly(dC), poly(dA-dI)·poly(dC-dT) and poly(dI-dC)·poly(dI-dC), which have a B-form that significantly differs from that of the other complementary deoxy polymers due to their different intramolecular packing arrangements and slightly changed value for the rise per residue. The helix parameters are presented in Table 1. The linear sequen-

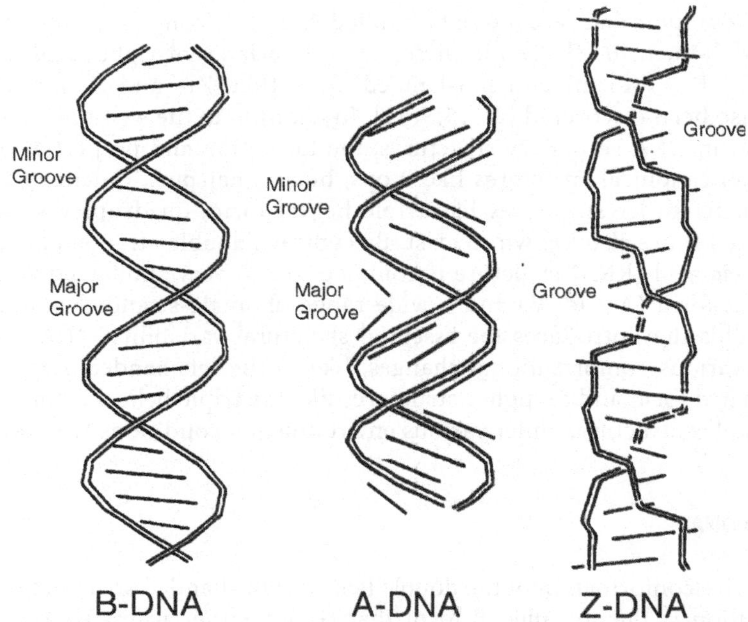

Fig. 1 Schematic representation of the various forms of DNA. Reprinted from [80] with permission from Wiley InterScience

Table 1 Conformational parameters of different polymorphic forms of DNA

DNA conformation	Occurrence	Axial rise per base (nm)	Base pair turn	Base pair repeat unit
A-form	Most natural and synthetic DNAs	0.26	11.0	1
B-form	Most natural and synthetic DNAs	0.34	10.0	1
C-form	Most synthetic DNA	0.33	7.9–9.6	1
D-form	Some synthetic DNA	0.30	8.0	1
Z-form	Few synthetic DNAs of poly purine-pyrimidine sequences	0.36–0.38	12	2

tial arrangement of the bases along the sugar phosphate backbone carries the genetic blueprint of the cell and is highly conserved in such processes as replication and transcription.

1.2
A-Form DNA

Double helical DNA takes on different conformations in different environmental conditions in the fibre. Depending on the type and amount of salt or amount of hydration, several forms of DNA structures including the A-form are observed [3, 5, 19]. Again, depending upon the environmental conditions, the intermolecular packing arrangement and helix parameters change as shown in Table 1. In the A-form DNA (Fig. 1) there are 11 residues per turn and the base pairs are displaced from the helix axis, thereby forming a very deep major groove and shallow minor groove. The sugar pucker is C3′-*endo*. All natural DNAs and most synthetic DNAs except poly(dA)·poly(dT), poly(dA-dG)·poly(dC-dT) can transform to the A-form structure [19]. Ivanov and colleagues [5] first reported the formation of A-form structure that can occur in B-DNA in ethanolic solution. A true A-form structure is observed in 60% (v/v) ethanol.

1.3
Z-DNA Structure

The existence of Z-DNA was first discovered using optical studies, demonstrating that a polymer of alternating deoxyguanosine and deoxycytidine residues, poly(dG-dC)·poly(dG-dC) produced a nearly inverted spectrum in a high salt solution [6]. The physical reason for this observation remained a mystery until an atomic resolution crystallographic study of [d(CG)$_3$] revealed a left-handed double helical structure maintaining Watson–Crick base pairing [7]. The Z-DNA helix is built from a dinucleotide repeat with the deoxycytidines in the *anti* conformation and deoxyguanosines in the unusual *syn* conformation. In Z-DNA, there is a single narrow groove that corresponds to the minor groove of B-DNA. The formation of Z-DNA from B-form DNA involves the "flipping" of the base pairs upside down. During this process, deoxycytidine remains in the *anti* conformation because both the sugar and the base rotate, while only the base of deoxyguanosine inverts, moving it into the *syn* conformation. As a consequence, the backbone follows a zigzag path; hence the name Z-DNA (Fig. 1). The Z-DNA can be formed from B-DNA under physiological salt conditions when deoxycytidine is 5′-methylated [20]. The helix parameters of Z-DNA are presented in Table 1.

Stabilization of Z-DNA by negative supercoiling illustrates a number of features. First, Z-DNA is a high energy conformation than the B-DNA and in plasmids it is formed only when they are torsionally stressed. The energy necessary to stabilize the Z-DNA can be determined by measuring the plasmid superhelical density at which Z-DNA formation occurs, and it is proportional to the square of the number of negative supercoils. Secondly, sequences other than the alternating purines and pyrimidines can form Z-DNA. These sequences are $d(TG)_n$ and $d(GGGC)_n$ repeats. It was observed that the occurrence of Z-DNA is best in $d(CG)_n$ and its methylated analogues, while the $d(GGGC)_n$ repeat is better than $d(TA)_n$. However, the functional significance of the Z-DNA has remained a provocative and controversial issue till now. Several observations raise the possibility that the left-handed motif might be involved in the regulation of biological process either directly by acting as potential template for DNA polymerase and by modulating strand exchange, thus affecting transcription and recombination process, or indirectly by altering the extent of DNA packaging and supercoiling [21–23]. Evidently, if the Z-DNA motif is indeed displaying regulatory effects, it should be characterized by a large sensitivity to cellular parameters. Such sensitivity has been indicated by the Z-DNA to B-DNA transition, which is affected by various DNA binding compounds [24–30] as well as osmotic stress [31]. It is known that Z-DNA unlike B-DNA is highly immunogenic and polyclonal. Also, monoclonal antibiotics can be raised against it that specifically recognize this conformation. Lafer et al. [32, 33] first demonstrated that Z-DNA formation is possible in human systems after analyzing the sera obtained from patients with autoimmune diseases. These patients produced antibodies which are highly specific for Z-DNA.

Identification of proteins that bind to Z-DNA added one further step to the establishment of the presence of Z-DNA in vivo and its possible biological role. Herbert and Rich [22] demonstrated an in vitro assay system where one type of double-stranded RNA adenosine deaminase, called DRAD-binding Z-DNA. There are evidences that topoisomerase II from *Drosophila*, human and calf thymus recognizes a number of DNA shapes, including Z-DNA [34, 35]. Bloomfield and coworkers [36] have found that the condensation of plasmids is enhanced by Z-DNA conformation in $d(CG)_n$ repeats. The information related to B–Z transition [31], the effect of ligands on it [28, 29] and X-ray crystal structure data [37, 38] appear to suggest that the possible biological role of this polymorphic form of DNA will be soon established.

1.4
H^L-DNA Structure

B-form duplex structures of poly(dG-dC)·poly(dG-dC) and its 5-methylated cytosine analogue can form a variety of unique structures depending on ionic strength, pH and various other environmental conditions [3, 6, 10, 20]. From extensive spectrophotometric, circular dichroic and thermal denaturation studies, Maiti and coworkers [39] advanced a model that the two GC polymers under controlled ionic strength, pH and temperature transformed to a unique and novel structure designated as the H^L-form DNA [10]. This model of left-handed structure with Hoogsteen base pairing was confirmed by FTIR studies [40, 41] and by high resolution Raman spectroscopy [42]. It was suggested that on protonation of the N3 of cytosines, guanines becomes unstacked and rotate out of the helix, reversing from the *anti* to *syn* conform-

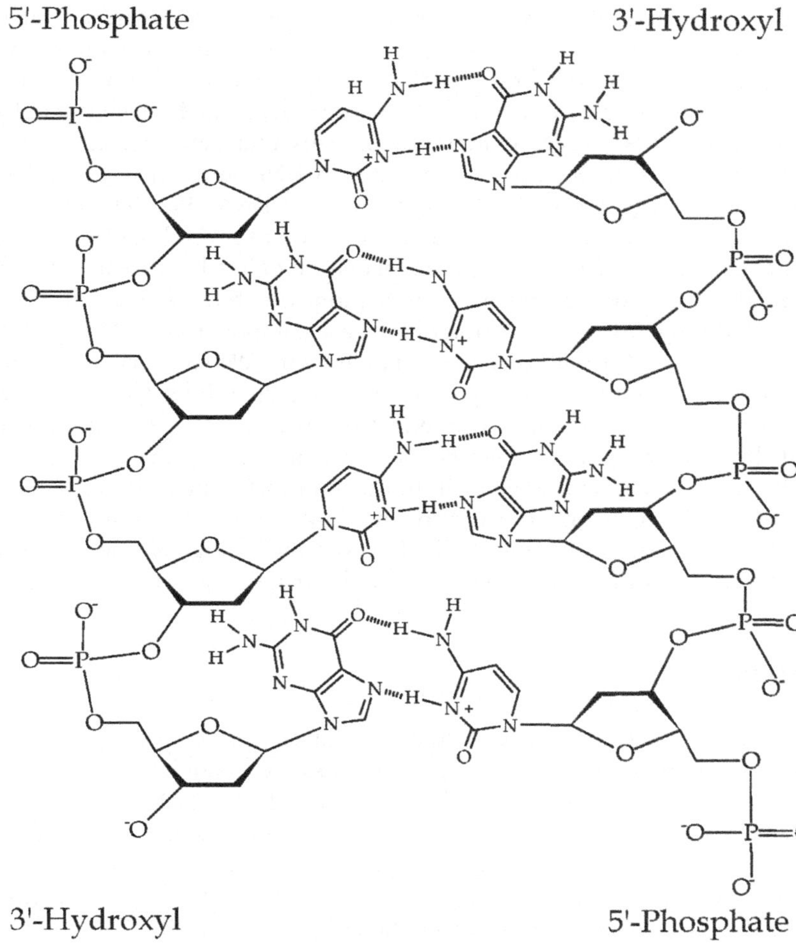

Fig. 2 Secondary structural representation of H^L-form DNA. Reprinted from [10] with permission from the publisher

ation to form Hoogsteen base pairing with N3 of cytosine. A representative secondary structure of the H^L-form DNA is shown in Fig. 2. However, the crystallographic data of H^L-form DNA and its biological role still remains to be elucidated and need further study.

1.5 Protonated DNA

In recent years there has been a great interest in the elucidation of non-conventional double helical conformations and their biological roles [43, 44].

Protonation of natural DNA has been studied for several years and it has been suggested that protonation leads to some conformational changes before acid denaturation [45–51]. However, the exact nature of these conformational changes remains obscure due to the controversy over which nucleic acid base gets protonated first. There is now a consensus that cytosines followed by guanines are the primary protonation sites in DNA and hence the GC-rich DNAs may undergo more changes than AT-rich DNAs [47–50]. Maiti and Nandi [47, 51] first demonstrated that protonation of several natural DNAs of different base composition leads to the formation of left-handed structures in certain DNAs, depending on base composition, which leads to an important probe facilitating the determination of the base composition of DNA [52]. But so far a true Z-conformation (i.e. left-handed with Watson–Crick base pairing) has not been reported in any polymer under the influence of pH. Tajmir Riahi et al. [40] have reported the effect of HCl on the solution structure of calf thymus DNA. There are reports involving the protonated structures, namely acid-induced exchange of imino-proton in GC pairs [53], helix–coil transition in DNA using pH variation methods [54] and i-motif structure formed by protonated cytosines [55]. However, the biological role of the protonated induced conformational polymorphic structure is yet to be identified.

1.6 Triplex DNA

The formation of three-stranded nucleic acid complexes was first demonstrated over five decades ago [56] but the possible biological role of an extended triplex was expanded by the discovery of the H-DNA structure in natural DNA samples [57–59]. H-DNA is an intermolecular triplex that is generally of the pyrimidine·purine × pyrimidine type ("dot"-Watson–Crick pairing and "cross" Hoogsteen base paring) and can be formed at mirror repeat sequences in supercoiled plasmids [59].

The earliest understanding of the conformation of the triple helices was largely based on evidence from X-ray diffraction data [60]. From the fibre diffraction data it was concluded that the third strand binds in the major groove of the duplex by Hoogsteen base pairing, that the sugar puckers were C3′-*endo* and that the duplex in the triplex has A-form DNA conformation. The first NMR studies [61–63] of DNA triplexes confirmed the proposed Hoogsteen and Watson–Crick base pairing of T·AxT and C·GxC$^+$ triplets. Subsequent studies [64–66] have revealed that the majority of sugars adopt a conformation close to the C2′-*endo* puckering, except for some cytosines, especially those in the third strand, which exhibit a proportion of C3′-*endo* conformation. Infrared (IR) studies [67, 68] of a (T·AxT)$_n$ (where n is any digit) triple helix has shown that its sugars are in the C2′-*endo* pucker and not in C3′-*endo*. Again from Raman and IR spectroscopic studies [69] of (C·GxC$^+$)$_n$ triplex it was concluded that sugars in the purine (dG) strand adopt a C2′-*endo* type pucker whereas those in the pyrimidine (dC) strands, have sugar puckers in the C3′-*endo* regions. Thus, vibrational spectroscopy (IR and Raman) has provided evidence that triplexes with base triplets T·AxT and C·GxC$^+$ are structurally and conformationally more similar to the B-form than to the A-form. A schematic representation of triplex nucleic acid structure along with base triplets is presented in Fig. 3.

Three classes of nucleic acid triple helices have been described for oligonucleotides containing only natural units. They differ according to the base sequences and the relative orientation of the phosphate-deoxyribose backbone of the third strand. All the three classes involve Hoogsteen or reverse Hoogsteen-like hydrogen bonding interaction between the triple helix form-

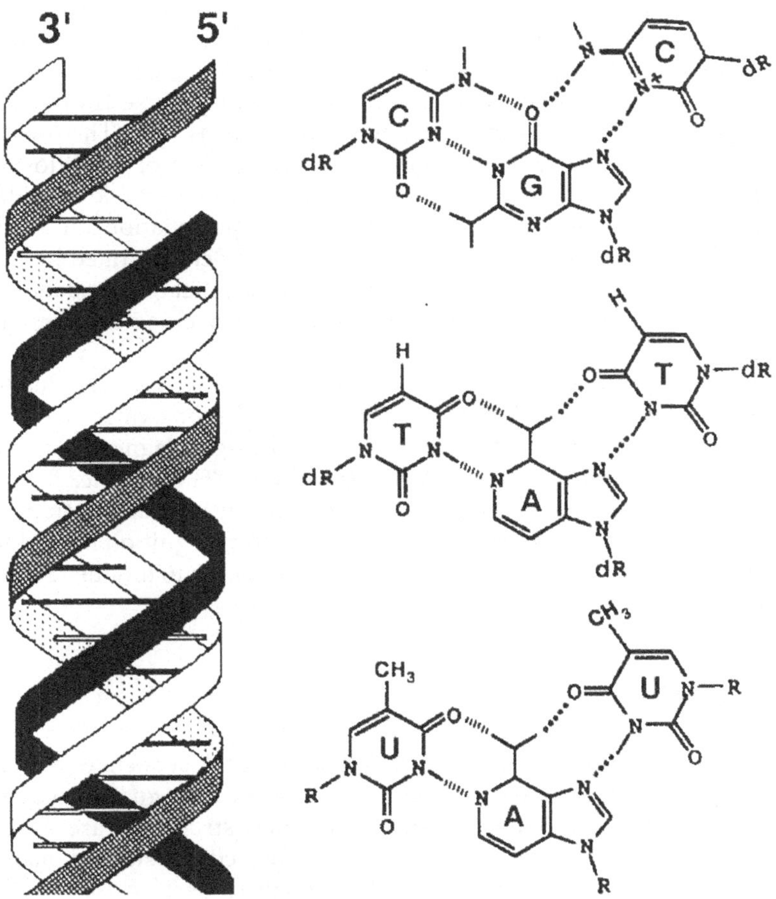

Fig. 3 Schematic representation of triplex form with chemical structures of base triplets of DNA and RNA triplexes. Reprinted from [80] with permission from Wiley InterScience

ing oligonucleotides (TFOs) and the purines of Watson–Crick base pairs. In the first class, the so called (C·T) or "pyrimidine motif" of the third strand binds in a parallel orientation with respect to the polypyrimidine strand of the duplex in the major groove, with all nucleotides in the *anti* glycoside torsional confirmation forming T·AxT and C·GxC$^+$ canonical isomorphous Hoogsteen base triplets [61–63, 70, 71]. In the second type of triplets, termed the (G·A) or "purine motif" the oligonucleotide binds in an antiparallel orientation with respect to the polypurine sequences of the Watson–Crick base pairing in the reverse Hoogsteen configuration to form non-isomorphous C·GxG and T·AxA base triplets [72, 73], with the occurrence of backbone distortion of the third strand. Such triplets are difficult to observe for some sequences. In the third class, the (G·T) or purine/pyrimidine mixed motif, the third strand binds to the oligopurine strand of the duplex by forming C·GxG and T·AxA base triplets through either Hoogsteen (parallel) or reverse Hoogsteen (antiparallel) hydrogen bond formation [74–76]. NMR data on an intramolecular triple helix with a (G·T) third strand have shown that all nucleotides adopt *anti* glycosidic torsion angles with reverse Hoogsteen hydrogen bonding of the C·GxG and T·AxA base triplets [77]. The formation of the parallel motif in C·GxC$^+$ requires the protonation of the cytosines of the third strand at the N3 position and it was found to be stable at acidic pH conditions [78]. The sequence-dependent cytosine protonation and methylation on thermal stability of DNA triplexes has been demonstrated by Leitner et al. [79] using NMR spectroscopy. They have shown that triplex stabilization by the methyl substituent primarily arises from the stacking energies and or hydrophobic effects.

The renewed interest in DNA triple helical structures has been due to the high selectivity of the third strand in duplex recognition, which may lead to a variety of potential applications in molecular biology, diagnostics and therapeutics [80–83]. To provide an in vivo application of TFOs, triplex formation has to occur at physiological conditions with a strictly regulated intramolecular pH of 7.0–7.4. The high specificity of TFO–DNA recognition has led to the development of an antigene strategy, the goal of which is to modulate the gene activity through TFOs [82, 83].

1.7 Quadruplex DNA

In addition to the familiar duplex DNA, certain DNA sequences can adopt non-duplex structures based on the association of four guanine (G) bases in a stable hydrogen-bonded arrangement; such structures are called G-quadruplexes. The core of G-quadruplex is formed by stacked planes of G-tetrads where four guanines associate through Hoogsteen-type hydrogen bonding in a head-to-tail manner. Monovalent cations such as Na^+ and K^+ have been shown to stabilize G-quadruplex structure, presumably by coordinating with eight carbonyl oxygen atoms between stacked tetrads [84], as detailed by crystallographic [85] and NMR studies [86]. A schematic representation of quadruplex structures with G-tetrads is shown in Fig. 4. In 1962 Gellert and coworkers [87] first proposed the G-tetrad structure for guanosine gels, and since then the unusual properties of G-quadruplex have been of interest to biophysists [88]. Similar structures are found at the ends of chromosomes, in the so-called telomeric regions, which may play an important role in telomerase maintenance and in transcriptional regulatory regions in several important oncogenes [89]. The transcriptional regulation of oncogene expression is an important target for drug design [89, 90] and a wide range of small molecules such as acridines, triazines, porphyrins and anthraquinones that are able to induce G-quadruplex structures have been intensively studied for their ability to inhibit teleomerase and thereby act as potential anticancer agents [91–93].

Telomeric DNA [94, 95], gene promoter regions [96] and immunoglobulin switch regions [97] have been the focus of attention because they contain continuous repeats of G-rich sequences that have the ability to form G-quadruplex structures [98, 99]. In addition, a number of proteins have been identified that exhibit specific recognition of G-quadruplex or promote G-quadruplex formation [100]. The telomeric sequences contain repeats of guanine-rich DNA sequences of several thousand bases that protect the ends from recombination, nuclease degradation and end-to-end fusion. Examples of such sequences are TTAGGG, TTGGGG, TTTT-GGGG and TTTTAGGG, which are found in telomers of human, *Tetrahymena*, *Oxytricha*, *Arabidopsin*, *Chrorella* and *Chlamydomona*, respectively [94]. Telomeric sequences have the potential to form structures held together by guanine tetrads by either the intramolecular folding of repetitive sequence, formation of a hairpin dimmer or by association of simple strands to form a tetramer. Thus

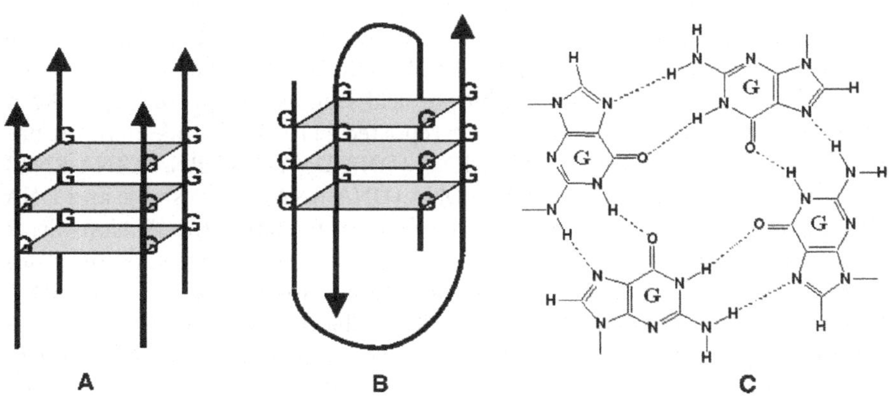

Fig. 4 Schematic representation of parallel **A** and antiparallel **B** quadruplex form of DNA with the chemical structure of a G-tetrad **C**

G-quadruplex formation may be involved in capping the chromosome end with a structure resistant to nucleases or to association of chromosomes. Recent studies suggest that the G-quadruplex structure plays a role in interfering with telomerase action, suggesting it as a potential therapeutic target for telomerase inhibition in cancer therapy [101–104]. Recently Gavathiotis and Seayle [105] determined that the structure of the intermolecular DNA quadruplex d(TTAGGG)$_4$ based on the human telomerase DNA sequence d(TTAGGG) provides a model system for exploitation in the design of novel telomerase inhibitors that bind to and stabilize G-quadruplex structure.

1.8 A-Form RNA

RNA differs from its deoxy counterpart by the presence of a 2′-OH group in the sugar residue, and the thymine CH_3 is replaced by an H in uracil [3, 106]. A topic of serious discussion for years was which structural variant amongst different forms of DNA would be appropriate to describe an RNA structure. Although the single-stranded forms of several RNA polymers do have a periodicity, considerable attention has been given to define a structural model to appropriately describe a double-stranded RNA structure. It is the 2′-OH group of RNA that stereochemically hinders the formation of a B-type structure [3]. On the other hand, an A-form structure can adequately fit a 2′-OH group of the ribose sugar into it. The A-DNA and the A-RNA double helices are quite similar and the related 11- to 12-fold helices of A-RNA are isostructural with those DNA–RNA hybrids [3, 19].

1.9 Triplex RNA

Besides DNAs, the RNAs are also potential candidates for triplex formation. Base triplets have an important role in RNA structure and function. They have been reported to occur in tRNA [107], 5SrRNA [108] and in *Tetrahymena* self-splicing intron [109]. Several biophysical studies and properties of RNA triplexes including hybrids have been reported [110–114]. In general, it has been found that RNA (DNA·RNA), RNA (DNA·DNA) and RNA (RNA·RNA) form the most stable triplexes, while DNA (DNA·RNA) is less stable and DNA (RNA·RNA) and DNA (RNA·RNA) are relatively unstable (the Watson–Crick strands are in parenthesis) [115]. These reports are mainly based on the studies involving the custom-made oligonucleotides. However, studies with RNA triplex structures involving homo-polyribonucleotides have also attracted recent attention. Chastain and Tinoco [116] have reported the formation of triple helical RNA structure upon binding of poly(rA) to double-stranded poly(rG)·poly(rC) from the measurements of UV absorbance mixing curves, melting curves and circular dichroic spectroscopy. They also reported that poly(rU) did not bind to poly(rG)·poly(rC) to form a triple helix. Buckin et al. [117] have reported the hydration effects accompanying the substitution of counter ions in the ionic atmosphere of poly(rA)·2poly(rU) triple helix (Fig. 3). Further, information regarding the structural features and stability of RNA triple helices as revealed from NMR studies [115] suggested that the Watson–Crick and Hoogsteen base pairings are disrupted simultaneously upon melting. The data are consistent with a structural model where Watson–Crick paired strands form an A-helix. RNA triple helical structures involving other homonucleotides such as poly(rI) have been studied and showed that poly(rA) does not form triplex with duplex poly(rI)·poly(rC). However, both poly(rC^+)·poly(rI)xpoly(rC) and poly(rI)·poly(rA)xpoly(rI) triple helices are formed in 0.1 M NaCl [116].

1.10 Quadruplex RNA

Matsugami et al. [118, 119], in their research on model oligonucleotides on hammerhead ribozymes, first observed that RNA oligomers containing repeating GGA sequences show properties characteristic of a G-quadruplex. NMR studies of $d(GGA)_n$ oligomers revealed that they form novel intrastrand parallel quadruplexes. Later Liu and coworkers [120, 121] determined an RNA quadruplex structure of r(GGAGGUUUUGGAGG) called R14, where two GGAGG segments are connected by UUU linker, as revealed from NMR studies. R14 forms an intrastrand parallel quadruplex that further dimerizes through stacking similar to the case of $d(GGAGG)_2$ [118].

2 Physical Techniques for Alkaloid–Nucleic Acid Interaction

Appropriately designed biophysical studies can elucidate the mode(s), the binding affinities and the nature of the ligand–nucleic acid interaction that give rise to the observed selectivity and specificities. Thus, it is worthwhile to describe briefly the various physical techniques used to study alkaloid–DNA/RNA interactions.

2.1 Spectrophotometry

This technique provides an easy and convenient method to evaluate the association of small molecules to various polymorphic forms of nucleic acid structures from the measurement of absorbance changes in the absorption maximum of the ligand, where the nucleic acid has no absorbance. Information about overall DNA/RNA base preference and nature of binding can also be obtained from this technique based on hypochromic and bathochromic effects and the presence of isosbestic points. [122].

2.2 Spectrofluorimetry

Where the binding ligand molecules are fluorescent, this method is very convenient for monitoring the association reaction from the enhancement or quenching of the fluorescence intensity on binding to DNA/RNA. Again, quantification in terms of binding constant, fluorescence quantum yield measurements, determination of Stern–Volmer quenching constant and fluorescence polarization measurements will provide a wealth of information about binding affinity, mode and specificity [123, 124].

2.3 Scatchard Analysis

Equilibrium binding constants can be determined for ligand binding to DNA or RNA, as revealed either from absorbance or fluorescence spectroscopy by keeping a constant concentration of ligand and varying the nucleic acid concentration [125], using the Scatchard equation:

$$r/C_f = k(n - r)\,, \tag{1}$$

where k is the binding constant, r is the number of molecules of ligand bound per mole of nucleotide or base pairs, n is the number of nucleotides occluded by the binding of a single molecule and C_f is the molar concentration of the free ligand. The results of absorbance and fluorescence titration can be expressed in the form of a Scatchard plot as r/C_f versus r. The parameters r and C_f were determined from the change in absorbance or fluorescence emission intensity. If A_i or F_i, A_f or F_f and A or F represent the absorbance or fluorescence of the initially, finally and partially titrated ligand, respectively, then the fraction of the bound molecules α_b would be given by:

$$\alpha_b = (A_f \quad \text{or} \quad F_f - A\text{or}F)/(A_f \quad \text{or} \quad F_f - A_i \quad \text{or} \quad F_i)\,. \tag{2}$$

The molar concentration of the free (C_f) and the bound (C_b) molecules and r could be evaluated from the following equations where D and P represent the total input ligand and the DNA nucleotide phosphate molar concentrations, respectively:

$$C_f = (1 - \alpha_b)D \tag{3}$$

$$C_b = \alpha_b D \tag{4}$$

$$r = C_b/P = (\alpha_b D)/P\,. \tag{5}$$

Usually non-cooperative and non-linear binding isotherms were observed in alkaloid–B-DNA complexation and the data were fitted to a theoretical curve drawn according to the excluded site model [126] developed by McGhee and von Hippel [127] for a non-linear non-cooperative ligand binding system using the following equation:

$$r/C_\mathrm{f} = K_\mathrm{i}(1 - nr)[(1 - nr)/\{1 - (n - 1)r\}]^{n-1} \,, \qquad (6)$$

where K_i is the intrinsic binding constant to an isolated binding site, and n is the number of nucleotides occluded by the binding of a single molecule. Binding data can be further analysed using any iterative computer programme (e.g. SCATPLOT version 1.2 developed by Ray et al. [128], which works on an algorithm [129]) to determine the best fit parameters of K_i and n to Eq. 6.

2.4 Thermal Stability

Nucleic acids undergo helix-to-coil transition with increasing temperature. The transition temperature T_m is defined as the temperature corresponding to which the hyperchromicity, H (absorbance at any temperature/absorbance at room temperature) is given by:

$$T_\mathrm{m} = (H_\mathrm{max} + 1)/2 \,, \qquad (7)$$

where H_max is the maximum value of hyperchromicity.

Since T_m of DNA is affected by the ligand, the stability or destability effect can be estimated by monitoring the 260 nm absorption profile as a function of temperature [130].

2.5 Fluorescence Quantum Yield

Steady state fluorescence quantum yield may be calculated using the equation of Parker and Rees [131] as described by Maiti et al. [132]:

$$\phi_\mathrm{s} = (F_\mathrm{s}\varepsilon_\mathrm{q}C_\mathrm{q}/F_\mathrm{q}\varepsilon_\mathrm{s}C_\mathrm{s})0.55 \,, \qquad (8)$$

where F denotes the integral area of the fluorescence in arbitrary units, ε represents the molar extinction coefficient at the wavelength of excitation. C represents molar concentration of the sample (s) and quinine sulfate (q) respectively. Quinine sulfate in 0.1 N H_2SO_4 was to be taken as reference standard for quantum yield measurements. The relative quantum yield of ligand–DNA or RNA complex can easily be obtained by calculating the total area of fluorescence spectrum for complex and free ligand.

2.6 Fluorescence Polarization Anisotropy

Fluorescence polarization anisotropy of a ligand and its complexes with DNA or RNA can be given by the following expression:

$$A' = (F_{\parallel} - F_{\perp})/(F_{\parallel} - 2F_{\perp}), \tag{9}$$

where $F_{\parallel}$ and $F_{\perp}$ are the fluorescence and intensities of the vertically ($_{\parallel}$) and horizontally ($_{\perp}$) polarized emission, when the sample is excited with vertically polarized light [133]. To account for the polarization bias of the L-format single channel detection system of the fluorescence spectrometer Eq. 8 has to be modified by introducing a polarization correction factor as follows:

$$A' = (F_{\parallel} - GF_{\perp})/(F_{\parallel} + 2GF_{\perp}), \tag{10}$$

where G is the polarization correction factor or G factor for the spectrometer and is defined in terms of a ratio between the vertical component ($F_{\parallel}$) (90°,0°) and horizontal component ($F_{\perp}$) (90°,90°) of fluorescence:

$$G = F_{\perp}/F_{\parallel}. \tag{11}$$

In practice, fluorescence anisotropy measurements are carried out as described by Larsson et al. [134] using:

$$A' = (F_{\mathrm{vv}} - F_{\mathrm{vh}}G)/(F_{\mathrm{vv}} + 2F_{\mathrm{vh}}G) \tag{12}$$

$$G = F_{\mathrm{hv}}/F_{\mathrm{hh}}, \tag{13}$$

used for instrumental correction, F_{vv}, F_{vh}, F_{hv}, F_{hh} represents the fluorescence signal for excitation and emission with the polarizer set at (0°,0°), (0°,90°), (90°,0°) and (90°,90°), respectively. This method provides valuable information regarding the orientation of ligand inside the nucleic acid structure in solution. For better results, the molecular weight of nucleic acid should be in the range of $2-4 \times 10^5$ Da.

2.7 Determination of Stern–Volmer Quenching Constant

This method is applicable when the fluorescence of a ligand is quenched in presence of DNA or RNA and provides base-dependent specificity [135]. In fluorescence quenching experiments the titration data is plotted according to the Stern–Volmer equation:

$$F_0/F = 1 + K_{\mathrm{sv}}[\mathrm{Q}], \tag{14}$$

where F_0 and F are the fluorescence emission intensities in the absence and in the presence of a quencher Q (square parenthesis represents concentration) and K_{SV} is the Stern–Volmer quenching constant, which is a measure of efficiency of quenching by the representative quencher (DNA or RNA).

2.8 Competition Dialysis Assay

This assay system developed by Chaires [136] is a new, powerful and effective tool based on the fundamental thermodynamic principle of equilibrium dialysis for the discovery of ligands that bind to nucleic acids with structural and sequence selectivity. Here, identical concentrations of various nucleic acid samples are dialysed in dispodialysers against a common ligand solution. At equilibrium, the contents of the ligand bound to each nucleic acid are determined and this is correlated directly to the ligand's specificity to a particular sequence.

2.9 Circular Dichroism Spectroscopy

Circular dichroism represents the difference in the absorption of left and right circularly polarized light by optically active chiral molecules. This technique provides information regarding conformational aspects of nucleic acid structures. Ligand binding to DNA or RNA by intercalation or by minor groove binding can also be elucidated. It is known that weak induction of circular dichroism is generally associated with rigid, planar polycyclic intercalating moieties, while asymmetric induction is due to minor grove binding ligands. The circular dichroic spectra can be expressed in terms of molar ellipticity by the following equation [137]:

$$[\theta] = 100/Cl\,, \qquad (15)$$

where $[\theta]$ is the molar ellipticity, C is the concentration expressed in moles per litre and l is the path length of the cuvette in centimetres. The unit of molar ellipticity $[\theta]$ is deg cm^2 dmole^{-1}. Usually the expressed molar ellipticity is based on either nucleic acid concentration for intrinsic circular dichroism or ligand concentration for induced circular dichroism.

2.10 NMR Spectroscopy

Nuclear magnetic resonance (NMR) spectroscopy is a powerful tool that can provide the binding sites of ligand–DNA interactions at the molecular level. A prerequisite for the examination of ligand–DNA complexes is assignment of all resonances in the NMR spectra of the free DNA and the ligand and of both components in the complex.

2.11 Electrospray Ionization Mass Spectroscopy (ESIMS)

This spectroscopy has been used as a sensitive and effective analytical technique for the characterization of specific ligand–oligonucleotide duplex non-covalent complexes [138], due to its specificity, sensitivity and quickness as well as its advantage in determining binding stoichiometry.

2.12 Viscosity

An increase in the viscosity of native rod-like duplex DNA or RNA is regarded as a diagnostic feature of the intercalative process. This technique provides information about the mode of ligand–DNA or RNA binding, i.e. intercalation or groove binding. In addition, this technique can provide information about the base preference in ligand binding to nucleic acid structure [122].

The relative viscosity is calculated using the equation:

$$\eta'_{sp}/\eta_{sp} = \{(t_{complex} - t_o)/t_o\}/\{(t_{control} - t_o)/t_o\}\,, \tag{16}$$

where η'_{sp} and η_{sp} are specific viscosities of the ligand nucleic acid complex and the nucleic acid respectively, $t_{complex}$, $t_{control}$ and t_o are the average flow times for the complex, free nucleic acid, and buffer respectively.

The relative increase L/L_o can be obtained from a corresponding increase in relative viscosity with the use of the following equation [139]:

$$L/L_o = (\eta/\eta_o)^{1/3} = 1 + \beta r\,, \tag{17}$$

where L and L_o are the contour lengths of nucleic acid in the presence and absence of the ligand, respectively. η and η_o are the corresponding values of intrinsic viscosity (approximated by the reduced viscosity), i.e. $\eta_{red} = \eta_{sp}/C$ with C being the DNA or RNA concentration expressed in grams per decilitre and η is defined as $[\eta] = \lim C \rightarrow 0\eta_{sp}/C$ solution, and β is the slope of a plot of L/L_o or $[\eta/\eta_o]$ versus r.

2.13 Determination of Thermodynamic Parameters

It is known that thermodynamic and structural studies are mutually complimentary and both are necessary for a complete elucidation of the molecular details of any binding process for the delineation of the molecular interaction involved at the interaction site. The Gibbs free energy change (ΔG) may be determined from the binding constant from the relation:

$$-\Delta G^{\circ} = RT \ln K\,. \tag{18}$$

The binding enthalpy change (ΔH) could be determined either from the plots of the temperature dependence of the binding constant according to the van't Hoff relationship:

$$[\delta \ln K/\delta(1/T)] = -\Delta H^\circ/R\,, \tag{19}$$

or directly from isothermal titration calorimetry and differential scanning calorimetry.

The entropy change ΔS may be estimated from the following equation:

$$\Delta S^\circ = -(\Delta G^\circ - \Delta H^\circ)/T\,. \tag{20}$$

There are several other techniques like the fluorescent dye displacement assays, footprinting, Fourier transform infrared spectroscopy, X-ray crystallography, electron microscopy, confocal microscopy, atomic force microscopy, surface plasmon resonance etc used for ligand–DNA interactions that are not discussed here.

3 Physicochemical Properties of Protoberberine Alkaloids

Alkaloids are nitrogen-containing bases produced mostly by plants during metabolism. They are regarded as reserve materials for protein synthesis, as protective substances discouraging animal or insect attacks, as plant stimulants or regulators or simply as detoxication products. They occupy an important position in applied chemistry and play an indispensable role in medicinal chemistry. There are several naturally occurring protoberberine alkaloids, among them berberine, palmatine and the synthetic coralyne (Scheme 1) are the most important for their medicinal and pharmacological utility. Their structure–activity relationships have been extensively investigated in the author's laboratory as well as in many other laboratories over the last 10 years. Recently Grycova et al. [140] have reviewed the isolation, biosynthesis, identification and biological activities of several protoberberine alkaloids.

3.1 Physical Properties

Spectrophotometric and spectrofluorimetric methods provide a wealth of information concerning structural determinations (identification, purity and precise measurement of concentration) and chemical changes in alkaloids. These techniques yield both quantitative and qualitative data on the effect of solvents, pH and other physiological conditions [141–143]. X-ray crystallography, ^{1}H and ^{13}C NMR spectroscopy, infrared spectroscopy (IR) and circular dichroic spectroscopy were also used to study the physical properties

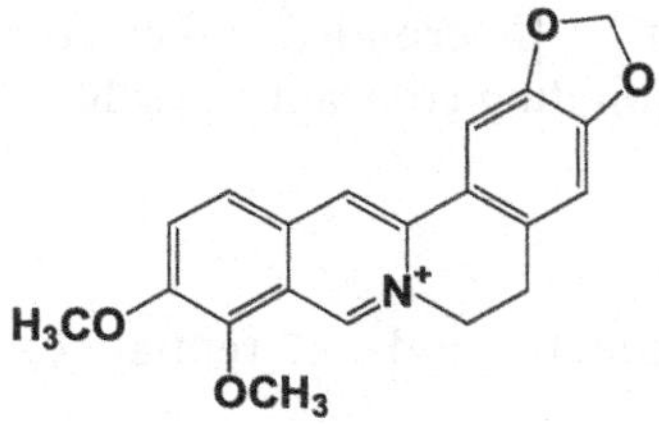

Berberine

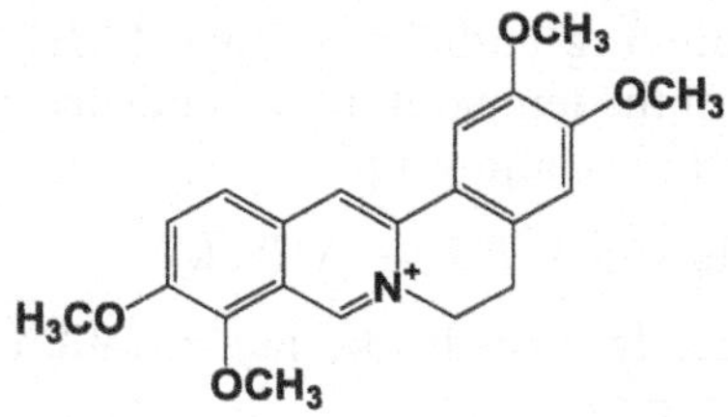

Palmatine

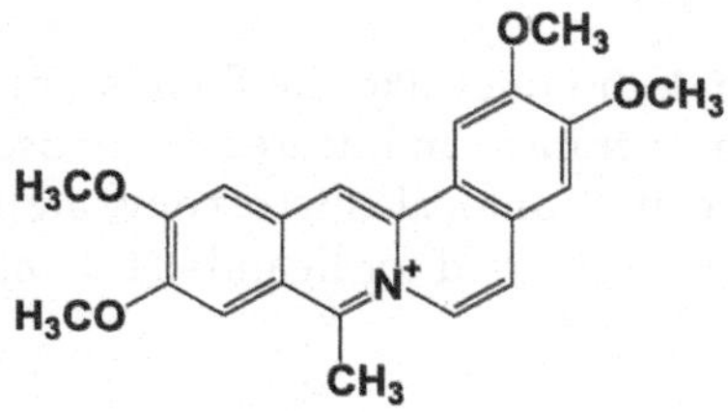

Coralyne

Scheme 1 Structures of protoberberine alkaloids

of alkaloids. Some physical properties of various protoberberine alkaloids are collated in Table 2.

Berberine (Scheme 1) is the most important member of the protoberberine group. Structurally, berberine has a polycondensate system with partial saturation in one of the rings that renders its polycyclic system slightly buckled. The characteristic UV-visible spectrum of berberine shows four absorption maxima centered around 230, 267, 344 and 420 nm in aqueous buffer at pH 7.0. It was observed that the absorption spectrum of berberine remained unaltered in the pH range 1.0–13.0 and at temperatures of 20–95 °C, which indicates its high stability in aqueous buffer. In aqueous conditions (pH 7.0) berberine has a weak fluorescence spectrum in the region 400–650 nm with a maximum around 550 nm when excited at 350 nm. This was again was unaffected in the pH range 1.0–13.0 [144]. Very recently, Lu et al. [145] measured the solubility of berberine chloride in ethanol, 2-propanol, 1-butanol, 1-octanol, ethanol + 2-propanol, ethanol + 1-butanol and ethanol + 1-octanol and showed that the solubility of berberine decreased in the order ethanol, 1-octanol, 2-propanol and 1-butanol while the same increased with increasing temperature.

Palmatine (Scheme 1) bears the same tetracyclic structure as berberine but differs in the nature of the substituents on the benzo ring; being methylene dioxy for berberine and dimethoxy for palmatine. In aqueous buffer, the UV-visible spectrum of palmatine shows maxima at 232, 268, 344.5 and 420 nm. Palmatine has a weak fluorescence spectrum in the range 400–650 nm with

Table 2 Some of the physical properties of isoquinoline alkaloids [144, 215]

Properties	Berberine	Palmatine	Coralyne
Empirical formulae	$C_{20}H_{18}O_4N^+$	$C_{21}H_{22}O_4N^+$	$C_{22}H_{22}O_4N^+$
Chemical name	7,8,13, 13a-Tetra-dehydro-9,10-dimethoxy-2,3-methylene-dioxy berberinium	7,8,13,13a-Tetra-dehydro-9,10-dimethoxy-berberinium	5,6,7,8,13,13a-Hexa-dehydro-8-methyl-2,3,10,11-tetra-methoxy berberinium
Crystal colour	Yellow	Canary yellow	Dark yellow
Solubility	Water	Water	Ethanol
Molecular weight (ion)	336.36	353.3	364.42
Melting point (°C)	210 (chloride salt)	221 (chloride salt)	250–252 (chloride salt)
Peak positions in absorption spectrum (nm)	230, 267, 344, 420 (in aqueous buffer)	232, 268, 344.5, 420 (in aqueous buffer)	219, 231, 300, 311, 326, 360, 405, 424 (30% ethanol)
Optical rotation $[\alpha]_D$ (solvent)	0° (H_2O)	0° (H_2O)	0° (30% ethanol)
IC_{50} value (μM) (in mice)	20.0	59.8	dna
Molar extinction coefficient (ε) (M^{-1} cm^{-1})	22 500 at 344 nm	25 000 at 344 nm	17 500 at 424 nm (30% ethanol)

dna data not available

an emission maximum around 530 nm when excited at 350 nm. In aqueous buffer of pH 7.0, the structural difference between berberine and palmatine was not reflected in their absorbance and fluorescence properties.

The synthetic alkaloid coralyne (Scheme 1) on the other hand is a planar molecule and is not readily soluble in aqueous buffers. It is highly soluble in ethanol and methanol. Coralyne is characterized by strong absorption maxima at 219, 300, 311, 326 and 424 nm with characteristic humps at 231, 360 and 405 nm in 30% (v/v) ethanol. It is highly fluorescent and gives an emission spectrum with a maximum at 460 nm when excitation was done either at 310 or 424 nm. It was observed that both absorbance and the fluorescence pattern of coralyne remained unaltered in buffer of various pH values ranging from 1.0 to 13.0 and also with salt concentration ranging from 4.0 to 500 mM. This implied that hydrophobic environment favoured the increment of their fluorescence properties [144].

3.2 Generation of Singlet Oxygen

The interaction of a compound (natural or synthetic) with light can lead to the production of singlet oxygen ($^1O_2$) and radical species. Such compounds are called photosensitizers. The singlet oxygen is the excited state of oxygen and is a very reactive species. Any such compound capable of promoting the conversion of molecular oxygen ($^3O_2$) (triplet ground state molecular oxygen is very inert) to singlet form ($^1O_2$) is potentially important in chemical and biological oxidation process. Many bioactive phytochemicals have been shown in recent years to be photosensitizers and their toxic activity against viruses, microorganisms, insects or cells have lead to the concept of therapeutic prospects in control of infectious diseases and cancer. If, in addition, the compound can intercalate with DNA and happens to be a natural product it becomes a good candidate for drug screening. In this context, several studies revealed the production of singlet oxygen by berberine and palmatine [146–149]. The singlet oxygen production of a compound can be generated as follows:

$$M \rightarrow M^* + {}^3O_{2-} \rightarrow M + {}^1O_2 , \qquad (21)$$

where M and M* represent the ground state and excited triple state of the photosensitizer molecule, respectively. The presence of fused aromatic rings evidently is responsible for its ability to function as photosensitizer for singlet oxygen production. The effectiveness of singlet oxygen generation has been quantified by analyzing absorbance spectral changes [146]. Brezova et al. [147] demonstrated the ability of berberine and palmatine to produce singlet oxygen upon photoexcitation in dimethyl sulfoxide using EPR spectroscopy and confirmed that the generation of $^1O_2$ during continuous irradiation of berberine, but the photochemical activities of palmatine were substantially lower. Hirakawa et al. [148] reported that the photoexcited berberine can act as functional photosensitizer enabling a switch in phototoxicity via singlet oxygen generation through DNA binding. Recently, it has been shown that the photoexcitation of berberine under aerated conditions resulted in the efficient electron and energy transfer to molecular oxygen ($^3O_2$) producing superoxide anion radical and singlet oxygen [149]. These reactive species may be involved in a variety of processes influencing the viability of cells after berberine/UVA treatment. However, the possible application of berberine in photodynamic therapy is yet to be confirmed.

4 Interaction of Protoberberine Alkaloids with Nucleic Acid Structures

A number of small organic molecules are endowed with attractive biological activities that are exerted through highly specific but non-covalent and reversible interactions with nucleic acid structures. The physical and molecular basis of interaction of natural alkaloids to nucleic acid has been a subject of extensive study in the recent past [140, 150]. The reversible interaction can be divided into three major classes: (i) binding by mechanism of intercalation, which was originally demonstrated by Lerman [151] as a process whereby the molecule is sandwiched between two adjacent base pairs of nucleic acids, (ii) binding to the major and minor grooves of DNA in a non-intercalative manner without inserting any part between the bound compound between the base pairs. Most DNA-interactive proteins bind in the major groove, while small molecules of molecular weight less than 1000 Da, including many antibiotics, bind in the minor groove [152], and (iii) binding simply to the outside of the helix involving electrostatic interaction between the anionic phosphate groups on the DNA/RNA backbone and cationic groups of the compound. Such outside binding may also be considered as a transient state before intercalation or groove binding. A schematic representation of the above three types of binding is given in Fig. 5. Elucidation of the nature, strength and energetics of these interactions may reveal the molecular basis of many antitumour and antiviral drugs and provide useful guidance for rational design of new potential chemotherapeutic agents [153]. The intercalating ligands constitute one of the most widely studied groups as they form important class of compounds for cancer therapy [154]. In this context, considerable attention has been paid to naturally occurring protoberberines for their low toxicity and wide range of biological activities that include antimicrobial, antileukemic, anticancer and topoisomerase inhibitory activities. The elucidation of nucleic acid binding properties has been the subject of extensive investigations over the last 10 years.

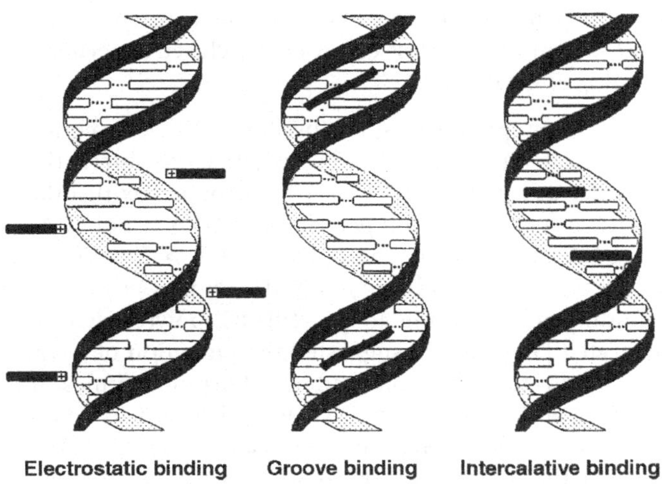

Fig. 5 Schematic representation of various binding modes of small molecules to DNA

4.1 Berberine–B-DNA Interaction

All earlier studies [155–158] reported the complexation of berberine with calf thymus DNA and suggested by a mechanism of intercalation. Maiti and coworkers [159–162] demonstrated first the base- and sequence-specificity of berberine from studies with several naturally occurring DNAs (*Clostridium perfringenes*, cholera bacteriophage $\phi 2$, calf thymus, *Escherichia coli*, *Micrococcus lysodeikticus*) and synthetic DNAs ((poly(dG-dC)·poly(dG-dC), poly(dG)·poly(dC), poly(dA-dT)·poly(dA-dT), poly(dA)·poly(dT)) using various physicochemical techniques. Several aspects of the interaction were reported:

1. Hypochromic and bathochromic effects of the absorption spectrum of berberine on binding of all B-form DNAs (Fig. 6a).
2. The remarkable enhancement of steady state fluorescence emission intensity and quantitative data on fluorescence quantum yield was sequence-dependent, being maximum with AT-rich DNA and alternating AT polymer (Fig. 6b).
3. Stabilization of all DNAs against thermal strand separation (Fig. 6c).
4. Binding of berberine caused changes in the circular dichroic spectrum of all B-DNAs with an increase of molar ellipticity of the 270 nm band; the molar ellipticity value at saturation depended strongly on the base composition of DNA being larger for the AT-rich DNA than the GC-rich DNA. However, the generation of berberine-associated extrinsic circular dichroism bands was not dependent on the base composition or sequence of base pairs.
5. Berberine increased the contour length of sonicated rod-like duplex of all B-DNA depending on base composition (Fig. 6d) and induced the unwinding–rewinding process of covalently closed superhelical DNA with an unwinding angle of 13°.

The various binding parameters for berberine–B-DNA complexation are shown in Table 3. It was observed [159–162] that the interaction of berberine with poly(dA-dT)·poly(dA-dT) was the strongest with other polymers. It varied in the order *Clostridium perfringenes* DNA > poly(dA)·poly(dT) > calf thymus DNA > cholera bacteriophage DNA > *E. coli* DNA > *Micrococcs lysodeikticus* DNA > poly(dA)·poly(dT) > poly(dG-dC)·poly(dG-dC). A strong binding of berberine to poly(dI-dC)·poly(dI-dC) structure was also demonstrated [163], which suggested the induction of a conformational change from B-form to A-form on binding. Thermodynamics of the interaction of berberine with calf thymus DNA was observed to be an exothermic and entropy-driven process favouring intercalation [164]. Saran et al. [165] first investigated the interaction of berberine with calf thymus DNA using high field ^{1}H NMR (Fig. 7) and all the proton resonances in the molecule

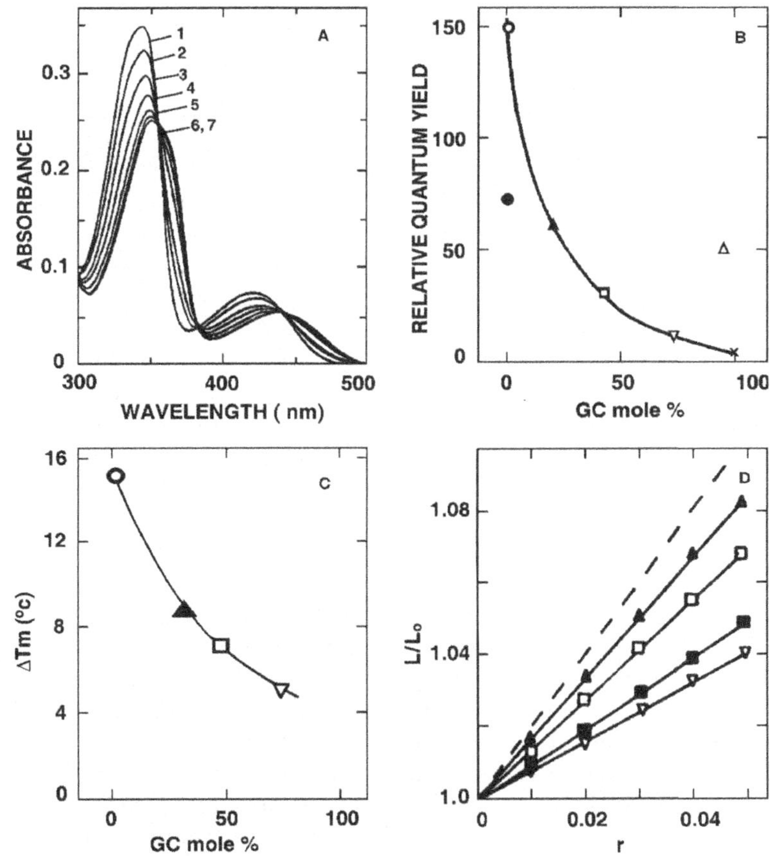

Fig. 6 Interaction of berberine with various B-DNAs as obtained from spectrophotometric (**a**), spectrofluorimetric (**b**), thermal melting (**c**) and viscometric (**d**) techniques. *Symbols*: *Clostridium perfringenes* (▲), calf thymus (□), *Escherichia coli* (■), *Micrococcus leisodeikticus* (▽), poly(dG)·poly(dC) (△), poly(dG-dC)·poly(dG-dC) (X), poly(dA)·poly(dT) (•) and poly(dA-dT)·poly(dA-dT) (∘). Reprinted from [161] with permission from the publisher

were assigned using a combination of 2D-COSY, NOESY and ROESY techniques. It was proposed that berberine binds to B-DNA by a mechanism of partial intercalation in which the planar isoquinoline moiety remained intercalated between the base pairs. This was subsequently supported by many other laboratories [166–170] using several analytical techniques that included absorption, fluorescence, electrospray ionization mass spectroscopy in combination with computer molecular modelling studies. However, a fluorescence spectral study of berberine with calf thymus DNA [171] and a high resolution ^{1}H NMR and UV spectroscopy [172] in conjunction with molecular modelling studies on berberine binding to various short oligonucleotide

Table 3 Various binding parameters for the interaction of berberine with several DNAs [144, 161]

DNA polymer	GC (mole %)	K (10^5 M^{-1})	n	ΔT_m at r_{max} (°C)	β	θ (degrees)
Clostridium perfringenes	30	4.0	3.4	7.5	1.65	
Calf thymus	42	3.8	4.0	7.0	1.37	
Escherichia coli	50	4.0	4.3	6.5	0.95	
Micrococcus lysodeikticus	72	3.9	5.5	6.0	0.75	
Poly(dG-dC)·poly(dG-dC)	100	3.5	3.0	dna	dna	
Poly(dG)·poly(dC)	100	3.5	4.0	14.0	dna	
Poly(dA-dT)·poly(dA-dT)	0	6.7	4.6	20.0	dna	
Poly(dA)·poly(dT)	0	4.6	8.8	16.5	dna	
CCS Col E1 DNA	54	–	–	–	–	13

dna data not available

duplexes favoured a groove binding geometry. This suggested that berberine is located in the minor groove with a preference for AT sequence, lying with the convex side on the helix groove where the positively charged nitrogen atom is close to the negative ionic surface of the oligonucleotide helices. However, the AT sequence preference of berberine binding to short oligonucleotides duplexes was contradicted by Chen et al. [169] based on results from electrospray ionization mass spectroscopy (ESIMS) and fluorescence titration experiments. The interaction of berberine with DNA was interpreted to be either by an intercalation or an external stacking parallel to the base pairs, as revealed from the results of electric linear dichroic spectroscopy [170].

Recently Chen et al. [173] have synthesized several dimeric berberines linked with alkyl chains of varying lengths (Scheme 2) and compared their affinity of binding to B-DNA with respect to monomeric berberine. They reported that dimers greatly enhanced DNA binding affinities up to approximately 100-fold with two double helical oligonucleotides, $d(AAGAATTCTT)_2$ and $d(TAAGAATTCTTA)_2$. Further, they demonstrated that these dimer's binding affinity to calf thymus DNA increased the same 100-fold compared to the binding of monomeric berberine [174]. Furthermore, these dimers linked by different spacers showed a prominent structure–activity relationship when bound to oligodeoxyribonucleotides, $d(AAGAATTCT)_2$, $d(AAGCATGCTT)_2$ and $d(TAAGAATTCTTA)_2$, as investigated by fluorescence titration and ethidium bromide displacement experiments. Among the dimers, a dimer linked with a propyl chain exhibited the largest binding affinity and this suggested that a propyl chain is the most suitable spacer for bridging the two berberine

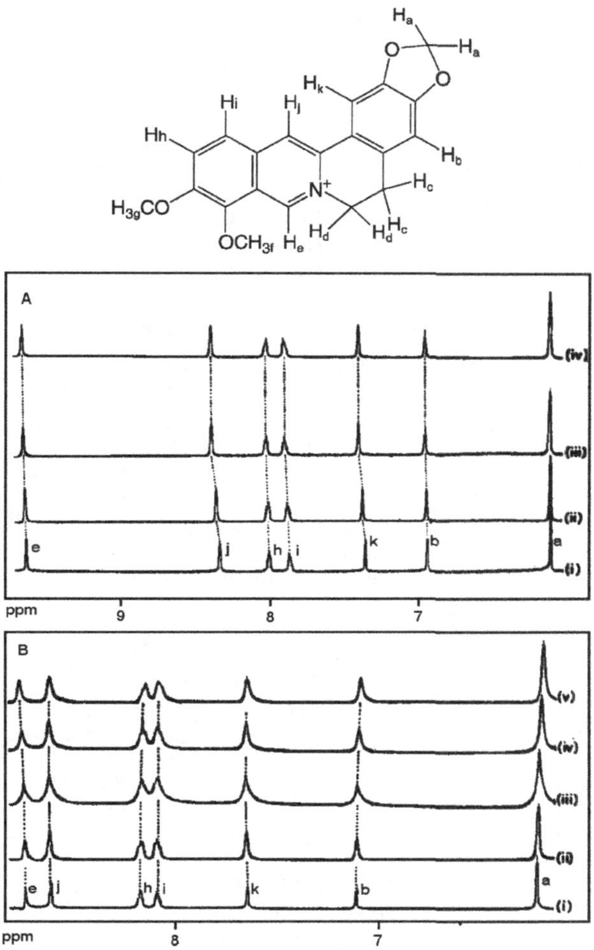

Fig. 7 Chemical structure of berberine with atom numbering and proton labels. ^{1}H NMR spectra of berberine titrated with increasing concentrations of calf thymus DNA at 298 K (**a**) and at 350 K (**b**). Reprinted from [165] with permission from the publisher

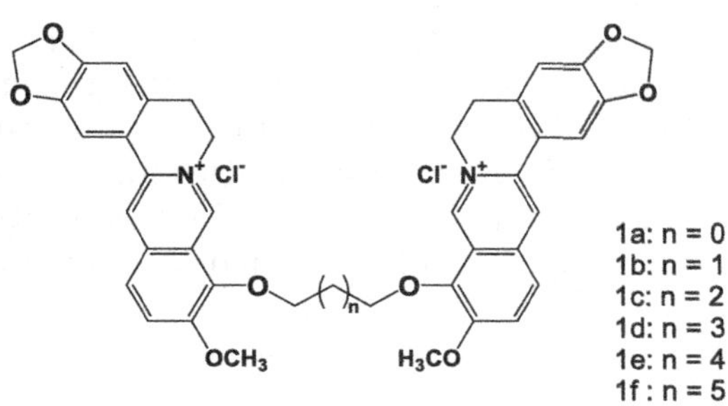

Scheme 2 Structures of berberine dimers

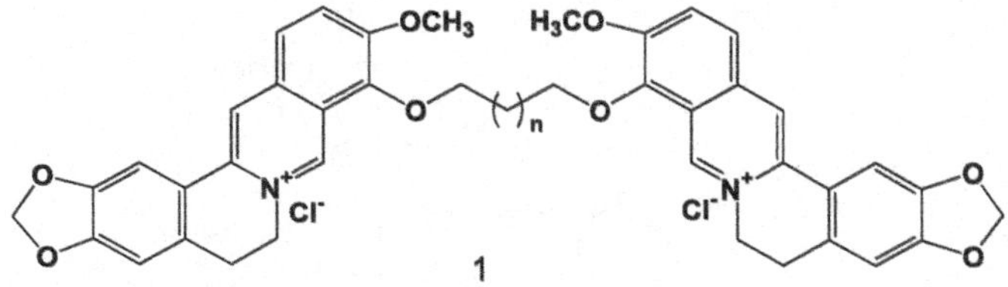

1

Ber-C_2-Ber (n = 0)
Ber-C_3-Ber (n = 1)
Ber-C_4-Ber (n = 2)

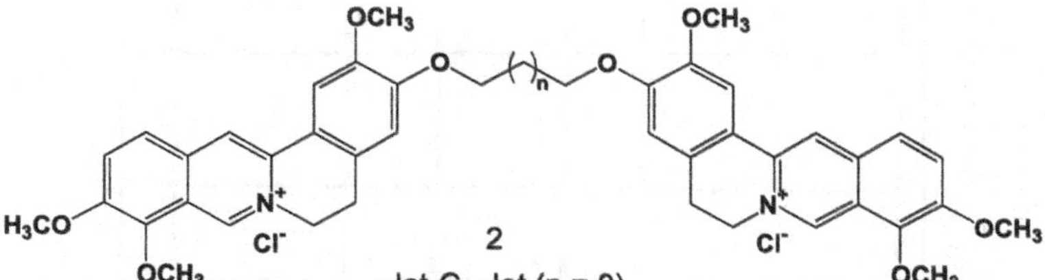

2

Jat-C_2-Jat (n = 0)
Jat-C_3-Jat (n = 1)
Jat-C_4-Jat (n = 2)

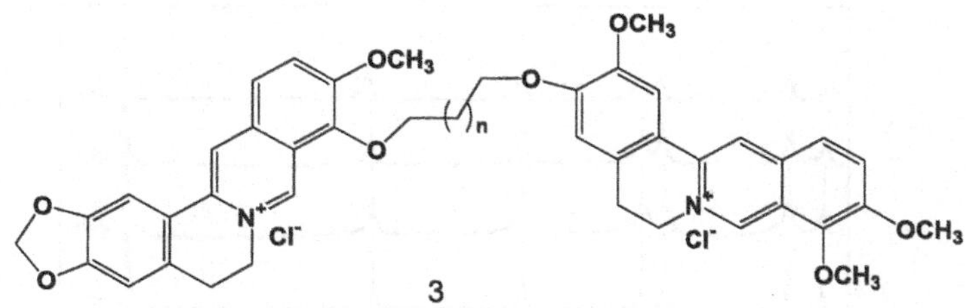

3

Ber-C_2-Jat (n = 0)
Ber-C_3-Jat (n = 1)
Ber-C_4-Jat (n = 2)

Scheme 3 Structures of berberine homodimers (Br-C_{2-4}-Ber), Jatrorrhizine homodimers (Jat-C_{2-4}-Jat) and berberine-Jatrorrhizine heterodimers (Ber-C_{2-4}-Jat)

units for DNA binding by intercalation. Recently, Long et al. [175] synthesized several jatzorrhizine homodimers and berberine–jatzorrhizine hetero dimers (Scheme 3) and studied their binding affinities towards calf thymus DNA and three double-stranded oligodeoxyribonucleotides, d(AAGAATTCTT)$_2$, d(TAAGAATAA)$_2$ and d(TTAAGAATTCTTAA)$_2$. It was suggested that spacer length and attaching position are of great importance on modulating DNA-binding affinities.

4.2
Palmatine–B-DNA Interaction

Palmatine is structurally similar to the well-known benzoquinolizine derivative berberine (vide supra). To evaluate the affinity and mode of binding to B-DNA, Kluza et al. [170] first reported that addition of calf thymus DNA induced marked changes in the absorption spectrum of palmatine, with strong hypochromic and bathochromic shifts, with well-resolved isosbestic points and with increased thermal melting temperature to form stable complexes with B-DNA. Further, from electric linear dichroism (ELD) studies it was interpreted that palmatine binds either by an intercalation or by an external stacking parallel to the base pairs. Recently, the author's laboratory [176] (Bhadra et al., unpublished results) first demonstrated the base- and sequence-dependent binding of palmatine from the studies with several natural and synthetic DNAs. Considerable specificity towards AT polynuclotides was evidenced by the competition dialysis experiment (Fig. 8a), increase of steady state fluorescence intensity (Fig. 8b), perturbation in the circular dichroic spectrum (Fig. 8c), stabilization against thermal denaturation (Fig. 8d), increase of contour length of sonicated rod-like duplex B-DNA

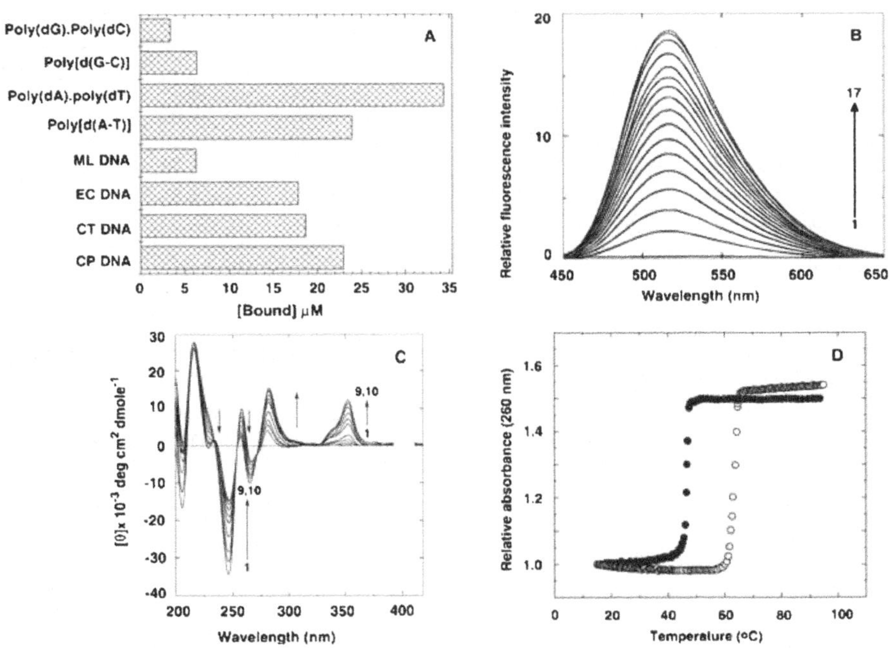

Fig. 8 Interaction of palmatine with various B-DNAs as obtained from competition dialysis (**a**), spectrofluorimetric (**b**), circular dichroism (**c**) and thermal melting (**d**) studies. (Reprinted from [176] with permission from Elsevier)

and unwinding–rewinding of covalently closed superhelical DNA with unwinding angle of 15° [176]. It was observed that the interaction of palmatine with poly(dA)·poly(dT) was the strongest, and with other DNAs it varied in the order, poly(dA-dT)·poly(dA-dT) > *Clostridium perfringenes* DNA > calf thymus DNA > *Escherishia coli* DNA > *Micrococcus lysodeikticus* DNA > poly(dG)·poly(dC) > poly(dG-dC)·poly(dG-dC). The binding parameters of palmatine complexation with various DNAs are presented in Table 4. The results established that palmatine formed an intercalation complex with all the polymers except the two GC polymers, which formed non-intercalative complex. The binding of palmatine to two AT polymers was further characterized by thermodynamic studies using isothermal titration calorimetry (ITC). The thermodynamic parameters (Table 5) of binding to the AT homopolymer showed a single strong endothermic binding event with a positive enthalpy (ΔH_1°) and positive entropy (ΔS_1°) values, while two distinct exothermic bind-

Table 4 Various parameters for the interaction of palmatine with several DNAs [176] (Bhadra et al., unpublished results)

DNA polymer	K (10^5 M^{-1})	n	ΔT_m at r_{max} (°C)	β	θ (degrees)
Calf thymus	0.19	4.0	11.0	1.40	
Poly(dG-dC)·poly(dG-dC)	0.13	7.9	dna	0.64	
Poly(dG)·poly(dC)	0.09	22.8	4.0	0.60	
Poly(dA-dT)·poly(dA-dT)	2.09	5.0	12.0	1.70	
Poly(dA)·poly(dT)	5.05	4.0	15.0	1.85	
CCS Col E1 DNA		–	–	–	15

dna data not available

Table 5 Binding of palmatine to homo and hetero polymer of AT as revealed from isothermal titration calorimetry [176]

Binding parameters	Poly(dA)·poly(dT)	Poly(dA-dT)·poly(dA-dT)
K_{b1} (10^5 M^{-1})	2.30	1.63
ΔH_1° (kcal mol^{-1})	+ 1.70	– 1.02
$T\Delta S_1^\circ$ (kcal mol^{-1})	+ 9.02	+ 6.04
ΔG_1° (kcal mol^{-1})	– 7.23	– 7.03
K_{b2} (10^3 M^{-1})		9.46
ΔH_2° (kcal mol^{-1})		– 36.77
$T\Delta S_2^\circ$ (kcal mol^{-1})		– 31.35
ΔG_2° (kcal mol^{-1})		– 5.36

ing events in the titration to the hetero AT polymer were observed and the first binding was overwhelmingly entropy-driven and the second binding is enthalpy-driven. However, an earlier study by Yu et al. [177] proposed a non-intercalative binding of palmatine.

4.3 Coralyne–B-DNA Interaction

Coralyne is an important planar synthetic protoberberine alkaloid that was found to be potentially capable of binding to DNA. Earlier works [178–180] established that coralyne formed complexation with DNA by a mechanism of intercalation, which led to speculation that the DNA binding may be correlated with the biological activity. It was reported that at lower alkaloid/DNA molar ratios coralyne intercalates into DNA by π-stacking along the DNA backbone, while at higher molar ratio, a DNA-induced aggregation occurred [178]. A more detailed analysis by Taira et al. [181] showed that coralyne caused aggregation of DNA due to the stacking arrangement of the self-aggregated coralyne along the DNA phosphate backbone. Although a wealth of data [178–182] is presently available regarding the complex formation between coralyne and DNA, the interpretation of these data needs to be performed with caution as coralyne exhibits a high tendency of aggregation in aqueous buffer even at very low concentrations. Maiti and co-workers [183, 184] first addressed this problem and showed that aqueous solution containing 30% (v/v) ethanol suppressed the aggregation of coralyne and allowed spectrophotometric titration with a fully monomeric form of coralyne (Fig. 9a) and the structure of DNA remained in the B-from conformation under these conditions [185]. The base- and sequence-selectivity

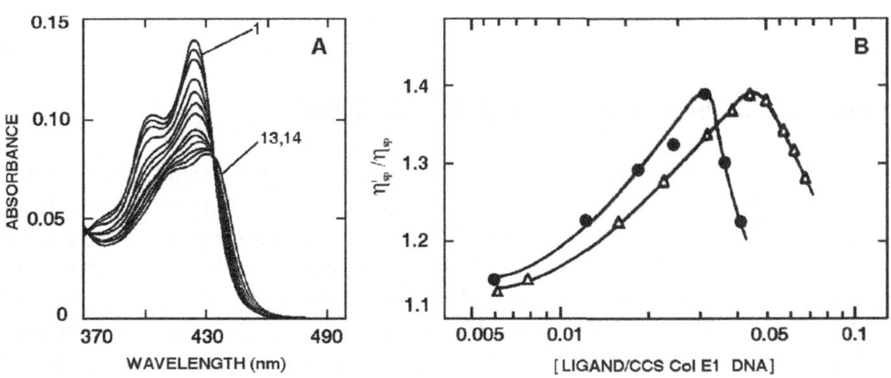

Fig. 9 Representative absorption spectra of coralyne with increasing concentrations of calf thymus DNA (**a**) and relative increase of the viscosities of CCS Col E1 DNA (**b**) by coralyne (△) and ethidium (•). Reprinted from [183] with permission from the publisher

Table 6 Various binding parameters for the interaction of coralyne with several DNAs [183, 184]

DNA polymer	K (10^5 M^{-1})	n	ΔT_m at r_{max} (°C)	β	θ (degrees)
Clostridium perfringenes	0.38	14.0	8.0	1.07	
Calf thymus	1.0	9.0	8.5	1.25	
Escherichia coli	1.55	8.5	dna	1.56	
Micrococcus lysodeikticus	2.87	7.0	9.0	2.13	
Poly(dG-dC)·poly(dG-dC)	9.8	3.9	dna	2.48	
Poly(dG)·poly(dC)	7.6	4.5	dna	dna	
Poly(dA-dT)·poly(dA-dT)	6.4	6.6	7.0	2.20	
Poly(dA)·poly(dT)	1.0	10.3	2.0	dna	
CCS Col E1 DNA					21

dna data not available

of binding of monomeric coralyne were studied by spectrophotometric and spectrofluorimetric titrations, viscometric and circular dichroism spectroscopic studies with various naturally occurring and synthetic DNAs of differing base composition and sequences. Coralyne increased the contour length of sonicated rod-like duplex DNA and induced the unwinding–rewinding process of covalently closed superhelical DNA with an unwinding angle of 21° (Fig. 9b). These experiments revealed that monomeric coralyne binds to B-DNA by a mechanism of intercalation with a relatively high specificity for GC-rich DNA and particularly for the heteropolymer of GC sequences. Varying binding parameters of coralyne–B-DNA complexes are depicted in Table 6.

5 Interaction of Protoberberine Alkaloids with Z-DNA

Poly(dG-dC)·poly(dG-dC) and its methylated analogue structures assume left-handed conformation (Z-DNA) in high molar sodium salt (Na^+, K^+), in low molar divalent cations (Ca^+, Mg^+, Ni^{2+}), micromolar concentrations of hexaamine cobalt chloride ($(Co(NH_3)_6)Cl_3$ and in millimolar concentrations of polyamines. In order to analyse the binding of berberine to Z-form DNA, Kumar et al. [186] reported that the Z-DNA structure of poly(dG-dC)·poly(dG-dC) prepared in either a high salt concentration (4.0 M) or in 40 mM ($Co(NH_3)_6)Cl_3$ remained invariant in the presence of berberine up to a nucleotide phosphate/alkaloid molar ratio of ~ 0.8 and suggested that berberine neither bound to Z-form DNA nor converted the Z-DNA to the bound B-form structure. To date there is no report on the binding of palmatine to the Z-form structure and study in this direction is awaited. It has been observed (Bhadra et al., unpublished results) that coralyne does not bind to high salt-induced Z-form but it induced a conformational switching from the left-handed form to the bound right-handed form. However, binding of coralyne to the Z-form structure has been revealed from competition dialysis experiments [187].

6 Interaction of Protoberberine Alkaloids with H^L-DNA

Both alternating GC polymer and its methylated analogue are known to be the ideal models for studying the B–H^L equilibrium transformation under low pH and low temperature as revealed from UV absorption, circular dichroism [39], FTIR [41], Raman [42] and force field microscopic studies [4]. This unique and stable left-handed structure with Hoogsteen base pairs (H^L-DNA) is distinctly different from the right-handed B-DNA and left-handed Z-DNA structures. Kumar et al. [186] demonstrated that the interaction of berberine with the H^L-DNA resulted in a remarkable increase of steady state fluorescence intensity, in intrinsic circular dichroism changes and in generation of extrinsic circular dichroic bands with opposite sign and magnitude compared to the B-form structure (Fig. 10). Based on these results [186], it was suggested that berberine could be used as a probe to detect the alteration of structural handedness due to protonation and may potentiate its use in regulatory roles for biological functions.

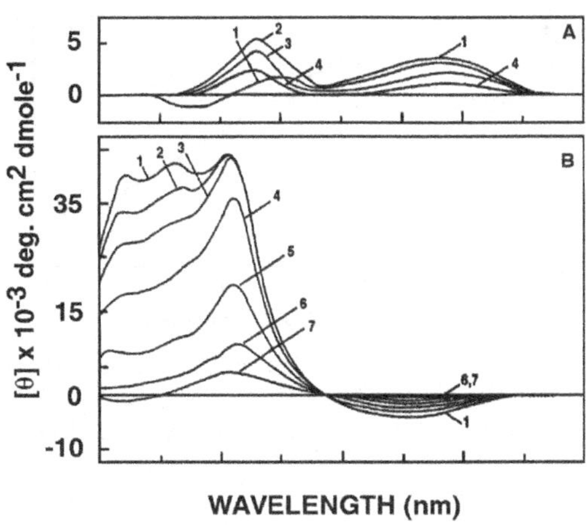

Fig. 10 Extrinsic circular dichroic spectra of berberine-B-DNA (**a**) and berberine-H^L-DNA (**b**) complexes. Reprinted in part from [186] with permission from Elsevier

7
Interaction of Protoberberine Alkaloids with Protonated DNA

Protonation-induced DNA structures have been very significant as pH is a critically important and sensitive factor in all biological processes. Studies on protonated structures of natural and synthetic DNAs have been reported by several authors [39–55]. Recently, binding of berberine to several protonated natural DNAs has been reported by Bhadra et al. [188] using various spectroscopic techniques. The results showed that in B-DNA, positive molar ellipticity at 270 nm decreased with increasing base composition while in the protonated form the negative molar ellipticity at 240 nm increased with increasing base composition of protonated DNA, revealing that berberine switches the specificity of binding from AT base pairs with B-form DNA to GC base pairs with protonated DNA (Fig. 11). Further, it was pointed out that berberine is the first example of an alkaloid molecule exhibiting different base pair specificity that is controlled by DNA conformation. Thus, the integrity of the structural conformation of DNA is necessary for alkaloid–DNA complexation.

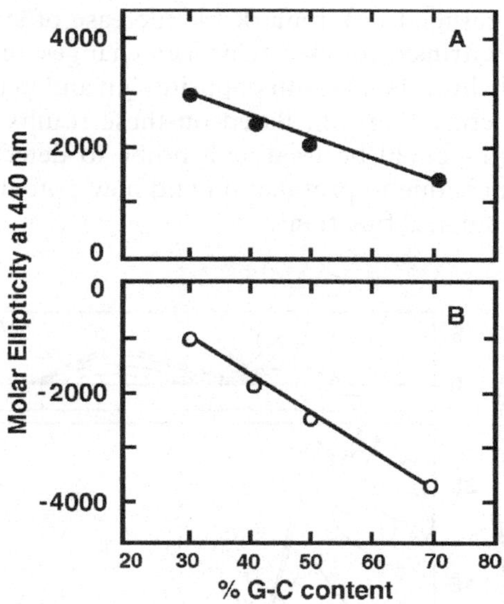

Fig. 11 Variation of extrinsic molar ellipticity at 440 nm of berberine–DNA complexes against GC content of DNA. **a** Berberine–B-DNA complexes and **b** berberine-protonated DNA complexes. Reprinted from [188] with permission from Elsevier

8 Interaction of Protoberberine Alkaloids with RNA

8.1 Protoberberine Alkaloid–Poly(A) Interaction

RNAs are versatile molecules that play an essential role in normal biological processes and in the progression of many diseases (like HIV and hepatitis C virus etc.), which has led to growing interest in the development of RNA binding agents [92, 189, 190, 192–194]. Polyriboadenylic acid [poly(A)] has been established to exist as a single-stranded helix stabilized by pair-wise stacking interaction between adjacent bases at physiological pH and temperature [195, 196]. Virtually all mRNAs in eukaryotic cells have a poly(A) tail at the 3′ end that is an important determinant in the maturation and stability of mRNA [197, 198]. Poly(A) structure plays an essential role in gene expression in eukaryotic cells [199–202]. Poly(A) polymerase (PAP), which catalyses 3′ end poly(A) synthesis, participates in an endonucleolytic cleavage step and is one key factor in the polyadenylation of the 3′ end mRNA [203]. Recent studies have identified NeoPAP, the human polyadenylic polymerase, to be significantly over-expressed in human cancer cells [204, 205]. It is thus likely that small molecules that could bind to the poly(A) tail of mRNA could interfere in the mRNA processing by PAP and would represent a new type of therapeutic agent. In this context, Maiti and coworkers [206] first observed a stronger affinity of berberine molecules to poly(A) structures over B-DNA and tRNA as revealed from absorption, fluorescence and circular dichroism studies (Fig. 12a). Employing various spectroscopic, thermal melting, viscometric and thermodynamic techniques it was further shown that berberine strongly binds to poly(A) in a non-cooperative manner by a mechanism of partial intercalation and that the binding process was endothermic and entropy-driven [207]. Very recently, Giri et al. [208, 209] studied the interaction of palmatine with poly(A) (Fig. 12b), tRNA, poly(U) and poly(C) using various biophysical techniques and their results also showed a strong affinity of palmatine to the poly(A) structure compared to B-DNA and tRNA while no binding was apparent with poly(U) and poly(C). The energetics of this strong binding of palmatine to poly(A) was also elucidated from the temperature dependence of the binding constant and ultra-sensitive isothermal titration calorimetry, both suggesting the binding to be exothermic and enthalpy-driven.

Coralyne–poly(A) binding studied by Xing et al. [187] revealed selective binding with poly(A) leading to a secondary self-structure formation in poly(A), as revealed from circular dichroic melting experiments. The binding was found to be very strong from circular dichroism changes (Fig. 12c) also. Calorimetric studies showed the binding to be predominantly enthalpy-driven with a stoichiometry of one coralyne per four adenine bases. A comparison of spectrophotometric and thermodynamic binding parameters of the interaction of berberine, palmatine and coralyne with poly(A) is presented in Table 7.

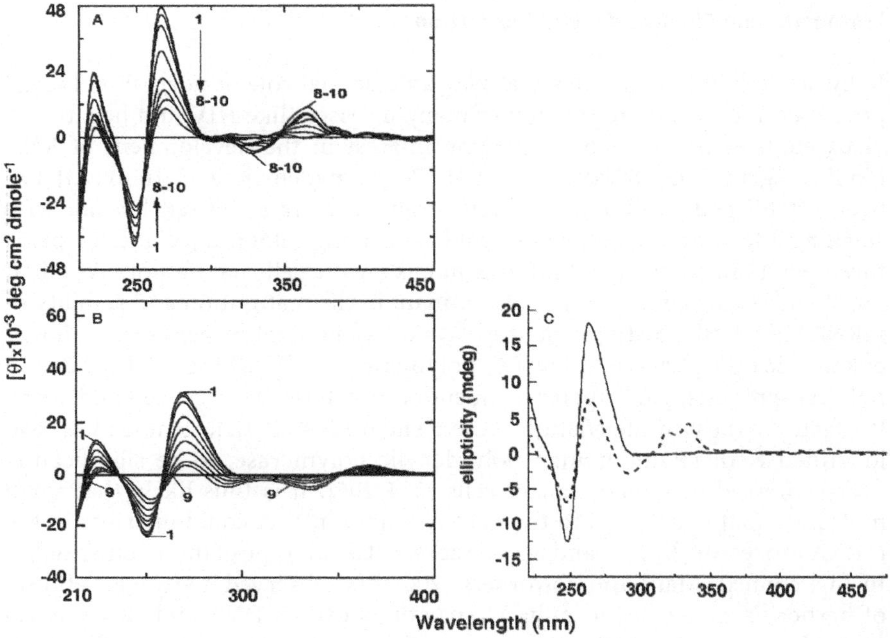

Fig. 12 Representative circular dichroic spectrum of poly(A) treated with various concentrations of berberine (**a**), palmatine (**b**) and coralyne (**c**). **a** Reprinted from [206], **b** reprinted from [208] and **c** reprinted from [187] with permission from Elsevier

Table 7 Comparative spectrophotometric and thermodynamic parameters for the interaction of berberine, palmatine and coralyne with poly(A) [187, 207–209]

Parameters	Berberine	Palmatine	Coralyne
K_i (10^5 M^{-1})	16.0	6.4	18.0
n	1.50	2.96	
$\Delta G°$ (kcal mol^{-1})	– 8.28	– 7.87 [a] – 7.99 [b]	– 8.40
$\Delta H°$ (kcal mol^{-1})	+ 8.57	– 16.63 [a] – 8.61 [b]	– 8.30
$\Delta S°$ (cal deg^{-1} mol^{-1})	+ 57.5	– 29.1 [a] – 2.27 [b]	+ 0.341

[a] van't Hoff analysis
[b] Isothermal titration calorimetry

8.2
Protoberberine Alkaloid–tRNA Interaction

Over the past few years significant advancement has taken place in the structural evaluation of various RNAs [210, 211]. Although all cellular RNAs have a single polynucleotide chain, they are highly versatile molecules that can fold into diverse structures and conformations and these structures can serve as receptors for specific drug recognition sites [212, 213]. Among the cellular RNAs, tRNAs are molecules with a cloverleaf structure that show a high degree of folding stabilized by base stacking, base pairing and other ternary interactions. tRNA (Fig. 13) is one of the most thoroughly characterized naturally occurring RNA molecules. It contains four stem regions and three loop regions. Very recently the binding of berberine and palmatine to $tRNA^{phe}$ was investigated [214] from multifaceted spectroscopic, thermal melting, viscometric, isothermal titration calorimetric and differential scanning calorimetric studies. The results revealed that both berberine and palmatine bind strongly to $tRNA^{phe}$ by mechanism of partial intercalation and that the binding was exothermic and driven by a moderately favourable enthalpy de-

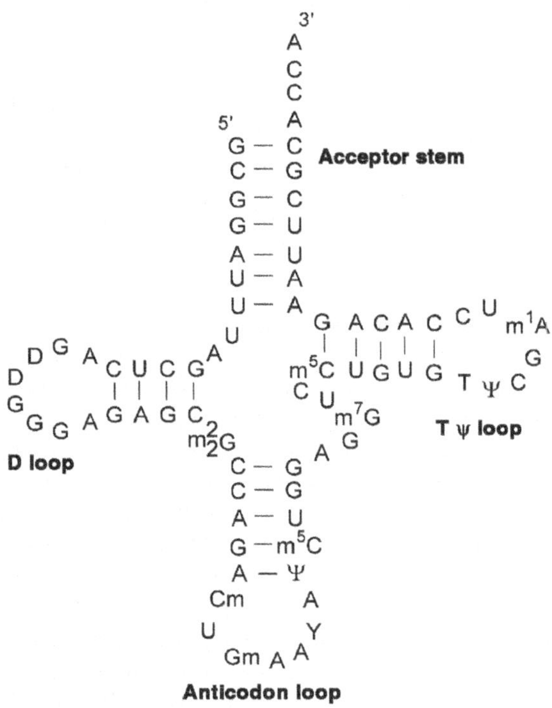

Fig. 13 Cloverleaf structure of yeast $tRNA^{phe}$. Reprinted from [214] with permission from Elsevier

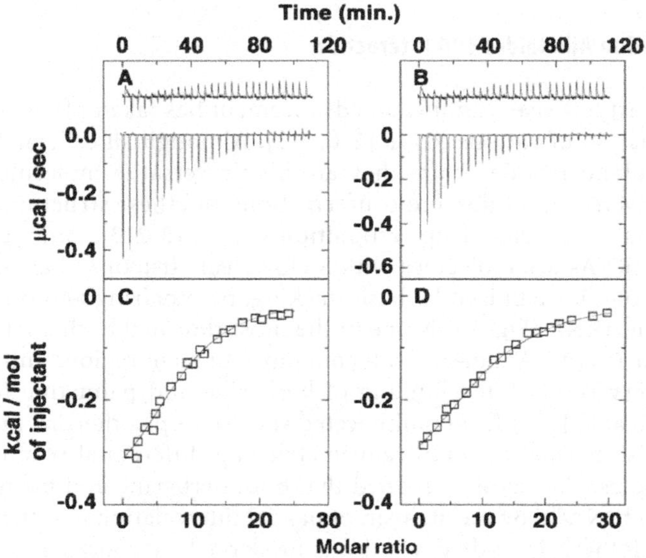

Fig. 14 Isothermal titration calorimetry for the binding of berberine (**a** and **d**) and palmatine (**b** and **e**) to tRNA. Reprinted from [214] with permission from Elsevier

crease and a moderately favourable entropy increase (Fig. 14). Comparative spectroscopic and thermodynamic data of the interaction of berberine and palmatine to tRNA are presented in Table 8. An interesting feature of the thermodynamic data is that despite the differences in the substituents of the two alkaloids there is little variation in apparent free energies.

8.3 Protoberberine Alkaloid–Duplex RNA Interaction

In order to develop compounds that can selectively target duplex RNA, Sinha et al. [194] studied the interaction of berberine with two different conformations of poly(rC)·poly(rG) structures. Poly(rC)·poly(rG) has been shown [15, 215] to exist in two conformations depending on the pH of the solution, the A-form at physiological pH and the protonated form at pH 4.3. These two conformations have been characterized to have clearly defined but distinctly different circular dichroic and absorption spectral characteristics. Both the A-form and the protonated form of the RNA induced moderate hypochromic change and bathochromic shifts in the absorption maxima peaks at 344 nm and 420 nm of the alkaloid with three isosbestic points centered around 357, 382 and 448 nm. Binding of berberine to both forms enhanced the fluorescence intensity, which was higher with the protonated form than with the A-from, suggesting clear differences in the nature of orientation

Table 8 Binding parameters of t-RNA complexation with berberine and palmatine obtained from spectrophotometric, spectrofluorimetric and isothermal titration calorimetric study [214]

Alkaloids	Spectrophotometry		Spectrofluorimetry		Isothermal titration calorimetry			
	K_i (10^5 M^{-1})	n	K_i (10^5 M^{-1})	n	K_b (10^5 M^{-1})	$\Delta G°$ (kcal mol^{-1})	$\Delta H°$ (kcal mol^{-1})	$\Delta S°$ (cal deg^{-1} mol^{-1})
Berberine	1.15	12.50	0.95	16.5	1.29	- 6.97	- 3.88	10.37
Palmatine	1.05	14.50	0.90	17.5	1.26	- 6.95	- 4.47	8.33

and/or environment of bound berberine molecules on these structures. Non-cooperative and non-intercalative mechanism of binding to the A-form and protonated form of poly(rC)·poly(rG) was proposed [194].

9 Interactions of Protoberberine Alkaloids with Higher Order Nucleic Acid Structures

9.1 Protoberberine Alkaloid–Triplex Nucleic Acid Interaction

The interest in triplex DNA structures has increased recently, intensified due to the realization that synthetic oligonucleotide-directed triple helix formation can be used in antigene agents for selectively targeting specific DNA sequences. Nevertheless, the instability of triple helical structures under normal physiological conditions is a critical limitation that restricts their uses in vitro. Several intercalating ligands have now been reported that can preferentially stabilize triplex structures [216–221]. Ren and Chaires [218] have shown that berberine preferentially and very strongly bound TxA·T triplex compared to AT duplex using novel competition dialysis and thermal melting experiments. Recently Das et al. [221] demonstrated a comparative study of berberine binding to TxA·T and UxA·U triplexes that showed that berberine specifically stabilized the Hoogsteen strand of TxA·T and UxA·U triplex structures without affecting the Watson–Crick strands (Fig. 15) The stability of the TxA·T triplexes at 210 nm [Na^+] was enhanced by 8 °C, while the UxA·U triplex stability at 35 mM [Na^+] was enhanced by 10 °C (Table 9) at saturation. The process of binding of berberine to DNA and RNA triplexes was also reported to be exothermic and entropy-driven (Table 9). The negative enthalpy and positive entropy values characterize the gross changes in the overall binding process and both favour tight binding at the intercalation site.

Earlier studies using thermal denaturation analysis and spectrophotometric titration with TxA·T and CxG·C^+ containing DNA triplexes showed that coralyne binds strongly to these triplexes by intercalation and does not exhibit a significant sequence-selectivity [222]. In a later study by Morau Allen et al. [217], employing DNase footprinting, thermal denaturation analysis, UV-visible spectrophotometric titrations, circular dichroism and NMR spectroscopy, showed that coralyne is fully intercalated into TxA·T triplex DNA whereas in C·GxC$^+$ triplex, it is partially intercalated due to electrostatic repulsion between the cationic alkaloid and the protonated cytosine [217]. Kepler et al. [223] demonstrated that coralyne intercalated to parallel triplex DNA but did not intercalate to antiparallel triplex DNA. Recently Hud and coworkers [219, 224] demonstrated that duplex poly(dA)·poly(dT) is trans-

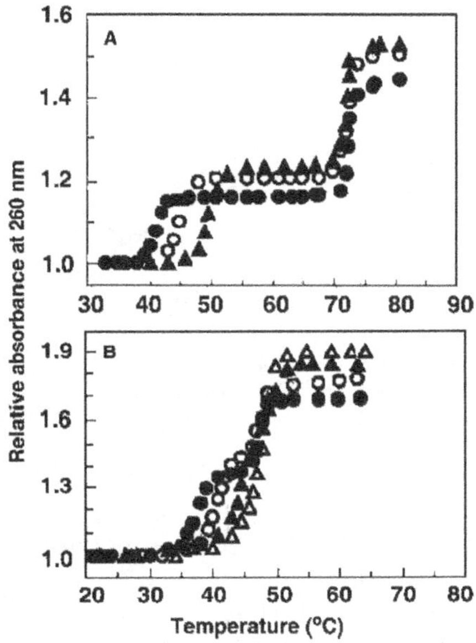

Fig. 15 Representative thermal melting profiles of **a** poly(dT)·poly(dA)xpoly(dT) and **b** poly(U)·poly(A)xpoly(U) triplexes with increasing concentration of berberine. Reprinted in part from [221] with permission from Adenine Press

Table 9 Various parameters for the interaction of berberine with DNA and RNA triplexes [221]

Parameters	Poly(dT)·poly(dA)xpoly(dT)	Poly(U)·poly(A)·poly(U)
K_i(10^5 M^{-1})	3.22	1.6
n	6.4	8.0
ΔT_m at r_{max} (°C)	7.8	10.0
$\Delta G°$ (20 °C) (kcal mol^{-1})	- 7.35	- 7.07
$\Delta H°$ (20 °C) (kcal mol^{-1})	- 4.61	- 6.05
$\Delta S°$ (20 °C) (cal deg^{-1} mol^{-1})	+ 9.52	+ 3.51

formed to a mixture of poly(dT)·poly(dA)xpoly(dT) and single-stranded poly(dA) in presence of coralyne at 35 °C. This disproportionation reaction of duplex DNA has been rationalized in terms of the formation of a stable complex between coralyne and DNA triplex [224]. Further, they demonstrated that coralyne may act as a template for the formation of an antiparallel DNA duplex from homo-adenine sequences [225]. Latimer et al. [226] prepared four derivatives of coralyne, (Scheme 4) ethylnorcoralyne, butylnorcoralyne and heptylnorcoralyne and studied their binding to

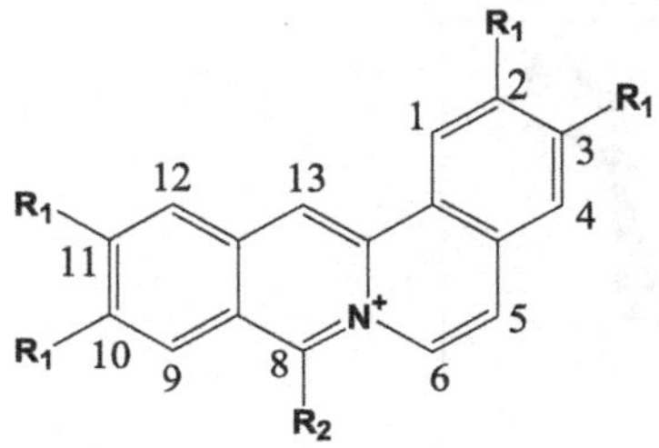

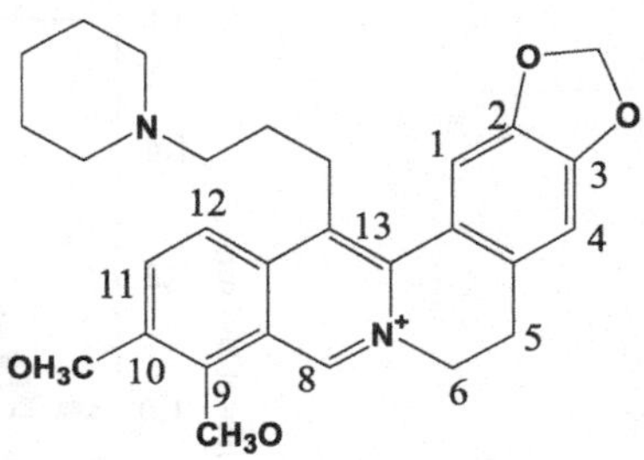

C1 Coralyne : R_1= OCH_3; R_2= CH_3
C2 Ethylnorcoralyne : R_1= OCH_3; R_2= C_2H_5
C3 Butylnorcoralyne : R_1= OCH_3; R_2= $CH_3CH_2CH_2CH_2$
C4 Heptylnorcoralyne : R_1= OCH_3; R_2= $CH_3(CH_2)_6$

C5 Piperidinoberberine

Scheme 4 Structures of coralyne derivatives and piperidinoberberine

poly(dT)·poly(dA)xpoly(dT) and poly[d(TC)]·poly[d(GA)]xpoly[d(C^+T)] as assessed from thermal denaturation profiles. A tetraethoxy derivative showed only weak binding to both type of triplexes. Analogues with extended 8-alkyl chains showed good binding to poly(dT)·poly(dA)xpoly(dT) but the preference for triplex poly[d(TC)]·poly[d(GA)]xpoly[d(C^+T)] was decreased in comparison with the duplex.

9.2 Protoberberine Alkaloid–Nucleic Acid Quadruplex Interaction

It is known that G-quadruplex structures at telomerases plays an important role in telomerase maintenance, both from the structural and the functional point of view [141–144, 227]. To identify small molecules that are able to induce G-quadruplex structures is an important area of current research for their ability to inhibit telomerase and thereby to act as potential anticancer agents. Recently berberine has been shown to inhibit telomerase elongation [228] and bind to G-quadruplex DNA [187]. The interactions of berberine with human parallel G-quadruplex structure indicated that berberine is stacked on the terminal G-tetrad of the quadruplex and that the side chain of the piperidinoberberine derivative can interact with one of the four grooves of the quadruplex [229]. Francechin et al. [230] also studied the interaction of berberine with various G-quadruplex DNA structures and its ability to inhibit telomerase have been examined and compared with those of a synthetic piperidinoberberine derivative (C_5) and the related compound coralyne. Their results showed that these alkaloids show selectivity for G-quadruplex compared to duplex DNA, which suggests that aromatic moieties play a dominant role in quadruplex binding. Coralyne has been found to have higher telomerase inhibitory activity than berberine and piperidinoberberine. It was observed that all berberine derivatives are selective in inducing intermolecular G-quadruplex compared to intramolecular G-quadruplex. In both cases, coralyne appears to be the most active compound of the series, probably due to its fully aromatic core.

10
Topoisomerase Activity of Protoberberine Alkaloids

DNA topoisomerases are nuclear enzymes that are able to break and rejoin the sugar–phosphate backbone bonds of duplex DNA and alter the topological state of the DNA helix [231, 232]. Topoisomerase I mediates transient breakage of one DNA strand, while topoisomerase II catalyses ATP-dependent relaxation of negative and positive supercoils, and causes knotting, unknotting, catenation and decatenation of DNA by passing the double-stranded DNA helix through a transient double-stranded break and then resealing the strand break [233, 234]. Developing compounds that are capable of poisoning these topoisomerases have emerged as an attractive approach for the search of novel cancer chemotherapeutic agents [231, 235–237]. Berberine, palmatine and several of their derivatives were shown to act as mammalian topoisomerase I and II inhibitors, although weakly [238], while coralyne and its derivatives were found to be strong inhibitors of topoisomerase I [239–241]. Both berberine and palmatine did not affect the DNA cleavage/relegation reaction catalysed by topoisomerases. This lack of effect most likely accounts for the presence of a methoxy substituent at the 9-position, which prevents hydrogen bonding interaction with the guanine residues. The 9-position in berberine analogues is an important determinant for DNA topoisomerase II inhibition [238].

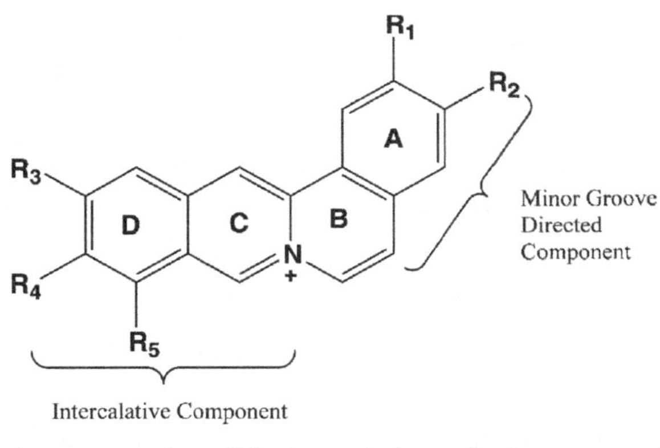

Fig. 16 Schematic representation of the intercalative and minor groove directed components of a protoberberine molecule as invoked by the "mixed mode" DNA binding model proposed by Pilch et al. [167]. Reprinted from [167] with permission from the American Chemical Society

The most important aspect of coralyne is its ability to inhibit DNA relaxation in a fashion significantly similar to the most potent antitumour alkaloid camptothecin, which is known to exert this property [242]. Presumably, the most notable biological action of these alkaloids appears to be topoisomerase inhibition [238–242], which has direct relevance to their DNA intercalating property. In this context, Pilch et al. [167] described a "mixed binding mode" model (Fig. 16) in which the protoberberine structure constitutes portions that can intercalate or bind to the minor groove of DNA. Wang et al. [240] demonstrated that coralyne (C_1) and several of its derivatives (C_6 to C_{14}) (Scheme 5), including the partial saturated

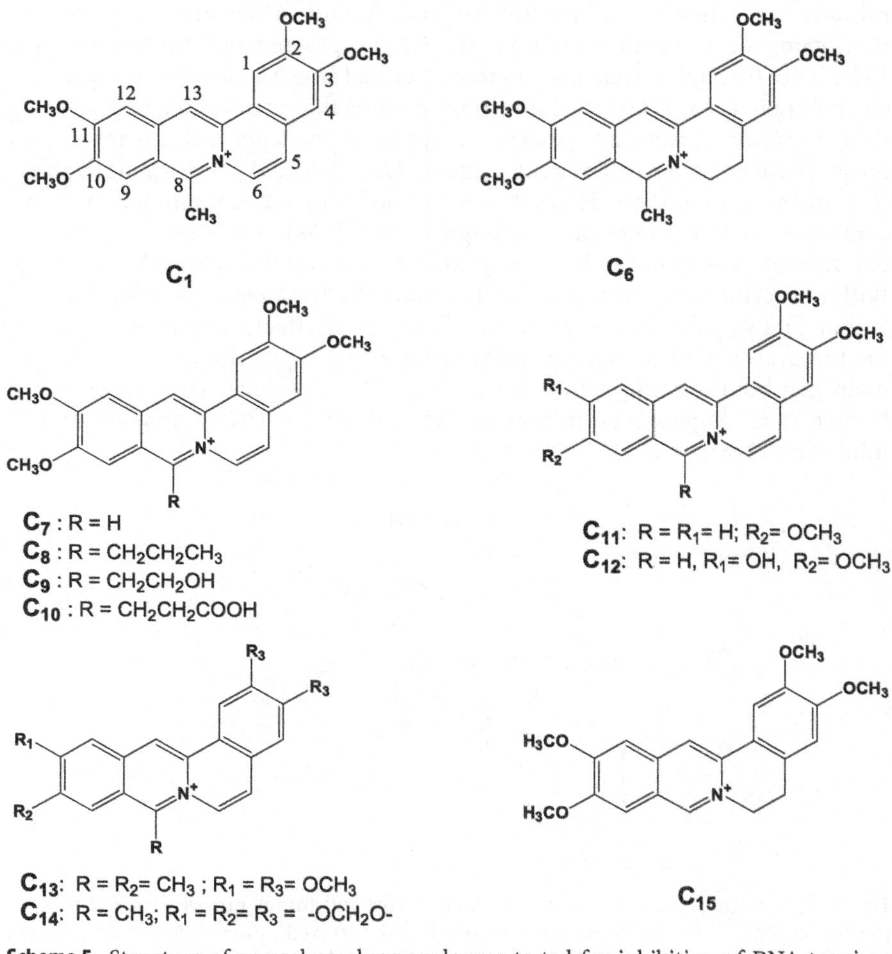

Scheme 5 Structure of several coralyne analogues tested for inhibition of DNA topoisomerase I

derivative 5,6-dihydrocoarlayne (C_6) were tested with respect to their ability to induce single-strand cleavage by formation of ternary complexes with plasmid DNA and topoisomerase I. They reported that all the derivatives induce cleavage in psp64 plasmid DNA, but are not as effective as coralyne (C_1) or 5,6-dihydrocoralyne (C_6), and that both stabilize the complex between topoisomerase I and DNA. However, coralyne was more effective than 5,6-dihydrocoralyne at lower concentration. In a subsequent study, Pilch et al. [167] confirmed that 8-desmethylcoralyne (C_7) and 5,6-dihydro-8-desmethylcoralyne (C_{15}) may also act as topoisomerase I poisons, whereas no topoisomerase II poisoning properties were observed. Further, to gain more insight into the correlation between the structure and activity of coralyne-based topoisomerase I poisons, Makhey et al. [239] studied the influence of the substitution pattern of coralyne derivatives (C_{16}–C_{20}) (Scheme 6) on their activity as topoisomerase poisons. They observed that those derivatives with a methylenedioxy functionality (i.e. C_{17} and C_{20}) exhibited the same activity as the known topoisomerase I poison of coralyne (C_1), whereas structures C_{16} and C_{19} did not exhibit significant activity towards topoisomerase I poisons. Most notably, the derivatives C_{17}, C_{18} and C_{20} induced topoisomerase II-mediated DNA cleavage, while the derivative with 2,3-dimethoxy substitution (C_{18}) did not exhibit topoisomerase I poisoning. The derivative C_{21} was the most efficient topoisomerase I poison with a similar activity to camptothecin. Makhey et al. [239] further investigated the cell toxicity of all derivatives (C_{16}–C_{20}). Interestingly, no correlation was observed between cytotoxicity of the coralyne derivatives and their activity as topoisomerase poisons.

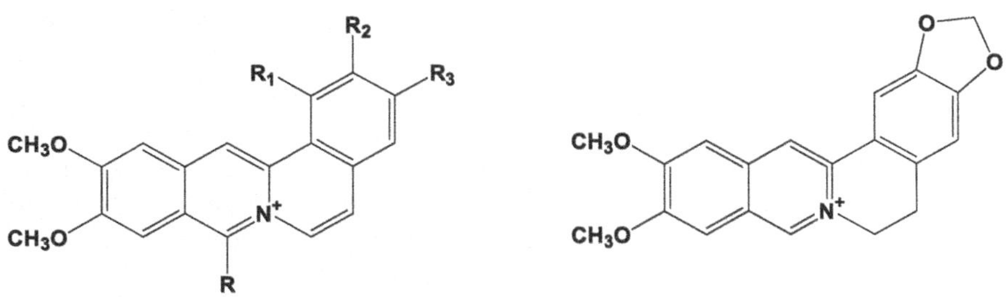

C_{16}: R = CH_3; R_1 = H; R_2=R_3 = OCH_3
C_{17}: R = CH_3; R_1 = H, R_2=R_3= -OCH_2O-
C_{18}: R = R_3 = H; R_1= R_2= OCH_3
C_{19}: R = R_1= H; R_2= R_3 = OCH_3
C_{20}: R = R_1= H; R_2= R_3= -OCH_2O-

C_{21}

Scheme 6 Structure of coralyne derivatives and a substituent in 8-position tested for cytotoxicity and topoisomerase I and II poisons

11
Biological Properties of Protoberberine Alkaloids

Berberine, a quaternary protoberberine alkaloid is a well-known naturally occurring medicine distributed widely in a number of clinically important medicinal plants such as *Captis japonica*, *Phellodendron aurense*, *Hydrastis Canadensis* (goldenseal), *Berberis aquifolium* (Oregone grape), *Berberis aristata* (tree turmeric), *Berberis vulgaris* (barbery), *Coptis chinesis* (coptis or golden thread) and a number of other plants. It occurs abundantly in the rhizomes, the outer bark of stems and roots of these plants. Berberine has a wide range of pharmacological [243] and biochemical effects [244, 245]. It has demonstrated a significant antimicrobial activity towards a variety of organisms including bacteria, fungi, protozoans, viruses, chlamydia and helminthes [246–251]. It has been used in the treatment of diarrhea and gastroenteritis for centuries and can be used as an antihypertension, antiarrhythmias and antiinflammatory agent [252–254]. In the clinics it has been shown that berberine inhibits the proliferation of all six types of esophageal cancer cell lines. The cell cycle analysis showed the accumulation of cells in the Go/G1 phase and a relative decrease of the S phase [255]. Chang et al. [256] reported that berberine induces inhibition of *N*-acetyltransferase activity and of 2-aminofluorene–DNA adduct formation in human leukemia cell in a dose-dependent manner. It was shown that berberine has antiproliferative and cytotoxic effects on the human cancer HL-60 cells and L12120 cells [257, 258]. Very recently, Jantova et al. [259] have investigated the effects of berberine on the cell morphology, cell cycle and apoptosis/necrosin in the human promonocytic U937 as well as in the murine melanoma B-16 cell lines and found that berberine acted cytotoxically on both tumour cell lines.

It has been reported that the compounds that exhibit photosensitizing properties and have a cytotoxic effect may be of interest for cancer research [259, 260]. In this context, Jantova et al. [261] demonstrated recently that the Ehrlich ascites carcinoma (EAC) cell line was more sensitive to these effects of non-photoactivated and photoactivated berberine than was the murine fibroblast NIH-373 cell line. The generation of superoxide anion radical and singlet oxygen ($^1O_2$) upon UVA irradiation of berberine increased the sensitivity of EAC cells while the sensitivity of NIH-373 cells was not changed. Moreover an anti-HIV [262] and anti-oxidative activity [263, 264] have recently been reported. The molecular mechanism underlying berberine-induced apoptosis through a mitochondria/caspases-dependent pathway has been demonstrated [265]. Bian et al. [266] synthesized several berberine derivatives and their antihyperglycemic activities were evaluated. Results indicated that only two compounds exhibited antihyperglycemic activity. Berberine has also been extensively studied as a promising antimalarial drug. Srivilaijareon et al. [267] demonstrated the presence of telomerase in *Plasmodium falciparium* and showed that the enzyme is inhibited by berberine.

Using human hepatoma-derived cell lines Kong et al. [268] showed that berberine increased mRNA and protein as well as the function of hepatic linear low density lipoprotein receptor (LDLR). It does not stimulate the transcription of LDLR, as the LDLR promoter activity was not increased by this compound. Post-transcriptional regulation appears to be the main working mechanism underlying the effect of this alkaloid on LDLR expression. It was proposed that berberine can be used as a monotherapy to treat hypercholesterolemic patients [268]. Very recently it was observed [269] that berberine reduces cholesterol and lipid accumulations in plasma as well as liver.

Palmatine, the protoberberine class of isoquinoline alkaloids, has been found in plants of various families and their name and occurrence has been reviewed recently [140]. Extracts containing palmatine and allied protoberberine alkaloids have been extensively used worldwide in folk medicine [270, 271]. Palmatine has been reported to be effective against experimental tumours by inhibiting the activity of reverse transcriptase [272, 273]. In addition, it was shown to possess antibiotic activity against bacteria, fungi and viruses [274–276]. It was found that palmatine exerts sedative effects by decreasing the levels of catecholamine in rat brain [277]. The cytotoxic potential of palmatine to HL-60 leukemia cells has been well documented [257]. Pharmacological activities of the alkaloid also include antipyretic, hepatoprotective and vasodilatory effects [140, 270].

Coralyne on the other hand has been shown to have low toxicity and very high antitumour activity compared to other protoberberines. It possesses significant activity in vivo in mice against P388 and L1210 leukemias at very low doses [277, 278]. Coralyne also exhibits selective toxicity against cultured tumour cell lines, including SF-268 glioblastoma cells, as compared with leukemia cells [140, 150]. Recently, Jiang's group [279] demonstrated inhibition of DNA topoisomerase I by several synthetic mono and dimeric protoberberine alkaloids (Schemes 2, 3) and showed that monomeric protoberberine alkaloids have little activity compared to the dimers, especially the compounds in Scheme 3 which are highly active at low concentrations with cleavage efficiency as effective as camptothecin.

12 Conclusion and Perspectives

In conclusion, protoberberine alkaloids represent a very interesting and significant group of natural products that exhibit a broad range of biological and pharmacological properties. In view of their extensive occurrence in various plants and significantly low toxicities, prospective development and use of these alkaloids as effective anticancer agents is a matter of great current interest. This chapter has focused on the physicochemical properties of berberine, palmatine, coralyne and their derivatives under various environmental conditions, on the general features of nucleic acid polymorphism and on complex formation between protoberberine alkaloids and several nucleic acid structures. These studies are based mainly on UV-visible spectroscopy, fluorescence spectroscopy, thermal melting, circular dichroism, NMR and molecular modelling, viscosity and thermodynamic analysis and provide detailed binding mechanisms at the molecular level for structure–activity relationships and for designing new and more specific compounds.

A large amount of published data suggest that berberine and palmatine exhibit strong binding affinity to the right-handed Watson–Crick base-paired B-form DNA by partial intercalation exhibiting AT base pair specificity, while they do not show any binding to the left-handed Watson–Crick base-paired Z-form DNA. On the other hand, both the alkaloids bind to the left-handed Hoogsteen base-paired H^L-form DNA suggesting the effective use of berberine and palmatine as sensitive probes to detect handedness change in DNA on protonation. Coralyne on the other hand exhibits high aggregation properties on binding to DNA, which has been effectively suppressed by the simple use of a mixed solvent system. This was a major breakthrough in understanding the molecular aspects of the interaction of monomeric coralyne and its derivatives to nucleic acid structures. The binding of monomeric coralyne to B-DNA has been established to be of intercalation with specificity towards GC base pairs. The diverse binding mode, mechanism and specificity of protoberberine alkaloids highlight the critical importance of these alkaloids in nucleic acid structure recognition. Again, berberine and coralyne have the ability to induce and stabilize DNA and RNA triplexes, which may lead to their use in triplex-targeted therapies. The ability of these compounds to induce and stabilize various G-quadruplex structures (including the human G-quadruplex) as well as their efficiency in inhibiting telomerase activity may potentiate their use as probes for G-quadruplex structures in biological systems as well as for their development as potential drugs. All protoberberine molecules bind selectively and with high affinity to single-stranded poly(A) structures, which may represent a new avenue for the development of RNA-specific therapeutic agents. Thus, the differential binding of these alkaloids to various polymorphic nucleic acid conformations may convey some specific meaning for their regulatory roles in biological systems and provide useful guidance for futuristic rational design of protoberberine derivatives with efficient and selective antitumour activity.

Acknowledgements Thanks are due to the Council of Scientific and Industrial Research (Govt. of India) for Emeritus Scientist award to M. Maiti. The work on protoberberine alkaloids and their interaction with polymorphic nucleic acid structures in the Biophysical Chemistry Laboratory was performed by a group of dedicated pre- and postdoctoral workers. The authors wish to express their indebtedness to the excellent contributions of Drs. Keya Chaudhuri, Ram Chandra Yadav, Ruma Nandi, Saswati Chakraborty, Dipanwita Debnath, Arunangshu Das, Anjana Sen, Arghya Ray, Suman Das, Sumana Pal, Anamika

Banerjee, Rangana Sinha, Ms. Kakali Bhadra, Mr. Md. Maidul Islam, Mr. Prabal Giri, Mr. Maidul Hossain and Mr. Anupam Adhikari.

References

1. Watson JD, Crick FHC (1953) Nature 171:737
2. Blackburn GM, Gait MJ (1990) Nucl Acid Chem Biol. IRL, Oxford
3. Saenger W (1984) Principles of nucleic acid structure. Springer, Berlin Heidelberg New York
4. Reddy SY, Leclerc F, Karplus M (2003) Biophys J 84:1421
5. Ivanov VI, Minchenkova LE, Schyolkina AK, Poletayev AI (1973) Biopolymers 12:89
6. Pohl FM, Jovin TM (1972) J Mol Biol 67:375
7. Wang AH, Quigley GJ, Kolpak FJ, Crawford JL, van Boom JH, van der Marel G, Rich A (1979) Nature 282:680
8. Wang AJ, Quigley GI, Kolpak FJ, van der Marel G, van Boom JH, Rich A (1981) Science 211:171
9. Maiti M (1986) Application of circular dichroic technique in nucleic acids research. In: Chatterjee SN (ed) Physical techniques in biology. Indian Physical Society, Calcutta, p 52
10. Maiti M (2001) Indian J Biochem Biophys 38:20
11. Stettler UH, Weber H, Koller T, Weissmann C (1979) J Mol Biol 131:21
12. Maiti M, Kumar GS, Das S, Ray A (2003) Biophysical aspects on the interaction of plant alkaloids with nucleic acid triplex structures. In: Maiti M, Kumar GS, Das S (eds) Recent trends in biophysical research. Double A Workstation, Kolkata, p 25
13. Feigon J (1993) Curr Biol 3:611
14. Gueron M, Leroy JL (2000) Curr Opin Struct Biol 10:326
15. Das S, Banerjee A, Sen A, Maiti M (2000) Curr Sci 79:82
16. Hall K, Kruz P, Tinoco I Jr, Jovin TM, van de Sande JH (1984) Nature 311:584
17. Felsenfeld G, Davies DR, Rich A (1957) J Am Chem Soc 79:2023
18. Liu H, Kanagawa M, Matsugami A, Tanaka Y, Katahira M, Uesugi S (2000) Nucleic Acids Symp Ser 44:65
19. Zimmerman SB (1982) Annu Rev Biochem 51:395
20. Behe M, Felsenfeld G (1981) Proc Natl Acad Sci USA 78:1619
21. Gruskin EA, Rich A (1993) Biochemistry 32:2167
22. Herbert A, Rich A (1996) J Biol Chem 271:11595
23. Wolfl S, Witting B, Rich A (1995) Bichim Biophys Acta 1264:294
24. Mirau PA, Kearns DR (1984) Biochemistry 23:5439
25. Chaires JB (1986a) J Biol Chem 261:8899
26. Chaires JB (1986b) Biochemistry 25:8436
27. Chen HH, Behe MJ, Rau DC (1984) Nucleic Acids Res 12:2381
28. Jimanez-Garcia E, Portugal J (1992) Biochemistry 31:11641
29. Ray A, Maiti M (1996) Biochemistry 35:7394
30. Das S, Kumar GS, Maiti M (1996) Biophys Chem 76:199
31. Preisler RS, Chen HH, Colombo MF, Choe Y, Short BJ Jr, Rau DC (1995) Biochemistry 34:14400
32. Lafer EM, Moller A, Noedheim A, Stolllar BD, Rich A (1981) Proc Natl Acad Sci USA 78:3546
33. Lafer EM, Valle RP, Moller A, Nordheim A, Schur PH, Rich A, Stollar BD (1983) J Clin Invest 71:314

34. Arndt-Jovin DJ, Udvardy A, Garner MM, Ritter S, Jovin TM (1993) Biochemistry 32:4862
35. Bechert T, Diekmann S, Arndt-Jovin DJ (1994) J Biomol Struct Dyn 12:605
36. Ma CL, Sun L, Bloomfield VA (1995) Biochemistry 34:3521
37. Ban C, Ramakrishnan B, Sundaralingam M (1996) Biophys J 12:1215
38. Ohishi H, Nakanishi I, Inubishi K, van der Marel G, van Boom JH, Rich A, Wang AHJ, Hakoshima T, Tomita K (1996) FEBS Lett 391:153
39. Kumar GS, Maiti M (1994) J Biomol Struct Dyn 12:183
40. Tajmir Riahi HA, Ahmed R, Naoui M, Diamantoglou S (1995) Biopolymers 35:493
41. Tajmir Riahi HA, Neault JF, Naoui M (1995) FEBS Lett 370:105
42. Seagers-Nolten GMJ, Sijtsema NM, Otto C (1997) Biochemistry 36:1324
43. Mirkin SM, Frank-Kamenetskii MD (1994) Annu Rev Biophys Biomol Struct 23: 541
44. Frank-Kamenetskii MD, Mirkin SM (1995) Annu Rev Biochem 64:65
45. Zimmer C, Venner H (1966) Biopolymers 4:1073
46. Guschlbauer W, Courtois Y (1968) FEBS Lett 1:183
47. Maiti M, Nandi R (1986) Indian J Biochem Biophys 23:322
48. Courtois Y, Fromageot P, Guschlbauer W (1968) Eur J Biochem 6:493
49. Hermann P, Frederiq E (1977) Nucleic Acid Res 4:2939
50. Smol'janinova TI, Zhidkov VA, Sokolov GV (1982) Nucleic Acid Res 10:2121
51. Maiti M, Nandi R (1986) Conformational changes of natural deoxyribonucleic acids: Evidence of formation of left handed helices in solution. In: Bawa SR (ed) Biopolymers and electron Microscopy. Punjab University Press, Chandigarh, p 107
52. Maiti M, Nandi R (1987) Anal Biochem 164:68
53. Nonin S, Leraoy JL, Gueron M (1996) Nucleic Acids Res 24:586
54. Natarajan G, Malathi R, Holler E (1996) Anal Biochem 237:152
55. Singh S, Hosur RV (1995) J Am Chem Soc 117:12637
56. Felsenfeld G, Davies DR, Rich A (1957) J Am Chem Soc 79:2023
57. Mirkin SM, Lyamichev VI, Drushlyak KN, Dobrynin VN, Filippov SA, Frank-Kamenetskii MD (1987) Nature 330:495
58. Hanvey JC, Klysik J, Wells RD (1988) J Biol Chem 263:7386
59. Wells RD, Collier DA, Hanvey JC, Shimizu M, Wohlrab F (1988) FASEB J 2:2939
60. Arnott S, Bond PJ, Selsing E, Smith PJ (1976) Nucleic Acids Res 3:2459
61. de los Santos C, Rosen M, Patel D (1989) Biochemistry 28:7282
62. Rajagopal P, Feigon J (1989) Biochemistry 28:7859
63. Rajagopal P, Feigon J (1989) Nature 339:637
64. Wang E, Malek S, Feigon J (1992) Biochemistry 31:4838
65. Macaya R, Wang E, Schultze P, Sklenar V, Feigon J (1992) J Mol Biol 225:755
66. Radhakrishnan I, Patel DJ (1993) Structure 1:135
67. Liquier J, Coffinier P, Firon M, Taillandier E (1991) J Biomol Struct Dyn 9:437
68. Howard FB, Miles HT, Liu KI, Frazier J, Raghunathan G, Sasisekharan V (1992) Biochemistry 31:10671
69. Ouali M, Letellier R, Adnet F, Liquier J, Sun JS, Lavery R, Taillandier E (1993) Biochemistry 32:2098
70. Moser HE, Dervan PB (1987) Science 238:645
71. Le Doan T, Perrouault L, Praseuth D, Habhoub N, Decout JL, Thuong NT, Lhomme J, Helene C (1987) Nucleic Acids Res 15:7749
72. Beal PA, Dervan PB (1991) Science 251:1360
73. Pilch DS, Levenson C, Shafer RH (1991) Biochemistry 30:6081
74. Cooney M, Czernuszewicz G, Postel EH, Flint SJ, Hogan ME (1988) Science 241:456

75. Sun JS, deBizemont T, Duval-Valentin G, Montenary-Garester J, Helene C (1991) CR Acad Sci III 313:585
76. de Bizemont T, Duval-Valentin G, Sun JS, Bisagni E, Montenary-Garestier J, Helene C (1996) Nucleic Acids Res 24:1136
77. Radhakrishnan I, de Los Santos C, Patel DJ (1991) J Mol Biol 221:1403
78. Bhaumik SR, Chary KV, Govil G (1998) J Biomol Struct Dyn 16:527
79. Leitner D, Schroder W, Weisz K (2000) Biochemistry 39:5886
80. Maiti M, Kumar GS (2007) Med Res Rev (in press) doi:10.1002/med.20087
81. Malvy C, Harel-Bellan A, Pritchard LL (1999) Triple helix forming oligonucleotides. Kluwer Academic, The Netherlands
82. Wells RD, Collier DA, Hanvey JC, Shimizu M, Wohlrab F (1988) FASEB J 2:2939
83. Helene C (1991) Anticancer Drug Des 6:569
84. Willamson JR, Raghuraman MK, Cech TR (1989) Cell 59:871
85. Parkinson GN, Lee MP, Neidle S (2002) Nature 417:876
86. Luu KN, Phan AT, Kuryavyi V, Lacroix L, Patel DJ (2006) J Am Chem Soc 128:9963
87. Gellert M, Lipset MN, Davies DR (1962) Proc Natl Acad Sci USA 48:2013
88. Guschlbauer W, Chantot JF, Thiele D (1990) J Biomol Struct Dyn 8:491
89. Han H, Hurley LH (2000) Trends Pharmacol Sci 21:136
90. Mergny JL, Helene C (1998) Nat Med 4:1366
91. Neidle S, Parkinson G (2002) Nat Rev Drug Rep Discov 1:383
92. Kelland LR (2005) Eur J Cancer 41:971
93. Fletcher TM (2005) Expert Opin Ther Target 9:457
94. Zakian VA (1995) Science 270:1601
95. Greider CW, Blackburn EH (1996) Sci Am 274:92
96. Simonsson T, Pecinka P, Kubista M (1998) Nucleic Acids Res 26:1167
97. Shimizu A, Honjo T (1984) Cell 36:801
98. Feigon J, Koshlap KM, Smith FW (1995) Methods Enzymol: Nuclear Magn Reson Nucleic Acids 261:225
99. Shafer RH (1998) Prog Nucleic Acids Res Mol Biol 59:55
100. Sundquist WI, Heaphy S (1993) Proc Natl Acad Sci USA 90:3393
101. Neidle S, Parkinson GN (2003) Curr Opin Struct Biol 13:275
102. Chou SH, Chin KH, Wang AH (2005) Trends Biochem Sci 30:231
103. Bertuch AA, Lundblad V (2006) Curr Opin Cell Biol 18:247
104. Autexier C, Lue NF (2006) Annu Rev Biochem 75:493
105. Gavathiotis E, Searle MS (2003) Org Biomol Chem 1:1650
106. Wilson WD, Ratmeyer L, Zhao M, Strekowski I, Boykin D (1993) Biochemistry 32:4098
107. Goddard JP (1977) Prog Biophys Mol Biol 32:233
108. Westhof E, Romby P, Romaniuk PF, Ebel JP, Ehresmann B (1989) J Mol Biol 207:417
109. Michel F, Westhof E (1990) J Mol Biol 216:585
110. Han H, Dervan PB (1993) Proc Natl Acad Sci USA 90:3806
111. Roberts RW, Crothers DM (1992) Science 258:1463
112. Escudee C, Franccois JC, Sun JS, Guunther O, Sprinzl M, Garestier T, Heelene JC (1993) Nucleic Acids Res 21:5547
113. Klinck R, Guittet E, Liquier J, Taillandier E, Gouyette C, Huynh-Dinh T (1994) FEBS Lett 355:297
114. Liquier J, Taillandier E, Klinck R, Guittet E, Gouyette C, Huynth-Dinh T (1995) Nucleic Acids Res 23:1722
115. Holland JA, Hoffman DW (1996) Nucleic Acids Res 24:2841
116. Chastain M, Tinoco I Jr (1992) Nucleic Acids Res 20:315
117. Buckin VA, Tran H, Morozov V, Marky LA (1996) J Am Chem Soc 118:7033

118. Matsugami A, Ouhasi K, Kanagawa M, Liu H, Kanagawa S, Uesugi S, Katahira M (2001) J Mol Biol 313:255
119. Matsugami A, Okuizumi T, Uesugi S, Katahira M (2003) J Biol Chem 278:28147
120. Liu H, Matsugami A, Katahira M, Uesugi S (2002) J Mol Biol 322:955
121. Liu H, Kuqimiya A, Sakurai T, Katahira M, Uesugi S (2002) Nucleosides Nucleotides Nucleic Acids 21:785
122. Nandi R, Chakraborty S, Maiti M (1991) Biochemistry 30:3715
123. Sen A, Ray A, Maiti M (1996) Biophys Chem 59:155
124. Ray A, Kumar GS, Maiti M (2003) J Biomol Struct Dyn 21:141
125. Maiti M, Nandi R, Chaudhuri K (1982) FEBS Lett 142:280
126. Crothers DM (1968) Biopolymers 6:575
127. McGhee JD, von Hippel PH (1974) J Mol Biol 86:469
128. Ray A, Maiti M, Nandy A (1996) Comput Biol Med 26:497
129. Nandy A, Kumar GS, Maiti M (1993) Indian J Biochem Biophys 30:204
130. Maiti M, Nandi R, Chaudhuri K (1984) Indian J Biochem Biophys 21:158
131. Parker CA, Rees WT (1960) Analyst (London) 85:587
132. Maiti M, Nandi R, Chaudhuri K (1983) Photochem Photobiol 38:245
133. Lokowicz JR (1983) Principles of fluorescence spectroscopy. Plenum, New York
134. Larsson A, Carlsson C, Jonsson M, Albinsson Bo (1994) J Am Chem Soc 116:8459
135. Sen A, Maiti M (1994) Biochem Pharmacol 48:2097
136. Chaires JB (2001) Methods Enzymol 340:3
137. Maiti M, Nandi R (1987) J Biomol Struct Dyn 5:159
138. Wan KX, Shibue T, Gross ML (2000) J Am Chem Soc 122:300
139. Cohen G, Eisenberg H (1969) Biopolymers 8:45
140. Grycova L, Dostal J, Marek R (2007) Phytochem 68:150
141. Simanek V (1985) The Alkaloids 26:185
142. Bhakuni DS, Jain S (1986) The Alkaloids 28:95
143. Bentley KW (2001) Nat Prod Rep 18:148
144. Debnath D (1992) Physico-chemical studies on the interaction of isoquinoline alkaloids with nucleic acids. PhD Thesis, Calcutta University
145. Lu YC, Lin Q, Luo GS, Cai YY (2005) J Chem Eng Data 1851:642
146. Maiti M, Chatterjee A (1995) Curr Sci 68:734
147. Brezova V, Dvoranova D, Kost'alova D (2004) Phytother Res 18:640
148. Hirakawa K, Kawanishi S, Hirano T (2005) Chem Res Toxicol 18:1545
149. Jantova S, Letasiova S, Brezova V, Cipak L, Labaj J (2006) J Photochem Photobiol B Biol 85:163
150. Ihmels H, Faulhaber K, Vedaldi D, Dall'Acqua F, Viola G (2005) Photochem Photobiol 81:1107
151. Lerman LS (1961) J Mol Biol 3:18
152. Reddy BS, Sondhi SM, Lown JW (1999) Pharmacol Therapeutic 84:1
153. Wolkenberg SE, Boger DL (2002) Chem Rev 102:2477
154. Denny WA (1989) Anticancer Drug Des 4:241
155. Klimek M, Hnilica L (1959) Arch Biochem Biophys 81:105
156. Krey AK, Hahn FE (1969) Science 166:755
157. Davidson MW, Lopp I, Alexander S, Wilson WD (1977) Nucleic Acids Res 4:2697
158. Cushman M, Dekow FW, Jacobsen LB (1979) J Med Chem 22:331
159. Maiti M, Chaudhuri K (1981) Indian J Biochem Biophys 18:245
160. Chaudhuri K, Nandi R, Maiti M (1983) Indian J Biochem Biophys 20:188
161. Debnath D, Kumar GS, Nandi R, Maiti M (1989) Indian J Biochem Biophys 26:201
162. Debnath D, Kumar GS, Maiti M (1991) J Biomol Struct Dyn 9:61

163. Kumar GS, Debanth D, Maiti M (1992) Anticancer Drug Des 7:305
164. Kumar GS, Debanth D, Sen A, Maiti M (1993) Biochem Pharmacol 46:1665
165. Saran A, Srivastava S, Coutinho E, Maiti M (1995) Indian J Biochem Biophys 32:74
166. Zhu JJ, Zhang JJ, Chen HY (1998) Spectroscopy Lett 31:1705
167. Pilch DS, Yu C, Makhey D, Lavoie EJ, Srinivasan AK, Olson WK, Sauers RR, Breslauer KJ, Geacintov NE, Liu LF (1997) Biochemistry 36:12542
168. Kim SA, Kwon Y, Kim JH, Muller MT, Chung IK (1998) Biochemistry 37:16316
169. Chen WH, Chan CL, Cai Z, Luo GA, Jiang ZH (2004) Bioorg Med Chem Lett 14:4955
170. Kluza J, Baldeyrou B, Colson P, Rasoanaivo P, Mambu L, Frappier F, Bailley C (2003) Eur J Pharm Sci 20:383
171. Lu WY, Lu H, Xu CX, Zhang JB, Lu ZH (1998) Spectroscopy Lett 31:1287
172. Mazzini S, Bellucci MC, Mondelli R (2003) Bioorg Med Chem 11:505
173. Chen WH, Pang JY, Qin Y, Peng Q, Cai Z, Jiang ZH (2005) Bioorg Med Chem Lett 15:2689
174. Qin Y, Pang JY, Chen WH, Cai Z, Jiang ZH (2006) Bioorg Med Chem 14:25
175. Long YH, Bai LP, Qin Y, Pang JY, Chen WH, Cai Z, Xu ZL, Jiang ZH (2006) Bioorg Med Chem 14:4670
176. Bhadra K, Maiti M, Kumar GS (2007) Biochim Biophys Acta - General Subjects 1770:1071
177. Yu Y, Long CY, Sun SQ, Liu JP (2001) Anal Lett 34:2659
178. Zee-Cheng KY, Cheng CC (1973) J Pharm Sci 62:1572
179. Wilson WD, Gough AN, Doyle JJ, Davidson MW (1976) J Med Chem 19:1261
180. Gough AN, Jones RL, Wilson WD (1979) J Med Chem 22:1551
181. Taira Z, Matsumoto M, Ishida S, Ichikawa T, Sakiya Y (1994) Chem Pharm Bull (Tokyo) 42:1556
182. Cho MJ, Repta AJ, Cheng CC, Zee-Cheng KY, Higuchi T, Pitman IH (1975) J Pharm Sci 64:1825
183. Pal S, Kumar GS, Debnath D, Maiti M (1998) Indian J Biochem Biophys 35:321
184. Pal S, Das S, Kumar GS, Maiti M (1998) Curr Sci 75:496
185. Albergo DD, Turner DH (1981) Biochemistry 20:1413
186. Kumar GS, Das S, Bhadra K, Maiti M (2003) Bioorg Med Chem 11:4861
187. Xing F, Song G, Ren J, Chaires JB, Qu X (2005) FEBS Lett 579:5035
188. Bhadra K, Kumar GS, Das S, Islam Md M, Maiti M (2005) Bioorg Med Chem 13:4851
189. Wilson WD, Li K (2000) Curr Med Chem 7:73
190. Hermann T (2000) Angew Chem Int Ed Engl 39:1890
191. Gallego J, Varani G (2001) Acc Chem Res 34:836
192. Tor Y (2003) Chem Biochem 4:998
193. Vicens Q, Westhof E (2003) Chem Biochem 4:1018
194. Sinha R, Islam Md M, Bhadra K, Kumar GS, Banerjee A, Maiti M (2006) Bioorg Med Chem 14:800
195. Holcomb DN, Tinoco I Jr (1965) Biopolymers 3:121
196. Leng M, Felsenfeld G (1966) J Mol Biol 15:455
197. Munroe D, Jacobson A (1990) Mol Cell Biol 10:3441
198. Tian B, Hu J, Zhang H, Lutz CS (2005) Nucleic Acids Res 33:201
199. Zarudnaya MI, Hovorun DM (1999) IUBMB Life 48:581
200. Bjork GR, Erickson JU, Gustafsson CE, Hagervall TG, Jonsson YH, Wikstrom PM (1987) Ann Rev Biochem 56:263
201. Alt FW, Bothwell AL, Knapp M, Siden E, Mather E, Koshland M, Baltimore D (1980) Cell 20:293
202. MeDevitt MA, Hart RP, Wong WW, Nevins JR (1986) EMBO J 5:2907

203. Dower K, Kuperwasser N, Merrikh H, Rosbash M (2004) RNA 10:1888
204. Topalian SL, Kaneko S, Gonzales MI, Bond GL, Ward Y, Manley JL (2001) Mol Cell Biol 21:5614
205. Topalian SL, Gonzales MI, Ward Y, Wang X, Wanq RF (2002) Cancer Res 62:5505
206. Nandi R, Debanath D, Maiti M (1990) Biochem Biophys Acta 1049:339
207. Yadav RC, Kumar GS, Bhadra K, Giri P, Sinha R, Pal S, Maiti M (2005) Bioorg Med Chem 13:165
208. Giri P, Hossain M, Kumar GS (2006) Bioorg Med Chem Lett 16:2364
209. Giri P, Hossain M, Kumar GS (2006) Int J Biol Macromol 39:210
210. Gallego J, Varani G (2001) Acc Chem Res 34:836
211. Vicens Q, Westhof E (2003) Biopolymer 70:42
212. Conn GL, Draper DE (1998) Curr Opin Struct Biol 8:278
213. Afshar M, Prescott CD, Varani G (1999) Curr Opin Biotechnol 10:59
214. Islam Md M, Sinha R, Kumar GS (2007) Biophys Chem 125:580
215. Das S (2000) Biophysical studies on the interaction of benzophenanthridine alkaloids with polymorphic nucleic acid structures. PhD Thesis, Jadavpur University
216. Scaria PV, Shafer RH (1991) J Biol Chem 266:5417
217. Moraru-Allen AA, Cassidy S, Asensio Alvarez JL, Fox KR, Brown T, Lane AN (1997) Nucleic Acids Res 25:1890
218. Ren J, Chaires JB (1999) Biochemistry 38:16067
219. Polak M, Hud NV (2002) Nucleic Acids Res 30:983
220. Ray A, Kumar GS, Das S, Maiti M (1999) Biochemistry 38:6239
221. Das S, Kumar GS, Ray A, Maiti M (2003) J Biomol Struct Dyn 20:703
222. Lee JS, Latimer LJP, Hampel KJ (1993) Biochemistry 32:5591
223. Keppler M, Neidle S, Fox KR (2003) Nucleic Acids Res 29:1935
224. Jain SS, Polak M, Hud NV (2003) Nucleic Acids Res 31:4608
225. Persil O, Santai CT, Jain SS, Hud NV (2004) J Am Chem Soc 126:8644
226. Latimer LPJ, Payton N, Forsyth G, Lee JS (1995) Biochem Cell Biol 73:11
227. Zaug AJ, Podell ER, Cech TR (2005) Proc Natl Acad Sci USA 102:10864
228. Naasani I, Seimiya H, Yamori T, Tsuruo T (1999) Cancer Res 59:4004
229. Parkinson GN, Lee MP, Neidle S (2002) Nature 417:876
230. Franceschin M, Rossetti L, D'Ambrosio A, Schirripa S, Bianco A, Ortaggi G, Savino M, Schultes C, Neidle S (2006) Bioorg Med Chem 16:1707
231. Chen AY, Liu LF (1994) Annu Rev Pharmacol Toxicol 34:191
232. Wang JC (1996) Annu Rev Biochem 65:635
233. Berger JM, Gamblin SJ, Harrison SC, Wang JC (1996) Nature 379:225
234. Roca J, Wang JC (1994) Cell 77:609
235. Corbett AH, Osheroff N (1993) Chem Res Toxicol 6:585
236. Froelich-Ammon SJ, Osheroff N (1995) J Biol Chem 270:21429
237. Convey JM, Jaxel C, Kohn KW, Pommier Y (1989) Cancer Res 49:5016
238. Krishnan P, Bastow KF (2000) Anticancer Drug Des 15:255
239. Makhey D, Gatto B, Yu C, Liu A, Liu LF, Lavoie EJ (1996) Bioorg Med Chem 4:781
240. Wang LK, Rogers BD, Hechst SM (1996) Chem Res Toxicol 9:75
241. Gatto B, Sanders MM, Yu C, Wu HY, Makhey D, LaVoie EJ (1996) Cancer Res 56:2795
242. Sanders MM, Liu AA, Li TK, Wu HY, Desai SD, Mao Y, Rubin EH, LaVoie EJ (1998) Biochem Pharmacol 56:1157
243. Kuo CL, Chi CW, Liu TY (2004) Cancer Lett 203:127
244. Creasey WA (1979) Biochem Pharmacol 28:1081
245. Schmeller T, Latz-Bruning B, Wink M (1977) Phytochemistry 44:257
246. Amin AH, Subbaiah TV, Abbasi KM (1969) Can J Microbiol 15:1067

247. Hwang BY, Roberts SK, Chadwick LR, Wu CD, Kinghorn AD (2003) Planta Med 69:623
248. Okunade AL, Hufford CD, Richardson MD, Petterson JR, Clark AM (1994) J Pharm Sci 83:404
249. Basha SA, Mishra RK, Jha RN, Pandey VB, Singh UP (2002) Folia Microbiol (Praha) 47:161
250. Nakamoto K, Sadamori S, Hamada T (1990) J Prosthet Dent 64:691
251. Yamamoto K, Takase H, Abe K, Saitio Y, Suzuki A (1993) Nippon Yakurigaku Zasshi 101:169
252. Tai YH, Feser JF, Marnane WG, Desjeux JF (1981) Am J Physiol 241:G253
253. Fukuda F, Hibiya Y, Mutoh M, Koshiji M, Akao S, Fujiwara H (199) Planta Med 65:381
254. Fukuda K, Hibiya Y, Mutoh M, Koshiji M, Akao S, Fujiwara H (1999) J Ethanopharmacol 66:227
255. Iizuka N, Miyamoto K, Okita K, Tangoku A, Hayashi H, Yosino S, Abe T, Morioka T, Hazama S, Oka M (2000) Cancer Lett 148:19
256. Chang JG, Chen GW, Huang CF, Lee JH, Ho CC et al. (2006) Am J Med 28:227
257. Kuo CL, Chou CC, Yung BYM (1995) Cancer Lett 93:193
258. Kettmann V, Kosfalova D, Jantova S, Cernakova M, Drimal J (2004) Pharmazie 59:548
259. Jantova S, Cipak L, Cernakova M, Kostalova D (2003) J Pharm Pharmacol 55:1143
260. Letasiova S, Jantova S, Cipak L, Mukova M (2006) Cancer Lett 239:254
261. Jantova S, Letasiova S, Brezova V, Cipak L, Labaj J (2006) J Photochem Photobiol B Biol 85:163
262. Vlietinck AJ, De Bruyne T, Apers S, Pieters LA (1998) Planta Med 64:97
263. Hwang JM, Wang CJ, Chou FP, Tseng TH, Hsieh YS, Lin WL, Chu CY (2002) Arch Toxicol 76:664
264. Yokozawa T, Ishida A, Kashiwada Y, Cho EJ, Kim HY, Ikeshiro Y (2004) J Pharm Pharmacol 56:547
265. Hwang JM, Kuo HC, Tseng TH, Liu JY, Chu CY (2006) Arch Toxicol 80:62
266. Bian X, He L, Yang G (2006) Biorg Med Chem Lett 16:1380
267. Sriwilaijareon N, Petmitr S, Mutirangura A, Ponglikitmongkol M, Wilairat P (2002) Parasitol Int 51:99
268. Kong W, Wei J, Abidi P, Lin M, Inaba S, Li C, Wang Y, Wang Z, Si S, Pan H, Wnag S, Wu J, Wang Y, Li Z, Liu J, Jiang JD (2004) Nature Med 10:1344
269. Abidi P, Chen W, Kraemer FB, Li H, Liu J (2006) J Lipid Res 47:2134
270. Nishiyama Y, Moriyasu M, Ichimaru M, Iwasa K, Kato A, Mathenge SG, Chalo Mutiso PB, Juma FD (2004) Phytochemistry 65:939
271. Slavik J, Slavikova L (1995) Collect Czech Chem Commun 60:1034
272. Sethy ML (1985) Phytochemistry 24:447
273. Gudima SO, Memelova LV, Borodulin VB, Pokholok DK, Mednikov BM, Tolkachev ON, Kochetkov SN (1994) Mol Biol (Mosk) 28:1308
274. Phillipson JD, Wright CW (1991) Trans R Soc Trop Med Hyg 85:18
275. Iwasa K, Kamigauchi M, Sugiura M, Nanba H (1997) Plant Med 63:196
276. Hsieh MT, Su SH, Tsai HY, Peng WH, Hsieh CC, Chen CF (1993) Jpn J Pharmacol 61:1
277. Zee-Cheng KY, Cheng CC (1976) J Med Chem 19:882
278. Zee-Cheng KY, Paull KD, Cheng CC (1974) J Med Chem 17:347
279. Qin Y, Pang JY, Chen MH, Zhao ZZ, Liu L, Jiang ZH (2007) Chem Biodiver 4:481

Top Heterocycl Chem (2007) 10: 211–238
DOI 10.1007/7081_2007_083

Published online: 29 June 2007

Polycyclic Diamine Alkaloids from Marine Sponges

Roberto G. S. Berlinck

Instituto de Química de São Carlos, Universidade de São Paulo, CP 780,
CEP 13560-970 São Carlos, Brazil
rgsberlinck@iqsc.usp.br

Abstract Marine sponges have been a source of a variety of polycyclic diamine alkaloids presenting different skeletal types, such as the halicyclamines, ‘upenamide, xestospongins, araguspongines, halicyclamines, haliclonacyclamines, arenosclerins, ingenamines, and madangamines. The occurrence of these alkaloids in sponges is quite possibly due to their biogenetic relatedness. Since these alkaloids also display different biological activities, they have been an interesting target in organic synthesis. The review includes the occurrence, isolation and structure determination, biological activities, and biogenetic relationships as well as the total synthesis of polycyclic diamine alkaloids isolated from marine sponges.

Keywords Alkaloids · Bisquinolizadine · Bis-1-oxaquinolizadine · Marine sponges · Piperidine

Abbreviations

2,4-DNB	2,4-Dinitrobenzyl
9-BBN	9-Borabicyclo[3.3.1]nonane
BINAP	2,2′-Bis(diphenylphosphino)-1,1′-binaphthyl
Boc_2O	*tert*-Butoxycarbonyl anhydride
COSY	Correlation spectroscopy
CSA	Camphorsulfonic acid
DCC	Dicyclohexylcarbodiimide
DEAD	Diethylazodicarboxylate
DIBAL-H	Diisobutylaluminum hydride
DMPU	*N*,*N*′-Dimethyl-*N*,*N*′-propyleneurea

ESI	Electrospray ionization
FAB	Fast atom bombardment
FGI	Functional group interconversion
gCOSY	Correlation spectroscopy with field gradient
gHMBC	^{1}H-detected multiple-bond correlation with field gradient
gHSQC	Heteronuclear single quantum correlation with field gradient
gNOE	Nuclear Overhauser enhancement with gradient field
HMBC	^{1}H-detected multiple-bond correlation
HMQC	Heteronuclear multiple quantum coherence
HOBT	1-Hydroxybenzotriazole
HOHAHA	Homonuclear Hartmann–Hahn spectroscopy
HRCIMS	High-resolution chemical ionization mass spectrometry
HRFABMS	High-resolution fast atom bombardment mass spectrometry
HSQC	Heteronuclear single quantum correlation
LC-MS/MS	Liquid chromatography coupled to tandem mass spectrometry
LDA	Lithium diisopropylamide
LR-COSY	Long-range correlation spectroscopy
m-CPBA	*meta*-Chloroperbenzoic acid
NOE	Nuclear Overhauser enhancement
NOESY	Nuclear Overhauser enhanced spectroscopy
PPTS	Pyridinium *p*-toluenesulfonate
ROESY	Rotating-frame Overhauser enhancement spectroscopy
TBDMS	*tert*-Butyldimethylsilyl
TMEDA	Tetramethylethylenediamine
TMS	Trimethylsilyl
TOCSY	Total correlation spectroscopy

1 Introduction

Among marine invertebrates, sponges have been widely investigated toward the discovery of biologically active secondary metabolites. Natural products isolated from marine sponges belong to every single structural class, from very simple, modified fatty acid derivatives to very complex polyfunctionalized macrolides, large non-ribosomal peptide synthase derived peptides, or compounds of mixed biosynthesis, which very often present potent biological activities [1–3].

Marine sponges are a source of an array of polycyclic diamine alkaloids of common biogenetic origin. This class of secondary metabolites has been the subject of four previous reviews [4–7]. Therefore, the present review will include literature reports previously not discussed, dealing with the isolation, structure determination, biological activities, and total synthesis of polycyclic diamine alkaloids isolated from marine sponges. This review will not include guanidine alkaloids [8, 9] or the manzamine alkaloids [10, 11], since these compounds have been recently reviewed elsewhere. Only polycyclic diamine alkaloids are included, but not related compounds such as alkylpyridine derivatives, which comprise another large class of sponge metabolites. Few of the marine sponge diamine alkaloids have been reviewed recently [8]. Since no attempt has been made to cover all synthetic approaches toward the total synthesis of marine sponge polycyclic diamine alkaloids, only the total syntheses of these compounds are reviewed herein. Finally, this review follows the biogenetic organization adopted by Andersen, van Soest, and Kong in the first review of this class of alkaloids [4].

2
Isolation and Biological Activities

2.1
Bis-3-alkylpiperidine Macrocycles

The EtOAc and *n*-BuOH extracts of the sponge *Haliclona viscosa* collected off the coast of Blomstrandhalvøya, Svalbard, Arctic Sea, yielded haliclamines C (**1**) and D (**2**) [12]. Both alkaloids have been isolated by gel permeation chromatography on Sephadex LH-20 and purified by C_{18} RP-HPLC. Analysis by FAB and ESI mass spectrometry indicated that haliclamine D was the higher homologue of haliclamine C. The structural elucidation of both **1** and **2** was challenging because of the large overlap of signals in the ^{1}H NMR spectrum. Analysis of the ^{1}H–^{15}N HSQC NMR spectrum indicated that both compounds were isolated as protonated salts. Due to the internal symmetry of **1** and **2**, only half of the ^{1}H and ^{13}C signals were observed in the NMR spectra. The length of each aliphatic chain in **1** and **2** was unambiguously established by LC-MS/MS analyses. The alkaloids halicyclamines C (**1**) and D (**2**) did not show feeding deterrence against the amphipod *Anonyx nugax*. However, both alkaloids displayed significant antibacterial activity. Interestingly, both **1** and **2** have been isolated from samples of *H. viscosa* collected during 1999, but these alkaloids were not detected in samples of the same sponge collected in 2000 and 2001 [12].

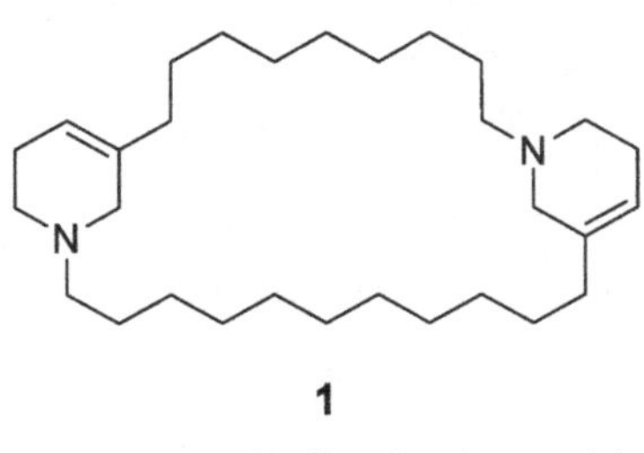

1

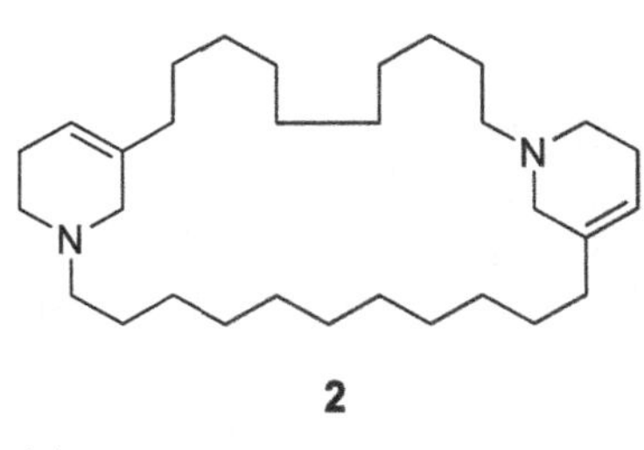

2

Fig. 1 Structures of halicyclamines C (1) and D (2)

2.2
Bisquinolizadine and Bis-1-oxaquinolizadine Macrocycles

The hexacyclic alkaloid upenamide (**3**) has been isolated from the Indonesian sponge *Echinochalina* sp. [13]. The sponge crude MeOH extract was subjected to a Kupchan liquid–liquid partition procedure, followed by C_{18} reversed-phase column chromatography of the least polar fraction. The alkaloid-containing fraction was further separated by gel permeation chromatography (LH-20), silica-gel column chromatography, and purified by C_{18} RP-HPLC. The structure of ‘upenamide (**3**) was established by a detailed analysis of spectroscopic data. The ^{13}C NMR spectrum displayed signals of eight *sp*^{2} methines accounting for four double bonds, which enabled the authors to start the construction of the structure **3** from the vinylic ^1H–^1H correlations observed in both ^1H–^1H COSY and HOHAHA spectra. Once ^1H spin systems were defined, long-range correlations observed in HMBC spectra enabled the authors to connect the structural fragments: (a) the triene chain to the cyclohexenol ring; (b) the cyclohexenol ring to the hexahydropyrido[2,1-*b*]-1,3-oxazin-4(6*H*)-one moiety; (c) construction of the *cis*-fused octahydro-2*H*-pyrano[2,3-*b*]pyridine bicyclic system; and (d) the final connection of structural fragments through the saturated alkyl chain. The relative stereochemistry of **3** was established by analysis of NOESY and ^1D gNOE NMR spectra. Additionally, ‘upenamide (**3**) did not present a Bohlmann absorption in the IR spectrum, therefore suggesting a *cis*-fused oxaquinolizidine system. Additional support for such a hypothesis was the unusually downfield signal observed for H-10 (δ 4.78) while C-10 is rather upfield (δ 88.7), in agreement with similar assignments of the structurally related alkaloid xestospongin C. The authors could establish the absolute configuration of the spiro-tricyclic moiety of **3** as 2*R*,9*S*,10*S*,11*R*,15*R* using the Mosher method. ‘Upenamide did not display cytotoxic activity against P388, A549, and HT29 cancer cell lines [13]. The authors proposed a biogenetic pathway for the formation of ‘upenamide (**3**) starting from two acrylaldehyde units, two molecules of ammonia, (2*E*,4*E*,6*E*,8*E*,10*Z*)-tetradeca-2,4,6,8,10-pentaenedial, and dodecanedial, through a highly oxidized intermediate (**4**).

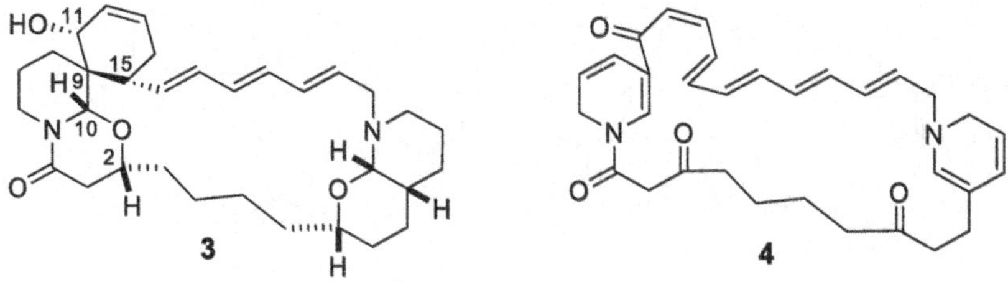

Fig. 2 Structures of ‘upenamide (3) and its putative biogenetic precursor (4)

The alkaloids (+)-xestospongin D (araguspongine A) (**5**) and 7-hydroxy-xestospongin A (**6**) have been isolated from a sponge of the genus *Xestospongia* collected in Western Australia [14], after extraction of the sponge with MeOH, partitioning of the crude extract with hexane, $CHCl_3$, and *n*-BuOH, further separation of the *n*-BuOH extract by silica-gel chromatography, and final purification by either normal-phase HPLC on silica gel or fractional crystallization. The structures of both alkaloids **5** and **6** were established by X-ray diffraction analysis and extensive analysis by HRCIMS and ^1H, ^{13}C, gCOSY, TOCSY, gHSQC, and gHMBC NMR spectra. The absolute configurations of (+)-xestospongin A (**5**) and 7-hydroxyxestospongin A (**6**) were assigned by anomalous X-ray scattering analysis and by the modified Mosher's ester method [14]. The results obtained enabled the authors to propose a series of stereochemical relationships, which confirmed the absolute stereochemistry of **5** and **6**, previously proposed after total synthesis [15], and questioned stereochemical assignments of related natural congeners based on insufficient experimental data [14]. Moreover, the occurrence of several stereoisomers in this series of alkaloids was proposed to be derived through an "incomplete stereochemical fidelity in the enantiospecific hydroxylation of the [putative] precursor (**7**) to the xestospongin/araguspongine family of alkaloids" [14]. This precursor may be converted with low stereoselectivity to either **8** or **9**, which are likely intermediates in this family of alkaloids, represented by either an enantiopure compound or by racemic mixtures co-occurring within a single sponge or in different sponge species (Scheme 1). While compound **5** displayed moderate antifungal activity against a fluconazole-resistant strain of *Candida albicans*, compound **6** was essentially inactive in this bioassay. The alkaloid 7-hydroxyxestospongin A (**6**) presented an in vitro blockage of Ca^{2+} release from IP3-dependent Ca^{2+} channels (ED_{50} 6.4 μM).

Further members of this class of alkaloids are the araguspongines K (**10**) and L (**11**), isolated from the marine sponge *Xestospongia exigua* collected at Bayadha, on the Saudi Arabian Red Sea coast [16]. After evaporation of the EtOH extract, it was partitioned between hexanes and MeCN. The polar fraction was subjected to a series of chromatographic separations by column chromatography on silica gel. The structures of both alkaloids **10** and

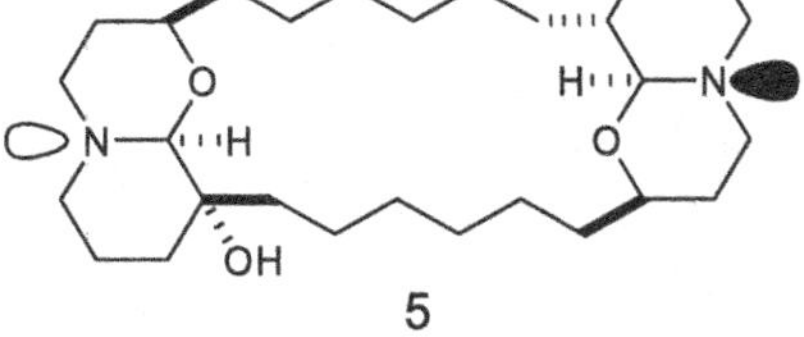

5

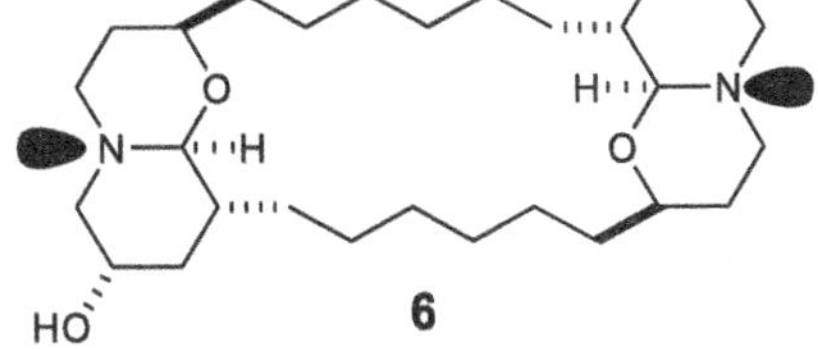

6

Fig. 3 Structures of (+)-xestospongin D (araguspongine A) (**5**) and 7-hydroxyxestospongin A (**6**).

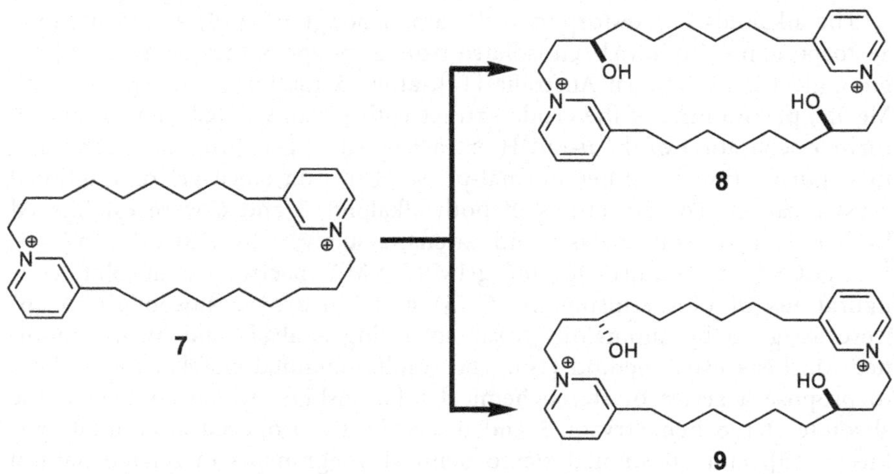

Scheme 1 Proposed biogenetic route for the formation of **8** and **9** as the putative precursors of xestospongin and araguspongine alkaloids

11 have been assigned by analysis of spectroscopic data and by comparison with NMR data reported previously for several related alkaloids. Since the IR spectrum of **10** showed a Bohlmann band at 2752 cm^{-1}, the compound presented a *trans*-bis-1-oxazoquinolizidine moiety. Furthermore, the presence of an *N*-oxide function was put in evidence by analysis of the ^1H–^{15}N HMBC spectrum, which showed a signal at δ 117.1 for an oxygenated nitrogen. The *N*-oxide bisquinolizidine moiety was also shown to have a *trans*-bicyclic system by analysis of the NMR data. Similar analysis enabled the authors to

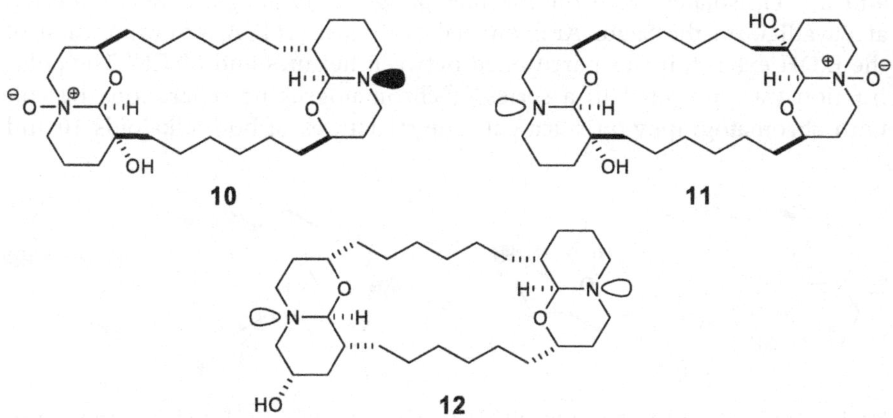

Fig. 4 Structures of araguspongines K (**10**), L (**11**), and M (**12**)

establish the structure of compound **11**. Both alkaloids were found to be inactive in antimalarial and antituberculosis bioassays [16].

An additional compound belonging to this structural class of alkaloids is araguspongine M (**12**), which has been isolated from the sponge *Neopetrosia exigua* [17]. The structure of **12** was established by analysis of spectroscopic data, symmetry properties, and also by comparison with data reported for araguspongines B and D, isolated from the same sponge [17]. Araguspongine M (**12**) displayed moderate cytotoxic activity against the human leukemia cell line HL-60.

2.3
Macrocycles with Conjoint Piperidine Rings

Haliclonacyclamine E (**13**) and arenosclerins A (**14**), B (**15**), and C (**16**) have been isolated from the marine sponge *Arenosclera brasiliensis*, endemic in Brazil. Crude extracts of this sponge displayed potent cytotoxic and antibiotic activities, and were subjected to fractionation by silica-gel flash chromatography, medium pressure chromatography on a SiOH cyanopropyl-bonded column, and reversed-phase C_{18} column chromatography to give compounds **13–16** [18]. The structure elucidation was based on spectroscopic analysis, including HRFABMS, ^1H–^1H COSY, HSQC, HSQC-TOCSY, and HMBC NMR experiments, along with the comparison of literature data previously reported for haliclonacyclamines A–D (**17–20**) [19, 20]. The relative stereochemistry of alkaloids **13–16** was established by analysis of NOESY and ROESY spectra,

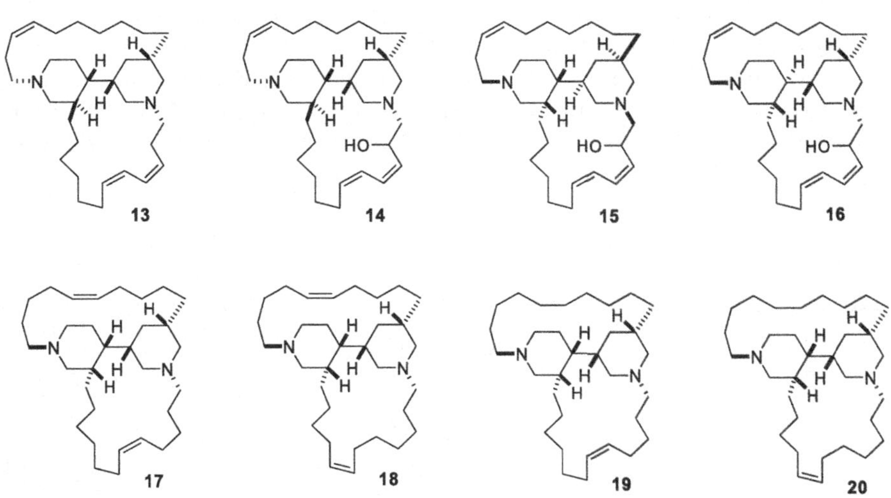

Fig. 5 Structures of haliclonacyclamine E (5), arenosclerins A–C (14–16), and haliclonacyclamines A–D (17–20)

and also by molecular modeling analysis. The alkaloids **13–16** displayed significant cytotoxic activity and antibiotic activity against several pathogenic bacterial strains, including antibiotic-resistant strains [21].

Additional alkaloids are halichondramine (**21**) isolated from the sponge *Halichondria* sp. collected at the Dahlak Archipelago in Red Sea, Eritrea [22], as well as tetrahydrohalicyclamine A (**22**) and 22-hydroxyhalicyclamine A (**23**) isolated from the sponge *Amphimedon* sp. collected in southern Japan [23]. Halichondramine (**21**) was isolated from the 5 : 5 : 1 EtOAc–MeOH–H_2O crude extract obtained from *Halichondria* sp., which was subjected to separations by gel permeation chromatography on Sephadex LH-20 and normal-phase chromatography on silica gel. The alkaloid **21** was identified by extensive analysis of spectroscopic data, including ^{1}H, ^{13}C, ^{1}H–^{1}H COSY, HMQC, HMBC, and HSQC-TOCSY NMR spectra. The relative stereochemistry of each of the piperidine rings in **21** was established by analysis of ^{1}H NMR and NOESY spectra. However, the authors have not obtained experimental data to relate the relative stereochemistry of each piperidine moiety to the other. They suggested the relative stereochemistry indicated in **21**, based on the Baldwin and Whitehead biogenetic proposal [24] (see below) and also by comparison with data for haliclonacyclamine E (**5**) and arenosclerin A (**14**).

The alkaloids **22** and **23** were obtained from the EtOH and 1 : 1 $CHCl_3$–EtOH extracts of the sponge *Amphimedon* sp. by a series of separations, including centrifugal partition chromatography and reversed-phase C_{18} column chromatography and HPLC [23]. The structure of tetrahydrohalicyclamine A (**22**) was established by analysis of HRFABMS, ^{1}H–^{1}H COSY, HSQC, HOHAHA, and HMBC spectra. The relative stereochemistry of **22** was proposed based on analysis of ^{1}H–^{1}H coupling constants between H-2 and H-3 ($J = 11.9$ Hz, *trans*-diaxial). The structure of the alkaloid **23** was proposed by comparison with NMR data previously reported for halicyclamine A (**24**), isolated from the sponge *Haliclona* sp. collected at Biak, Indonesia [25]. Tetrahydrohalicyclamine A (**22**) and 22-hydroxyhalicyclamine A (**23**) dis-

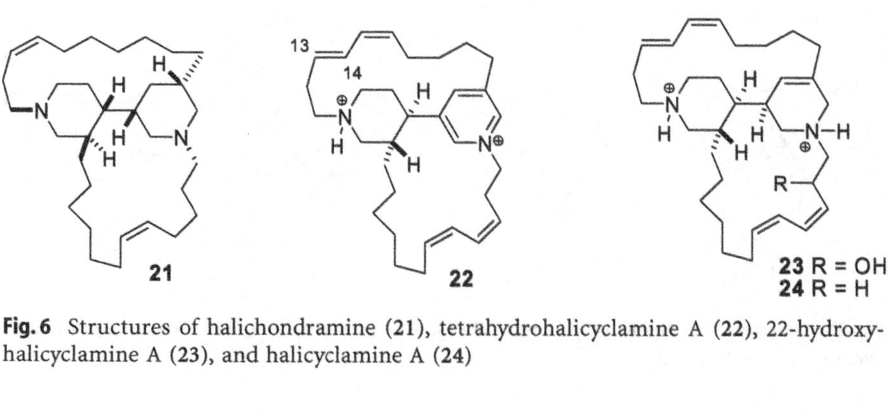

Fig. 6 Structures of halichondramine (**21**), tetrahydrohalicyclamine A (**22**), 22-hydroxyhalicyclamine A (**23**), and halicyclamine A (**24**)

played cytotoxic activity against P388 cells with IC_{50} at 2.2 and 0.45 μg/mL, respectively.

Haliclonacyclamine F (**25**), arenosclerin D (**26**), and arenosclerin E (**27**) have been recently isolated from the sponge *Pachychalina alcaloidifera* endemic in Brazil [26]. The alkaloids **25–27** were isolated from the cytotoxic, antibiotic, and antituberculosis MeOH crude extract of *P. alcaloidifera* by a series of separations on silica-gel and cyanopropyl-bonded silica-gel columns. The structures of compounds **25–27** were established by the same approach employed for the structural elucidation of haliclonacyclamine E (**13**) and arenosclerins A–C (**14–16**) [18], as well as by comparison with NMR data for this last series of alkaloids. The alkaloids **25–27** displayed moderate cytotoxic activity against SF295 (human CNS), MDA-MB435 (human breast), HCT8 (colon), and HL60 (leukemia) cancer cell lines.

The alkaloids **13–27** evidently share a common biogenetic origin, based on an extension of the Baldwin and Whitehead proposal [24]. Jaspars et al. [25] and Clark et al. [20] devised a biogenetic pathway for the formation of the 4-(piperidin-3-yl)piperidine moiety of these alkaloids, starting from the condensation of two acrylaldehyde units, two ammonia units, and two dialdehydes with a variable number of carbon atoms, probably formed from the condensation of acetate/malonate precursors (Scheme 2). The condensation of these units would give a "haliclamine-framework" intermediate (**28**), which may suffer an intramolecular Diels–Alder cyclization to yield an ingenamine/ingamine/xestocyclamine intermediate (**29**). After redox exchange with simultaneous skeleton rearrangement, a putative halicyclamine/haliclonacyclamine/arenosclerin precursor (**30**) is formed, presumably with a defined stereochemistry at C-3 and C-9. Selective reduction of the didehydro- and/or tetradehydropyridine moieties in **30** gives either the halicyclamine (**31**, 7,8-didehydro) or the haliclonacyclamine/arenosclerin (**32**) skeletons. Two structural variations are noteworthy in the alkaloids **13–26:** the relative stereochemistry at C-2, C-3, C-7, and C-9 as well as the positions and geometry of double bonds. The position and geometry of unsaturations can

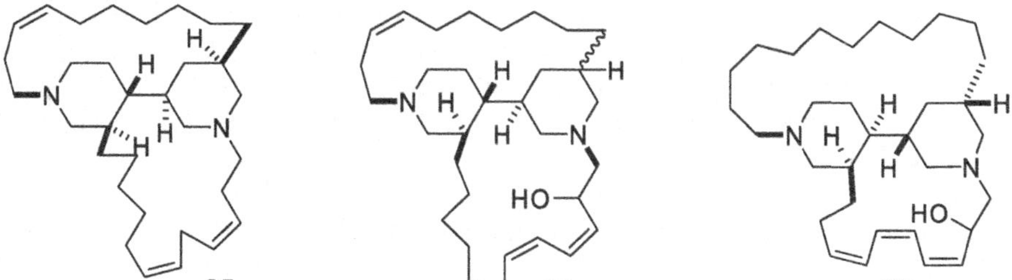

25 **26** **27**

Fig. 7 Structures of haliclonacyclamine F (**25**), arenosclerin D (**26**), and arenosclerin E (**27**)

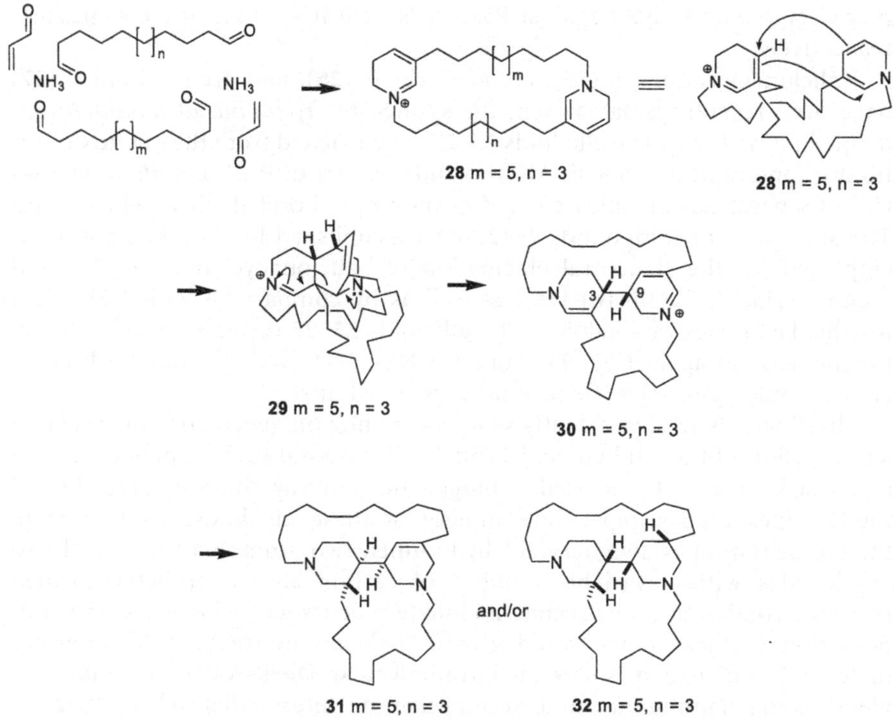

Scheme 2 Biogenetic pathway proposed by Crews et al. [25] and Garson et al. [19, 20] for the formation of halicyclamine and haliclonacyclamine/arenosclerin alkaloid skeletons

be easily explained in terms of polyketide biosynthesis. However, both alkaloids **22** and **23** have an $E - \Delta^{13,14}$ double bond, a unique structural feature in this class of secondary metabolites.

On the other hand, the relative stereochemistry within the bispiperidine system of the alkaloids is highly variable, and constitutes an interesting biogenetic problem. The relative stereochemistry of haliclonacyclamines A–D (**17–20**) was established by X-ray crystallographic analysis of haliclonacyclamine A (**17**), followed by hydrogenation of the alkaloids **17–19**, a procedure which gave an identical, totally saturated alkaloid with $[\alpha]_D + 23.9°$ (*c* 0.45, CH_2Cl_2) for tetrahydro-**17**, $[\alpha]_D + 24.9°$ (*c* 0.45, CH_2Cl_2) for tetrahydro-**18**, and $[\alpha]_D + 12.7°$ (*c* 0.28, CH_2Cl_2) for dihydro-**19**. Due to the insufficient amount of **20**, it was not subjected to the hydrogenation. Instead, its relative stereochemistry was compared with those of **17–19** by analysis of NOESY data [20]. Therefore, the relative stereochemistry of haliclonacyclamines A–D was defined as in **A**, in which both piperidine rings assume boat conformations contrary to each other. In the case of haliclonacyclamine E (**13**) and arenosclerins A–C (**14–16**), the relative stereochem-

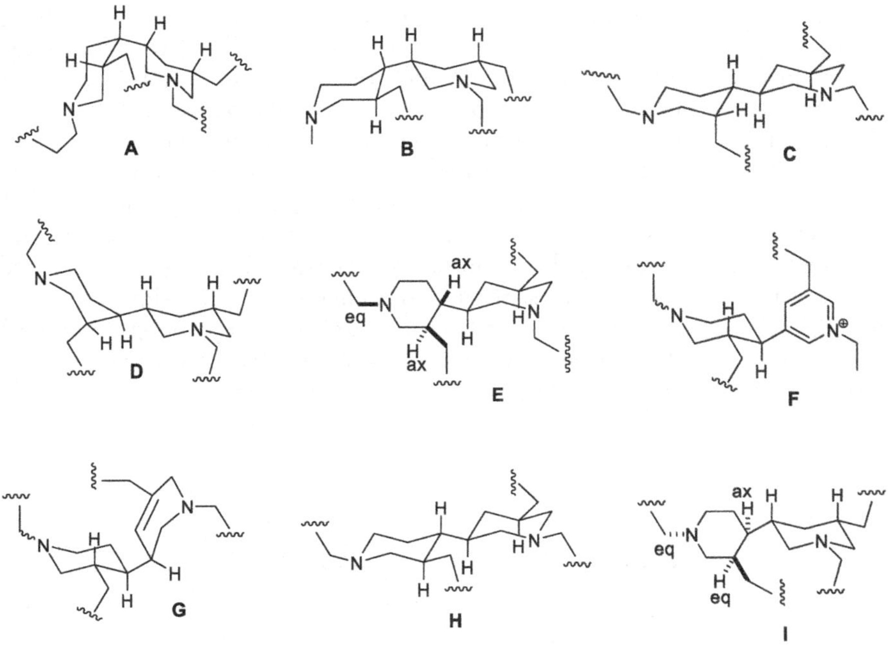

Fig. 8 Relative stereochemistry of haliclonacyclamine, arenosclerin, and halicyclamine alkaloids

istry of the bispiperidine moiety was established by analysis of ^{1}H, NOESY, and ROESY NMR spectra as **B** for compounds **14** and **15**, as **C** for compound **16**, and as **D** for compound **17** [18]. The relative stereochemistry of halichondramine (**13**) was tentatively suggested as **E** by analysis of NOE spectra and by comparison with NMR data of haliclonacyclamine E and arenosclerin A. According to the authors, the conformation **E** should have both H-3 and H-9 in a 90° dihedral angle, since no NOE coupling has been observed between these protons or between protons of the two piperidine moieties [22]. The relative stereochemistry of tetrahydrohalicyclamine A (**22**) was defined as **F** by analysis of its ^{1}H NMR spectrum. Therefore, the relative stereochemistry of 22-hydroxyhalicyclamine A (**23**) is likely to be **G** [23]. The relative stereochemistry of haliclonacyclamine F (**25**), arenosclerin D (**26**), and arenosclerin E (**27**) have been proposed by analysis of NOESY spectra to be as depicted in **H** for **25** and **26**, and as in **I** for the alkaloid **27** [26]. The structure of arenosclerin E (**27**) is unique, since it is the only alkaloid of this series in which the length of the alkyl bridges is reversed. Moreover, the ^{1}H NMR and NOE data for **27** suggest that the dihedral angle between H-3 and H-9 must be close to 90°, in order to support the strain promoted by the presence of the conjugated triene system in the C_{10} alkyl bridge [26].

Considering the number of halicyclamine/haliclonacyclamine/arenosclerin stereoisomers isolated so far, it is tempting to suggest that if the biosynthesis of these compounds proceeds through the intermediate **29**, formed by a highly stereospecific Diels–Alder reaction, there might be some sort of equilibration processes during the redox exchange and hydrogenation from **29** to **31**/**32**, leading to distinct stereoisomers. Although no biosynthetic studies have yet been performed on this class of alkaloids, Garson's group obtained indirect evidence that metabolites **17**–**20** are likely to be biosynthesized by the sponge *Haliclona* sp., since the compounds have been detected in the sponge cells spongocytes, choanocytes, and larger archaeocytes [27]. However, since no Diels–Alderase enzyme has yet been isolated from any sponge source of such metabolites, there is still no experimental support for the very elegant biogenetic pathway proposed by Baldwin and Whitehead.

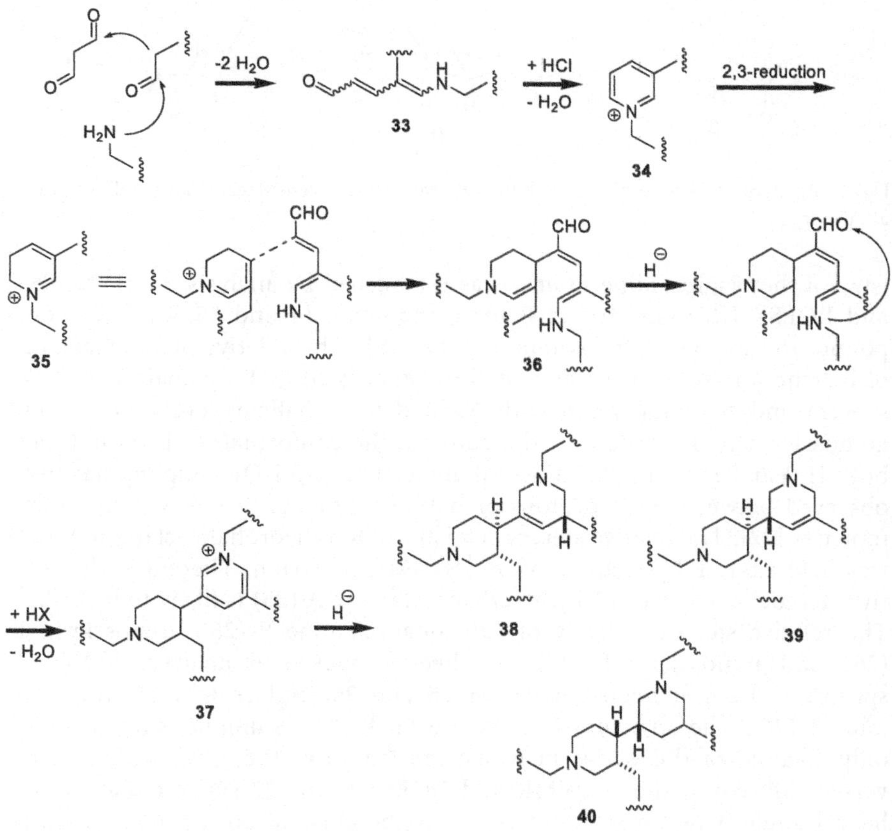

Scheme 3 Biogenetic pathway proposed by Marazano et al. [28, 29] for the formation of halicyclamine and haliclonacyclamine/arenosclerin alkaloid skeletons

An alternative biogenetic pathway proposed by Marazano's group may explain the formation of different stereoisomers of the halicyclamine/haliclonacyclamine/arenosclerin bispiperidine alkaloids (Scheme 3) [28]. Based on the biomimetic synthesis of model compounds, it seems likely that substituted 5-amino-2,4-pentadienal derivatives (**33**) may be biosynthesized by sponges (or associated microorganisms) as precursors of pyridine and pyridinium salts. The formation of such intermediates was hypothesized from the condensation of malodialdehyde with a long-chain aldehyde in the presence of a long-chain substituted primary amine. The 5-amino-2,4-pentadienal intermediates can easily be converted to 3-substituted pyridinium (**34**) or 2,3-dihydropyridinium salts (**35**) [28]. The latter would be prone to condense with a second 5-amino-2,4-pentadienal moiety, to give an intermediate (**36**) which would be subjected to an intramolecular cyclization, leading to the formation of a substituted 4-(piperidin-4-yl)pyridinium (**37**). In principle, such an intermediate may be reduced to a maximum of 32 different stereoisomers, if one considers the relative stereochemistry of the alkyl substituents at the nitrogen atoms. These include the stereoisomers of halicyclamines **38** and **39** as well as those of haliclonacyclamines A–D (**40**) [28]. According to Marazano's proposal [28], the intermediates such as **34**, **35**, and **37** are much less susceptible to oxido-reduction reactions than intermediates such as **28** (Scheme 2). Further experimental support for such a hypothesis was obtained, and showed the validity of Marazano's proposal for the biogenetic pathway of the halicyclamine, haliclonacyclamine, and arenosclerin alkaloids [29].

2.4
Condensed Bis-3-alkylpiperidines with Unrearranged Skeletons

Misenine (**41**), an alkaloid with a new carbon skeleton, was isolated from the marine sponge *Reniera* sp. [30]. The structure elucidation of compound **41** was complicated by the fact that the NMR spectra suffered from the influence of the concentration of the sample, the pH of the solvent in which the spectra were acquired, and traces of inorganic impurities, leading to irreproducible spectra. Spectra of good quality have been obtained in acid-free $CDCl_3$. The structure **41** was proposed by extensive analysis of $^1H-^1H$ COSY, TOCSY, HMQC, HMBC, and NOESY spectra. However, the spectroscopic data did not enable the authors to establish unambiguously the length of two alkyl bridges, as well as between **41** and two other possible structures, **42** and **43**. The structure **41** was proposed by considering a biogenetic proposal (Scheme 4), in which the first intermediate **44**, formed after condensation of two dialdehyde units, two ammonia molecules, and two acroleins, suffers intramolecular cyclization to give **45**. This second intermediate would be exposed to an oxidative retro-imine bond cleavage, leading to the aldehyde **46**. Nevertheless, the cleavage of the C-8/C-g bond is less understandable, but is postulated for the formation of the direct precursor (**47**) of misenine, after

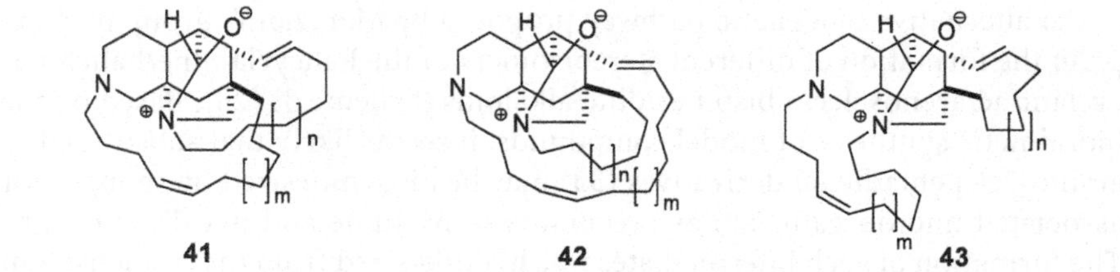

Fig. 9 Structure of misenine (41) and two alternative structure proposals 42 and 43 for misenine

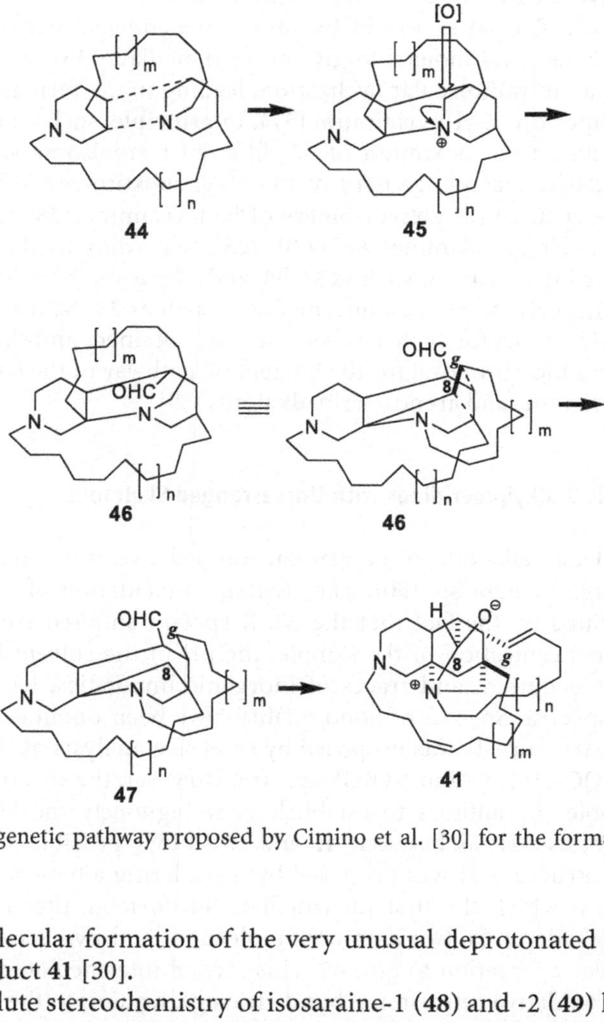

Scheme 4 Biogenetic pathway proposed by Cimino et al. [30] for the formation of misenine (41)

the intramolecular formation of the very unusual deprotonated hemiaminal natural product **41** [30].

The absolute stereochemistry of isosaraine-1 (**48**) and -2 (**49**) has been established using the modified Mosher's method on the reduced derivatives **50** and **51** obtained after reaction of the natural products with $NaBH_4$ [31].

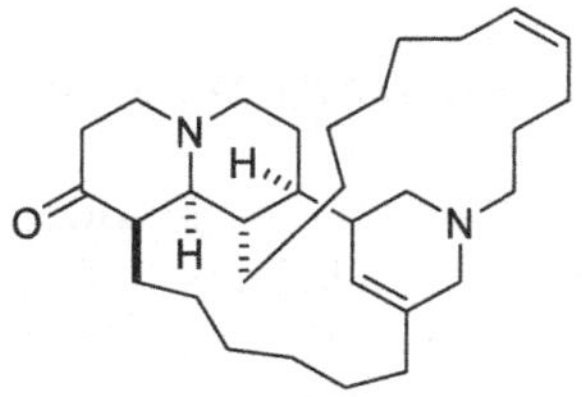

48

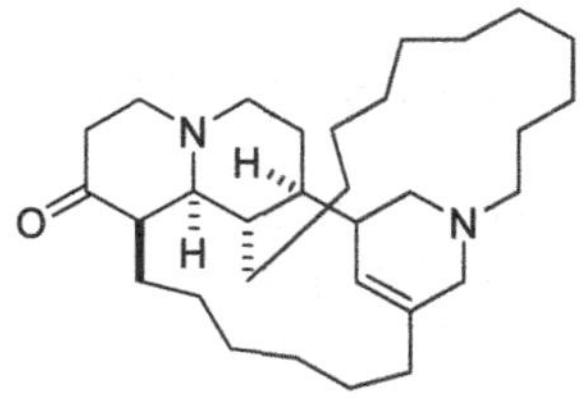

49

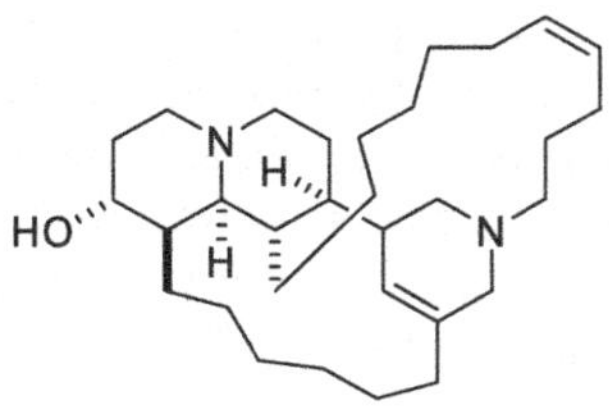

50

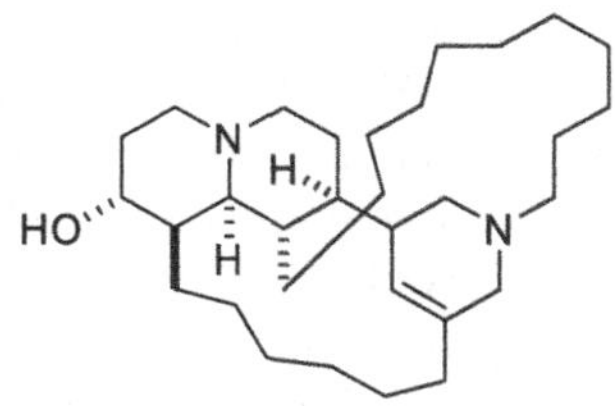

51

Fig. 10 Structures of isosaraine-1 (**48**) and -2 (**49**) and the respective products of reduction with $NaBH_4$, **50** and **51**

After the initial reports on the xestocyclamines, ingamines, and ingenamines [4], a new alkaloid belonging to this class, namely ingenamine G (**52**), was isolated from *Pachychalina* sp. (= *P. alcaloidifera*) [32]. After evaporation, the MeOH crude extract of the sponge was subjected to an acid–base partitioning procedure, which yielded both apolar and polar complex alkaloid fractions. These fractions were further subjected to a series of column chromatography separations, using silica-gel and cyanopropyl-bonded silica-gel columns. The separations yielded a mixture of bispyridine alkaloids (cyclostelletamines) as well as ingenamine G (**52**). The alkaloid **52** was identified by analysis of spectroscopic data (HRFABMS, COSY, HSQC, HMBC, HSQC-TOCSY, and NOESY) and also by comparison with data reported for previously isolated xestocyclamines, ingamines, and ingenamines [4]. Ingenamine G (**52**) was the first member of this series presenting a diene allylic alcohol moiety in the C_8 alkyl bridge. It also displayed cytotoxic, antibacterial, and antituberculosis activities [32].

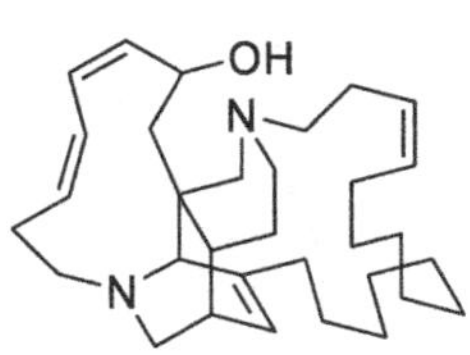

52

Fig. 11 Structure of ingenamine G (**52**)

2.5 Condensed Bis-3-alkylpiperidines with Seco or Rearranged Skeletons

The initial isolation of madangamine A (**53**) [33] was followed by the report of a series of five additional madangamine alkaloids B–E (**54–57**) from the same sponge species, *Xestospongia ingens* [34]. The MeOH crude extract was subjected to liquid–liquid partition between hexanes/MeOH/H_2O and EtOAc/H_2O. The hexanes fraction was separated by normal-phase chromatography on silica gel and normal-phase HPLC to give the alkaloids **54–57**, which were identified by extensive analysis of spectroscopic data, including COSY, LR-COSY, NOE, HMQC, and HMBC. A common structural feature of all madangamines isolated from *X. ingens* is a tricyclic alkaloid central core with a C_8 alkyl bridge connecting N-1 to C-3. The structures of the alkaloids **53–57** differ by the oxidation level of a C_{10}, C_{11}, or C_{12} alkyl chain connecting N-7 to C-9. Analysis of the ^1H signals of the N-7 to C-9 alkyl chain was particularly difficult due to signal broadness, which suggested a slow conformation exchange of the methylene groups within this alkyl bridge. Compounds **56** and **57** were isolated as an inseparable mixture. Madangamine D (**56**) has an odd number of carbon atoms in the N-7/C-9 alkyl chain, an unusual feature within this alkaloid class. As pointed out by the authors, the central tricyclic core of the alkaloids **53–57** has three six-membered rings in chair conformations, in which H-4ax and H-10ax are very close to each other, indicated by the strong NOE correlation observed in the NOESY spectra of these alkaloids. Moreover, the low polarity of these alkaloids (soluble in hexane and benzene) is probably due to the madangamine structural rigidity, in which N-7 cannot suffer amine inversion and, therefore, the N-7 lone pair is locked into the tricyclic central core and unavailable for protonation [34].

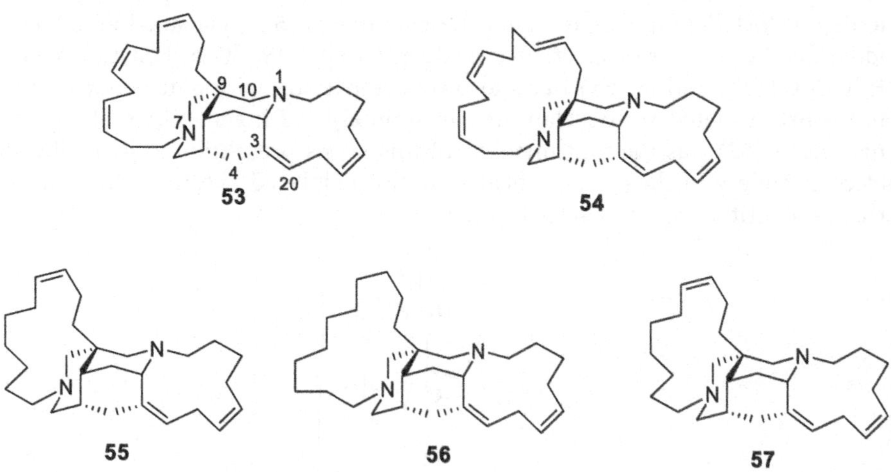

Fig. 12 Structures of madangamines A–E (53–57)

The authors also proposed a biogenetic route for the formation of the madangamine skeleton (Scheme 5) from a putative ingenamine precursor (**58**), which may suffer an enzyme-catalyzed skeleton rearrangement, leading to a spiro tetracyclic intermediate (**59**). The intermediate **59** may present a redox exchange to **60** which, after further oxidation, may give **61**, an immediate precursor of the madangamine tricyclic core **62** [33, 34].

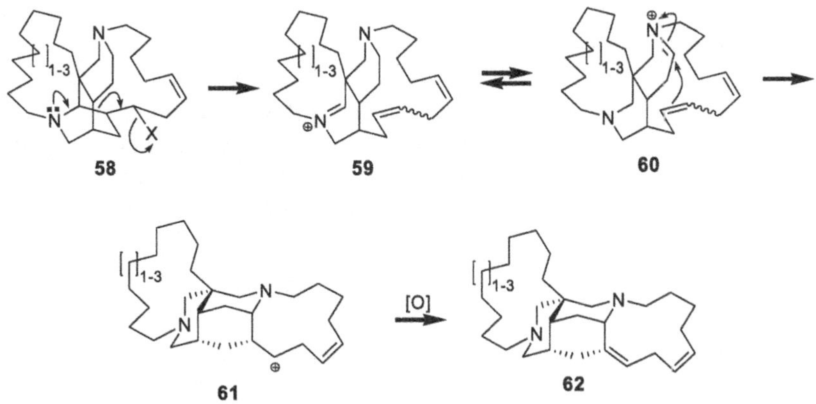

Scheme 5 Biogenetic pathway proposed by Andersen et al. [33, 34] for the formation of the madangamine skeleton

The biogenetic route proposed by Andersen et al. [4] for the biosynthesis of madangamines was supported by the recent isolation of madangamine F (**63**) from the marine sponge *Pachychalina alcaloidifera* [26]. Madangamine F was identified by extensive analysis of spectroscopic data, and also by comparison with data previously reported for the madangamines [33, 34]. Differently from madangamines A–E (**53–57**) isolated from *X. ingens*, madangamine F (**63**) has a conjugated polyunsaturated C_{10} alkyl chain connecting N-1 to C-3. Additionally, the geometry of the C-3/C-22 double bond is the inverse of the C-3/C-20 double bond in alkaloids **53–57**. However, the stereochemistry of this double bond is *Z* in all madangamine alkaloids, since the nomenclature priority rule changes in madangamine F due to the presence of the hydroxyl group at C-4.

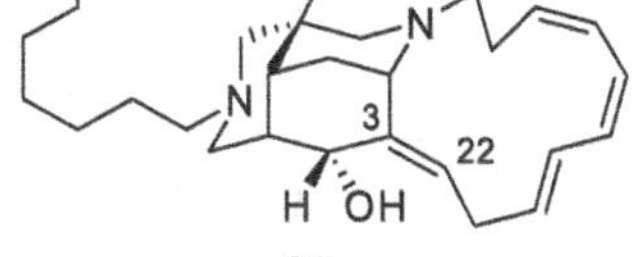

63

Fig. 13 Structure of madangamine F (**63**)

3
Total Synthesis

Although several synthetic approaches have been developed toward the total synthesis of polycyclic diamine sponge alkaloids, only a few members of this class have been synthesized.

3.1
Bis-3-alkylpiperidine Macrocycles

The synthesis of both haliclamines A (**64**) and B (**65**) has been achieved (Scheme 6). The first synthesis of both **64** and **65** [35, 36] was a convergent approach starting from the monoprotection and functional group interconversions (FGIs) of 1,4-butanediol (**66**) to the monoprotected iodide (**67**), which

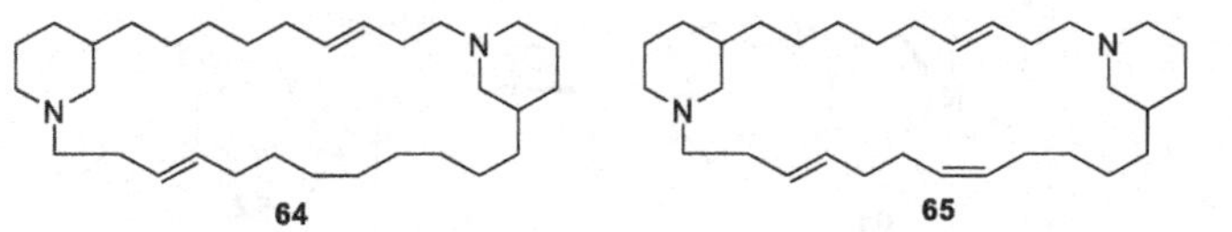

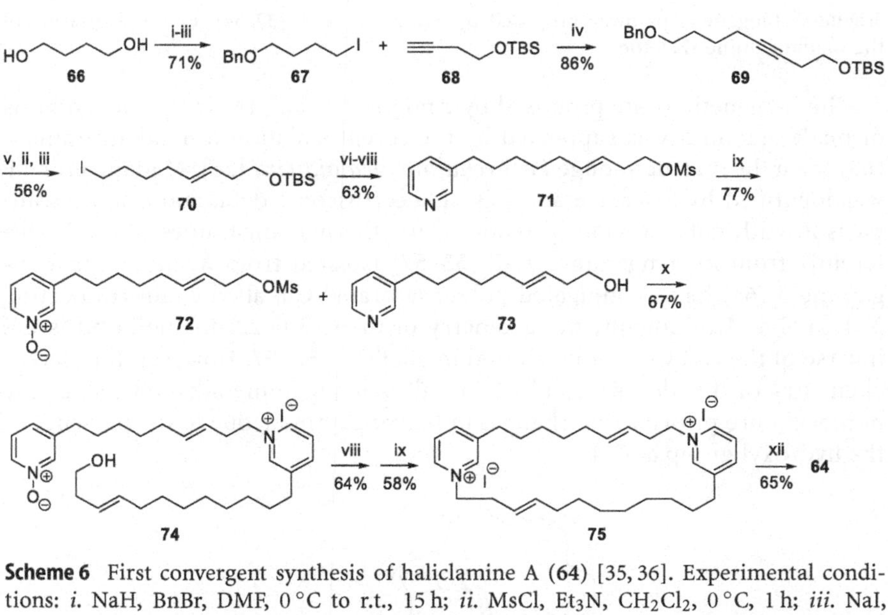

Scheme 6 First convergent synthesis of haliclamine A (**64**) [35, 36]. Experimental conditions: *i.* NaH, BnBr, DMF, 0 °C to r.t., 15 h; *ii.* MsCl, Et_3N, CH_2Cl_2, 0 °C, 1 h; *iii.* NaI, acetone, reflux, 3–4 h; *iv.* Li, 1 : 1 DMPU–THF, – 15 °C, 30 min then r.t. overnight; *v.* excess Na, *t*-BuOH, NH_3–Et_2O, – 40 °C, 3–4 days; *vi.* LDA, 3-picoline, THF, – 78 °C, 20 min then r.t., 5–6 h; *vii.* 3 : 2 AcOH–H_2O, r.t., 2–3 h; *viii.* MsCl, Et_3N, CH_2Cl_2, 0 °C, 1 h; *ix. m*-CPBA, CH_2Cl_2, 0 °C, 2 h then r.t., 10 h; *x.* KI, MeCN, reflux, 4 days; *xi.* KI, MeCN, reflux, 5 days; *xii.* $NaBH_4$, 3 : 2 MeOH–H_2O, 0 °C to r.t., 11 h

was condensed with the lithium acetylide of **68**. Further FGI to **70** provided the necessary protected acetylene to be condensed with 3-picoline. After additional FGIs, the *N*-oxide **72** was condensed with the 3-substituted pyridine **73**, in order to obtain the complete carbon framework of haliclamine A. Macrocyclization proceeded through the mesylate of **74** to give the bispyridinium diiodide **75**, which yielded natural **64** after reduction with $NaBH_4$. The synthesis of haliclamine B (**65**) was developed by essentially the same approach, using (3*E*,8*Z*)-12-(pyridin-3-yl)dodeca-3,8-dien-1-ol as the analogous pyridine of **73**, prepared by *Z*-selective Wittig olefination.

A second convergent synthesis of haliclamine A (**64**) was achieved in a stepwise sequence from cyclopropyl(thiophen-2-yl)methanone (**76**) (Scheme 7) [37]. The protected thiophene **77** was condensed with formylpiperidine to give **78**, suitable for a Wittig olefination with **79**. After desulfurization of the product **80**, the deprotected alcohol **82** was subjected to homoallylic rearrangement using Me_3SiBr in the presence of $ZnBr_2$. The re-

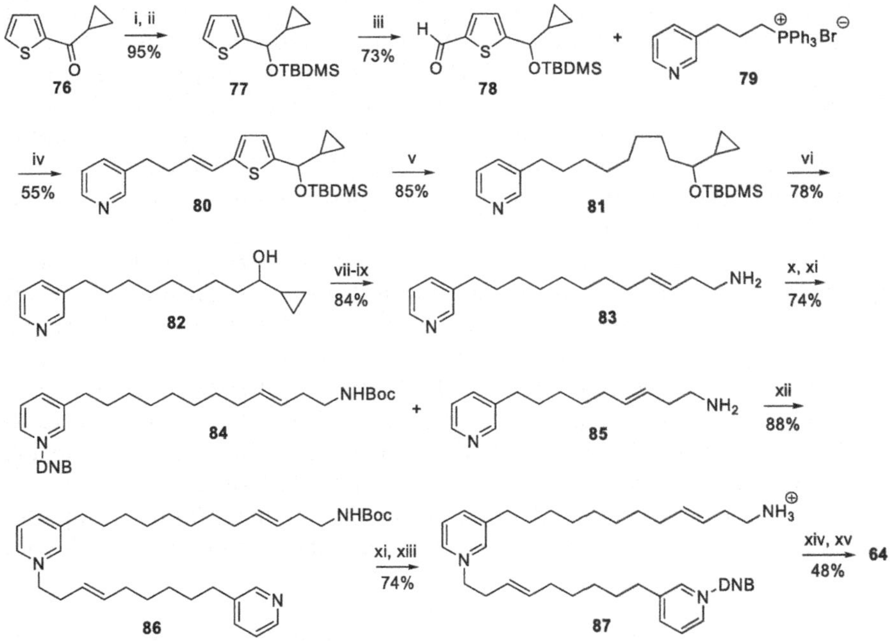

Scheme 7 Second convergent total synthesis of haliclamine A (**64**) [37]. Experimental conditions: *i.* $NaBH_4$, MeOH, r.t.; *ii.* TBDMSCl, imidazole, DMF, r.t.; *iii. n*-BuLi, THF, – 78 °C, then formylpiperidine (2.5 eq.); *iv.* **83**, NaH, THF, 0 °C, then **82**, –78 °C; *v.* Na_2CO_3, Ni (Raney), EtOH, Δ; *vi.* 6 eq. Bu_4NF, THF, r.t.; *vii.* 0.5 eq. $ZnBr_2$, 2.5 eq. TMSBr, CH_2Cl_2, – 20 °C; *viii.* 5 eq. NaN_3, DMF, 60 °C; *ix.* 1.7 eq. PPh_3, 0 °C, r.t., overnight, then NH_2OH, 0 °C; *x.* Boc_2O; *xi.* 3 eq. 2,4-DNBCl, MeOH, Δ, 48 h; *xii. n*-BuOH, Δ, 15 min.; *xiii.* 1 eq. TFA, CH_2Cl_2, 1 h; *xiv.* Et_3N, *n*-BuOH, Δ; *xv.* $NaBH_4$, MeOH–H_2O

spective bromide was substituted with sodium azide before reduction to the corresponding amine **83**. Condensation of the homologous aminopyridine **85** similarly obtained with the deprotected substrate **84** gave the dimeric pyridine **86**, which could be cyclized under alkaline (Et_3N) conditions, before usual reduction to yield the natural haliclamine A (**64**).

3.2 Bisquinolizadine and Bis-1-oxaquinolizadine Macrocycles

A linear synthesis of petrosin (**88**) was achieved [38, 39], starting from methyl 8-formyloctanoate (**89**) obtained from the ozonolysis of methyl oleate (Scheme 8). The substrate **89** was stepwise converted to the nitrile **90**, which was alkylated with the dianion of propionic acid. Half of the product was subjected to ester methylation and hydrogenation to the primary amine **92**. The amine **92** was coupled with the acid previously prepared to give the amide **93**, which was further stepwise converted to the protected amine **94**. Macrolactamization was performed through the pentafluorophenol ester **95**. For the single-step formation of the pentacyclic skeleton, the amides were reduced, the amines Boc-protected, the hydroxyl groups oxidized, the protecting groups removed, and the cyclization reaction was performed in refluxing dilute acid. Under such conditions, petrosin (**88**) was crystallized in 23% yield. Additional stereoisomers could be obtained by chromatographic separation of the mother liquor. Further equilibration of the stereoisomer mixture in butylamine improved the yield of petrosin up to 33%. The overall synthesis was achieved in 20 steps and 4.6% total yield.

Petrosins C (**98**) and D (**99**) have also been synthezised by Heathcock's group, but in a completely distinct approach (Schemes 9 and 10) [40]. Use of Makosza's "vicarious nucleophilic substitution" of 2-methyl-4-nitropyridine (**100**) with phenyl chloromethyl sulfone gave **101** which was converted to the methyl ether **102**. Alkylation of **102** with the TBDMS ether of (*Z*)-4-chlorobut-2-en-1-ol gave the corresponding sulfone **103**, which was desulfonylated, lithiated, and treated with the corresponding primary chloride of **104** to give the bispyridine **105**. Thiolation of **105** required special care and was performed by initial lithiation of the methylpyridine group followed by transmetalation to the corresponding magnesium derivative before

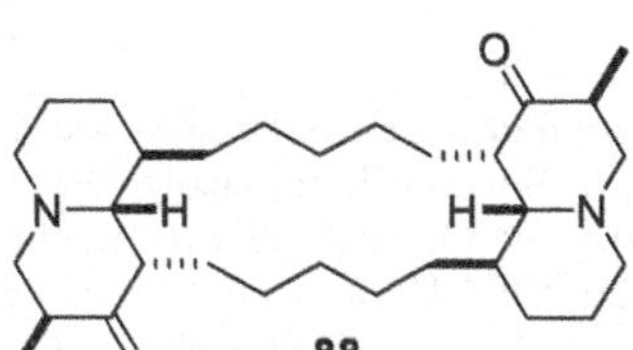

88

Fig. 14 Structure of petrosin (**88**)

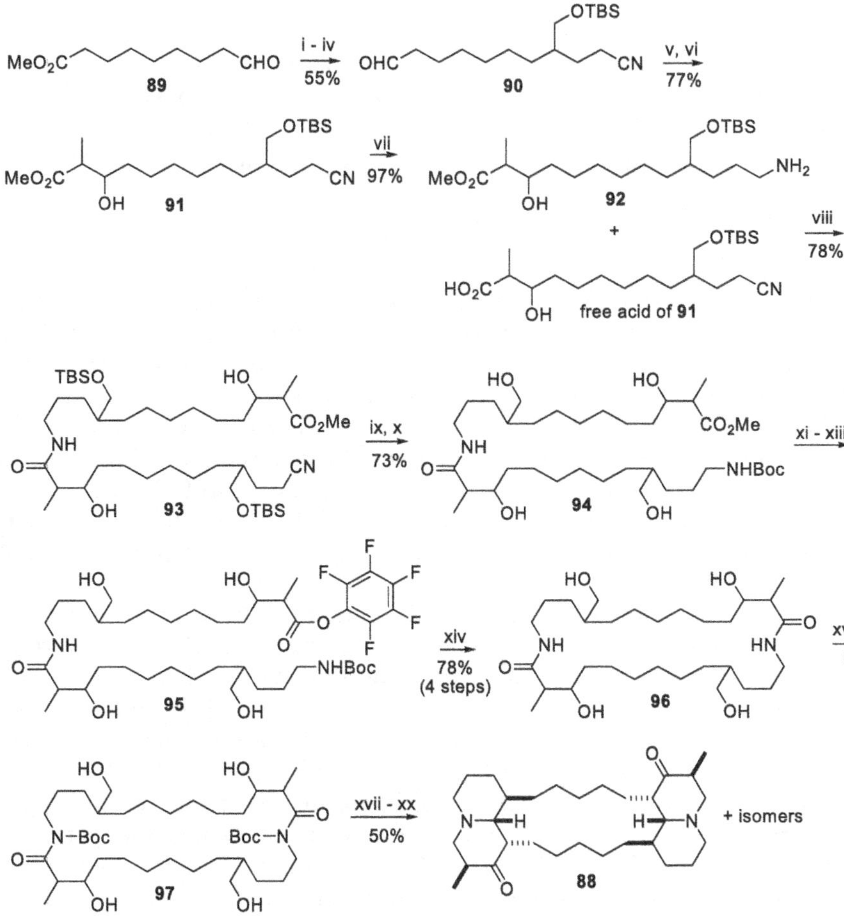

Scheme 8 Total synthesis of petrosin (**88**) [38, 39]. Experimental conditions: *i.* (a) pyrrolidine, K_2CO_3, (b) $H_2C=CHCN$, MeCN, (c) H_2O; *ii.* $NaBH_4$, MeOH; *iii.* TBSCl, imidazole, DMF; *iv.* DIBAL-H, CH_2Cl_2, – 95 °C; *v.* LDA, propionic acid (dianion), THF; *vi.* CH_2N_2, Et_2O; *vii.* H_2, PtO_2, EtOAc, HOAc; *viii.* DCC, HOBT, THF; *ix.* H_2, PtO_2, HCl, EtOH; *x.* Boc_2O, dioxane, H_2O; *xi.* NaOH, MeOH, THF; *xii.* DCC, C_6F_5OH, THF; *xiii.* 6N HCl, dioxane; *xiv.* high dilution into dioxane/pyridine, 90 °C; *xv.* $LiAlH_4$, THF; *xvi.* Boc_2O, dioxane, H_2O; *xvii.* Dess–Martin periodinane, CH_2Cl_2; *xviii.* 1 M HCl, EtOH, H_2O; *xix.* 0.2 M HOAc, EtOH; *xx.* (a) *n*-butylamine, 3-Å molecular sieves, (b) propylammonium acetate, $Cl(CH_2)_2Cl$, (c) H_2O

addition of diphenyl sulfide. The product **106** was oxidized to the corresponding sulfone before deprotection to the alcohol **107**. Conversion of **107** to **109** was achieved via the unstable primary chloride **108**. Further transformations included reduction of the two double bonds, the difficult alkylation of the sulfone, which required 2,6-lutidine as a cosolvent, reductive desul-

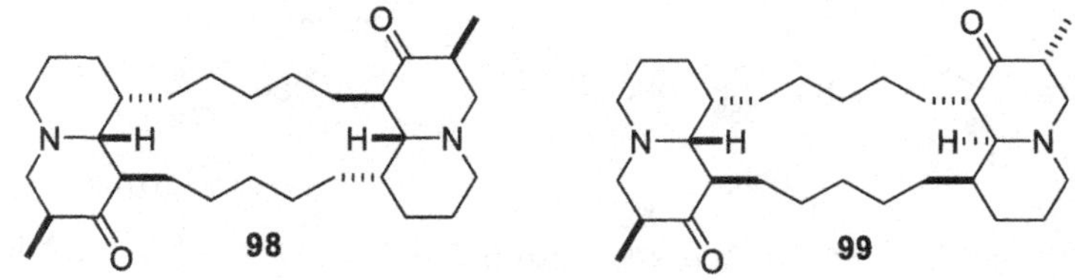

Fig. 15 Structures of petrosins C (**98**) and D (**99**)

fonylation, and allylation of the macrocyclic bispyridine to give a mixture of **111** (14%) and **112** (42%), along with undesired products of N alkylation, 2-pyridine bromination, and 2-pyridine alkylation. Further transformations employed identical conditions for the transformation of both stereoisomers **111** and **112** into **98** and **99**, respectively. These included hydroboration of the allylic double bonds, cyclization through the corresponding dimesylates followed by the reduction of the pyridine rings, and hydrolysis of the enol ether groups. Finally, methylation of the ketone **115** and equilibration under alkaline conditions afforded **98** from **111** and **99** from **112** (not shown in Scheme 10).

The stereoselective synthesis of xestospongin A (**116**), also known as (+)-araguspongine D, has been achieved (Scheme 11) [41], starting from **117** through a series of FGIs to prepare the thiophene **119**. At this point, the thiophene group was lithiated and treated with DMF in order to obtain the corresponding thiophenecarbaldehyde, which was converted to **120** after reaction with lithioacetonitrile and acetylation. The resulting ester was kinetically "resolved" with Amano P-30 lipase derived from *Pseudomonas fluorescens*. The "enantiopure" (at C-3) alcohol **121** obtained was partially converted to the aldehyde **122** and the amino alcohol **123**. Both thiophenes **122** and **123** were condensed under equilibrating conditions in CH_2Cl_2 and 5% aqueous NaOH to give 42% of the *trans* isomer (**124**) and 29% of the corresponding *cis* isomer (not shown). The latter could be further equilibrated to **124** in Et_3N–$CDCl_3$ at 80 °C over 3 h. After reduction of the nitrile and conversion of the dimethylketal to the corresponding aldehyde **125**, the second cyclization was performed in very diluted conditions (0.02 M of **125** in DMSO was diluted to 0.02 mM in 1 : 1 CH_2Cl_2–H_2O) at pH higher than 12. Final desulfurization of **126** gave xestospongin A (**116**). The results obtained enabled the authors to expand the synthetic approach on a less rigid substrate (**127**), which gave both xestospongins A (**116**) and C (**128**) in a neutral buffered (AcOH/NaOAc, 0.05 M, pH 7.0) MeOH solution, in 2.1–2.5 : 1 ratio [42].

A biomimetic synthesis of (–)-xestospongin A (**116**), (+)-araguspongine B (**129**), and (+)-xestospongin C (**130**) was developed (Scheme 12) [43], starting from a Weiler's alkylation of ethyl acetoacetate (**131**) with 1-bromo-4-chlorobutane (**132**) to give the ethyl ester of 8-chloro-3-oxooctanoate

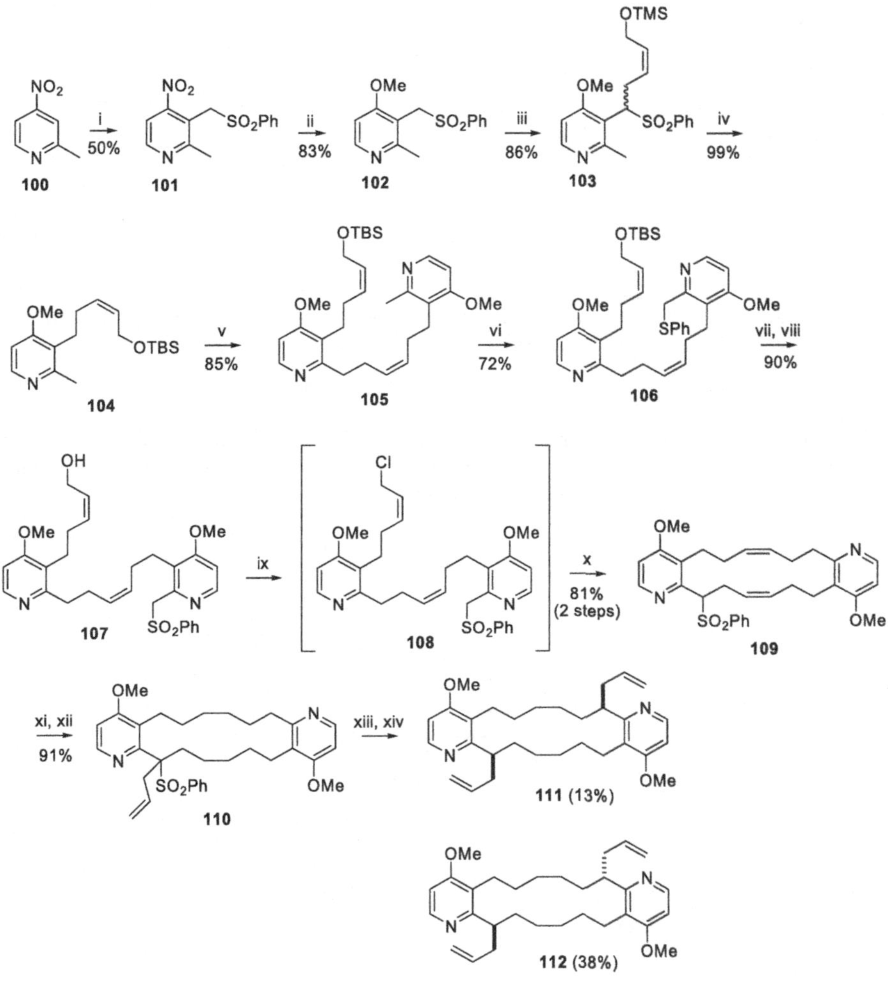

Scheme 9 Total synthesis of petrosins C (**98**) and D (**99**) [40]. Experimental conditions: *i.* $PhSO_2CH_2Cl$, NaOH, DMSO; *ii.* NaOMe, MeOH, rt; *iii.* (a) *n*-BuLi, THF; (b) (*Z*)-$ClCH_2CH = CHCH_2OTBS$; *iv.* 5% Na/Hg, MeOH; *v.* (a) *n*-BuLi, THF; (b) TBOS replaced by chloride in **104**; *vi.* (a) *n*-BuLi, THF; (b) $MgBr_2 \cdot Et_2O$; (c) PhSSPh; *vii.* $(NH_4)_6Mo_7O_{24}$, H_2O_2; *viii.* 1N HCl, THF; *ix.* MsCl, LiCl, 2,6-lutidine, DMF; *x.* NaHMDS, THF; *xi.* *p*-$TsNHNH_2$, NaOAc, THF, H_2O; *xii.* (a) 3 eq. *n*-BuLi, 2,6-lutidine, THF; (b) $CH_2 = CHCH_2I$; *xiii.* Na/Hg, THF, MeOH; *xiv.* (a) *s*-BuLi, TMEDA, ether, (b) $CH_2 = CHCH_2Br$

(**133**), which was subsequently transformed into the dimethylketal iodide **135**. Alkylation of 3-picoline with **135**, followed by FGI of the diol moiety, gave the pyridine **136** which was dimerized in a refluxing solution of NaI in butanone. Partial reduction of the dimerized pyridine gave the bistetrahydropyridine **137**, which was subjected to intramolecular cycliza-

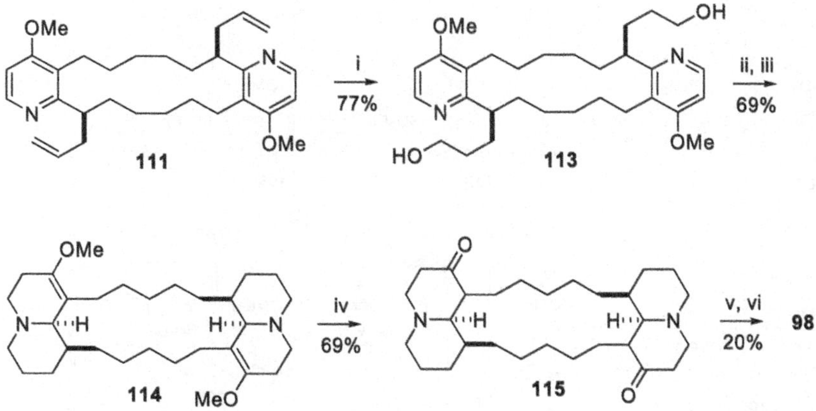

Scheme 10 Total synthesis of petrosins C (**98**) and D (**99**) [41] (contd.). Experimental conditions: *i.* (a) 9-BBN, THF, (b) H_2O_2; *ii.* MsCl, Et_3N, CH_2Cl_2; *iii.* $NaBH_4$, EtOH; *iv.* aqueous HCl; *v.* LDA, THF, MeI; *vi.* K_2CO_3, MeOH, H_2O

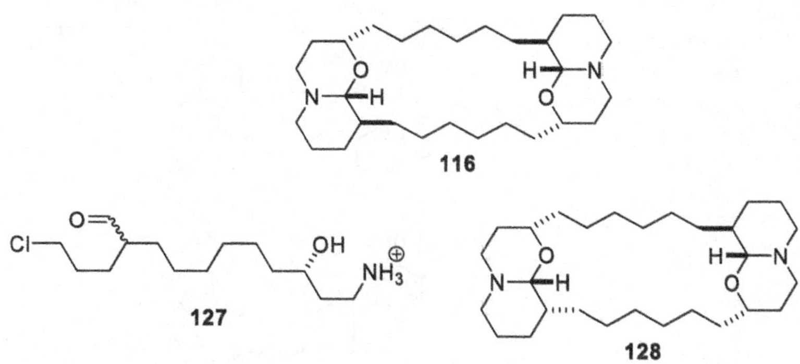

Fig. 16 Structures of xestospongin A (**116**) and of a less rigid substrate employed for an alternative synthesis of both **116** and xestospongin C (**128**)

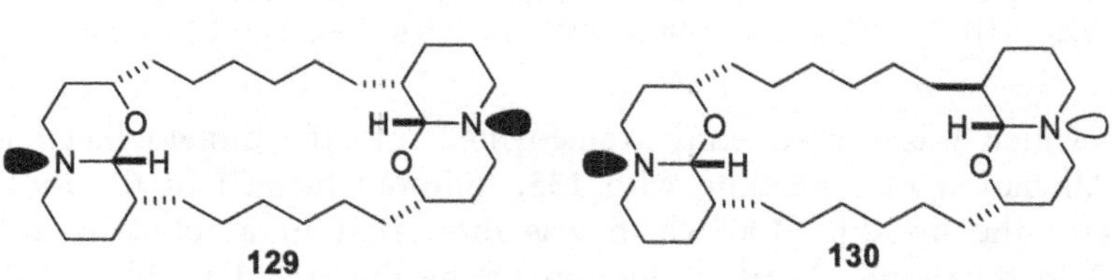

Fig. 17 Structures of (+)-araguspongine B (**129**) and (+)-xestospongin C (**130**)

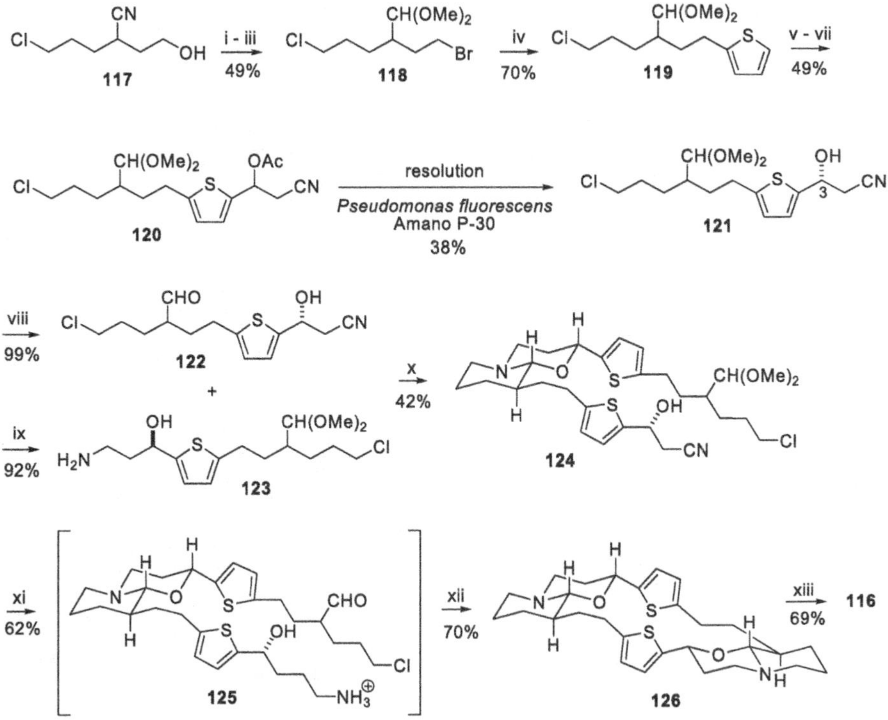

Scheme 11 Total synthesis of xestospongin A (**116**) [41]. Experimental conditions: *i.* NBS, PPh_3; *ii.* DIBAL-H, CH_2Cl_2, – 78 °C; *iii.* $HC(OMe)_3$, CSA, MeOH; *iv.* (a) thiophene, Li, THF then **118**, 0 °C to r.t.; *v.* (a) *n*-BuLi, – 78 °C, (b) DMF; *vi.* $LiCH_2CN$, THF, – 78 °C; *vii.* Ac_2O, pyridine; *viii.* TFA, H_2O, DMSO, 65 °C; *ix.* $LiAlH_4$, Et_2O, 0 °C to r.t.; *x.* (a) CH_2Cl_2, (b) 5% aq. NaOH; *xi.* (a) $LiAlH_4$, Et_2O, 0 °C to r.t., (b) TFA, H_2O, DMSO, 80 °C; *xii.* CH_2Cl_2, H_2O, 5% NaOH, pH > 12; *xiii.* H_2, Ni (Raney), EtOH

tions. The stereochemistry of the reaction product **138** was defined by X-ray diffraction analysis. Hydrogenation of **138** with Raney nickel in MeOH afforded (+)-araguspongine B (**129**) as a major stereoisomer, along with (+)-xestospongin C (**130**). On the other hand, hydrogenation of **138** with rhodium on alumina gave (–)-xestospongin A (**116**), (+)-xestospongin C (**130**), and (+)-araguspongine B (**129**). By comparison of the specific rotation values of the synthetic compounds with values of the corresponding natural products, the absolute stereochemistry of the natural compounds could be revised as shown.

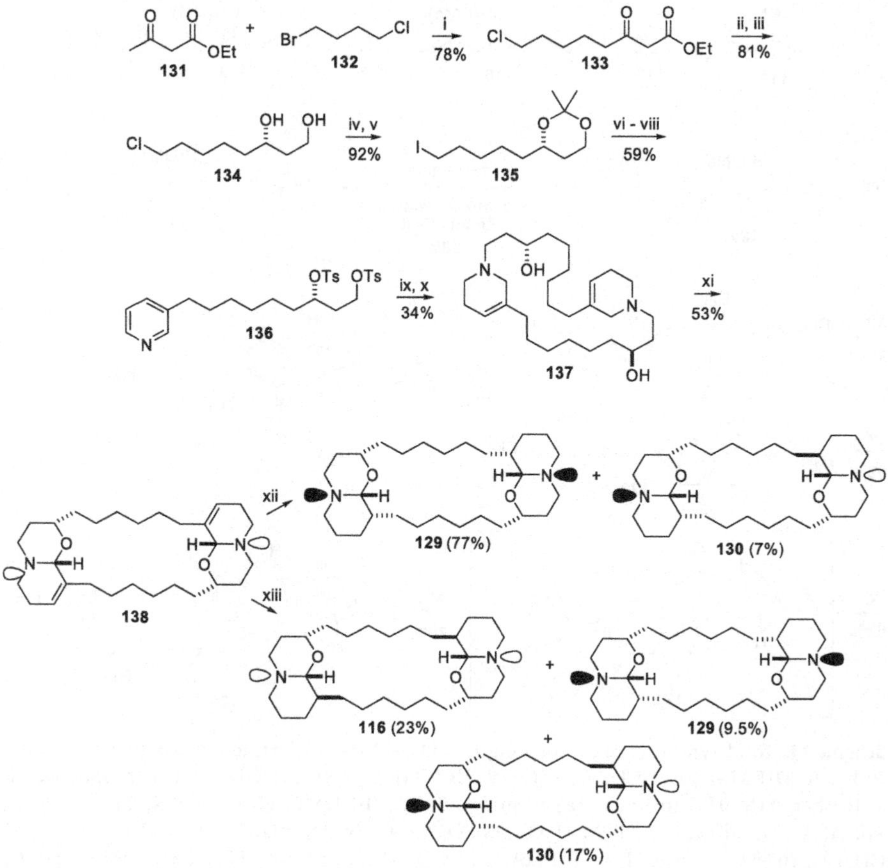

Scheme 12 Total synthesis of (–)-xestospongin A (**116**), (+)-araguspongine B (**129**), and (+)-xestospongin C (**130**) [41]. Experimental conditions: *i.* (a) NaH, THF, (b) *n*-BuLi, (c) **132**; *ii.* Ru(II)-*S*-BINAP, H_2, EtOH; *iii.* $LiBH_4$, Et_2O; *iv.* PPTS, 2,2-dimethoxypropane, acetone; *v.* NaI, acetone, reflux; *vi.* 3-picoline, LDA, THF; *vii.* HCl(aq.), EtOH; *viii.* TsCl, Et_3N, CH_2Cl_2; *ix.* NaI, butanone, reflux; *x.* $LiBH_4$, MeOH, *i*-PrOH; *xi.* DEAD, CH_2Cl_2; *xii.* H_2, Ni (Raney), MeOH; *xiii.* Rh on alumina, MeOH, H_2, then add alumina, reflux

References

1. Blunt JH, Copp BR, Hu W-P, Munro MHG, Northcote PT, Prinsep MR (2007) Nat Prod Rep 24:31
2. Faulkner DJ (2002) Nat Prod Rep 19:1
3. Mayer AMS, Rodríguez AD, Berlinck RGS, Hamann MT (2007) Comp Biochem Physiol 145(C):553
4. Andersen RJ, van Soest RWM, Kong F (1996) 3-Alkylpiperidine alkaloids isolated from marine sponges in the order Haplosclerida. In: Alkaloids: chemical and biological perspectives. Pergamon, New York, p 301

5. Almeida AMP, Berlinck RGS, Hajdu E (1997) Quim Nova 20:170
6. Specic K (2000) J Toxicol Tox Rev 19:139
7. Rodríguez J (2000) Stud Nat Prod Chem 24E:573
8. Bienz S, Bisegger P, Guggisberg R, Hesse M (2005) Nat Prod Rep 22:647
9. Berlinck RGS, Kossuga MH (2005) Nat Prod Rep 22:516
10. Magnier E, Langlois E (1998) Tetrahedron 54:6201
11. Hu JF, Hamann MT, Hill R, Kelly M (2003) Alkaloids Chem Biol 60:207
12. Volk CA, Lippert H, Lichte E, Köck M (2004) Eur J Org Chem 3154
13. Jiménez JI, Goetz G, Mau CMS, Yoshida WY, Scheuer PJ, Williamson RT, Kelly M (2000) J Org Chem 65:8465
14. Moon SS, MacMillan JB, Olmstead MM, Ta TA, Pessah IN, Molinski TF (2002) J Nat Prod 65:249
15. Baldwin JE, Melman A, Lee V, Firkin CR, Whitehead RC (1998) J Am Chem Soc 120:8559
16. Orabi KY, El Sayed KA, Hamann MT, Dunbar DC, Al-Said MS, Higa T, Kelly M (2002) J Nat Prod 65:1782
17. Liu H, Mishima Y, Fujiwara T, Nagai H, Kitazawa A, Mine Y, Kobayashi H, Yao X, Yamada J, Oda T, Namikoshi M (2004) Mar Drugs 2:154
18. Torres YR, Berlinck RGS, Magalhães A, Schefer AB, Ferreira AG, Hajdu E, Muricy G (2000) J Nat Prod 63:1098
19. Charan RD, Garson MJ, Brereton IM, Willis AC, Hooper JNA (1996) Tetrahedron 52:9111
20. Clark RJ, Field KL, Charan RD, Garson MJ, Brereton IM, Willis AC (1998) Tetrahedron 54:8811
21. Torres YR, Berlinck RGS, Nascimento GGF, Fortier SC, Pessoa C, Moraes MO (2002) Toxicon 40:885
22. Chill L, Yosief T, Kashman Y (2002) J Nat Prod 65:1738
23. Matsunaga S, Miyata Y, van Soest RWM, Fusetani N (2004) J Nat Prod 67:1758
24. Baldwin JE, Whitehead RC (1992) Tetrahedron Lett 33:2059
25. Jaspars M, Pasupathy V, Crews P (1994) J Org Chem 59:3253
26. Oliveira JHHL, Nascimento AM, Kossuga MH, Cavalcanti BC, Pessoa CO, Moraes MO, Macedo ML, Ferreira AG, Hajdu E, Pinheiro UP, Berlinck RGS (2007) J Nat Prod 70:538
27. Garson MJ, Flowers AE, Webb RI, Charan RD, McCaffrey EJ (1998) Cell Tissue Res 293:365
28. Kaiser A, Billot X, Gateau-Olesker A, Marazano C, Das BC (1998) J Am Chem Soc 120:8026
29. Jakubowicz K, Abdeljelil KB, Herdemann M, Martin M-T, Gateau-Olesker A, Mourabit AA, Marazano C, Das BC (1999) J Org Chem 64:7381
30. Gui Y, Trivellone E, Scognamiglio G, Cimino G (1998) Tetrahedron 54:541
31. Gui Y, Trivellone E, Scognamiglio G, Cimino G (1998) Tetrahedron Lett 39:463
32. De Oliveira JHHL, Grube A, Köck M, Berlinck RGS, Macedo ML, Ferreira AG, Hajdu A (2004) J Nat Prod 67:1685
33. Kong F, Andersen RJ, Allen TM (1994) J Am Chem Soc 116:6007
34. Kong F, Graziani EI, Andersen RJ (1998) J Nat Prod 61:267
35. Morimoto Y, Yokoe C (1997) Tetrahedron Lett 38:8981
36. Morimoto Y, Yokoe C, Kurihara H, Kinoshita T (1998) Tetrahedron 54:12197
37. Michelliza S, Mourabit AA, Gateau-Olesker A, Marazano C (2002) J Org Chem 67:6474
38. Scott RW, Epperson JR, Heathcock CH (1994) J Am Chem Soc 116:8853
39. Scott RW, Epperson JR, Heathcock CH (1998) J Org Chem 63:5001

40. Heathcock CH, Brown RCD, Norman TC (1998) J Org Chem 63:5013
41. Hoye TR, North JT, Yao LJ (1994) J Am Chem Soc 116:2617
42. Hoye TR, Ye Z, Yao LJ, North JT (1996) J Am Chem Soc 118:12074
43. Baldwin JE, Melman A, Lee V, Firkin CR, Whitehead RC (1998) J Am Chem Soc 120:8559

Top Heterocycl Chem (2007) 10: 239–263
DOI 10.1007/7081_2007_065

Published online: 4 July 2007

Catechins and Proanthocyanidins: Naturally Occurring *O*-Heterocycles with Antimicrobial Activity

Pietro Buzzini[1] (✉) · B. Turchetti[1] · F. Ieri[2] · M. Goretti[1] · E. Branda[1] · N. Mulinacci[2] · A. Romani[2]

[1]Dipartimento di Biologia Vegetale e Biotecnologie Agroambientali, Section of Applied Microbiology, University of Perugia, Borgo XX Giugno 74, 06121 Perugia, Italy
pbuzzini@unipg.it

[2]Dipartimento di Scienze Farmaceutiche, Polo Scientifico, University of Firenze, Via U. Schiff 6, 50019 Sesto Fiorentino (FI), Italy

Abstract Pharmacological interest in drug discovery has increased in recent years. Accordingly, an increasing part of the scientific body has thus far studied several thousands of compounds of natural and synthetic origin to ascertain their biological activity. The occurrence in plant tissues of secondary metabolites characterized by a *O*-heterocyclic structure and exhibiting antimicrobial properties is a well-known phenomenon. Among them, catechins and proanthocyanidins are two classes of compounds that exhibit antiviral or antimicrobial properties towards prokaryotic and eukaryotic microorganisms. Yet, despite the profusion of studies published so far, the real potentialities and limitations presented by the use of the above-mentioned class of molecules as antiviral or antimicrobial agents have not been critically evaluated. The present chapter represents an overview of the recent literature regarding the antiviral and antimicrobial properties exhibited by this class of compounds. Their mode of action as well as their synergy with currently used antibiotic molecules are also reviewed. In addition, their potentialities and the causes that have hampered their pharmaceutical exploitation so far are discussed.

Keywords Antibacterial activity · Antimycotic activity · Antiprotozoal activity · Antiviral activity · Cathechins · Proanthocyanidins

Abbreviations

EC	Epicatechin
ECG	Epicatechin-3-*O*-gallate
EGC	Epigallocatechin
EGCG	Epigallocatechin-3-*O*-gallate
GC	Gallocatechin
GCG	Gallocatechin-3-*O*-gallate
SAR	Structure-activity relationships

1 Introduction

The increasing interest of a growing part of the scientific body (academic or industrial) in discovering novel antimicrobial compounds is essentially justified by the expectation that a few of them could play a role in supporting or even in substituting some commercial molecules currently used in clinical practice. It has been estimated that although a number of novel antimicrobial molecules (either of biological or synthetic origin) are currently launched each year, their downturn is becoming very rapid [1, 2]. Among the causes, the increasing development of resistant microbial genotypes is undoubtedly one of the most relevant. This phenomenon, first described in bacteria, is becoming a looming threat also for eukaryotic microorganisms. Although many factors may determine such changes in microbial genomes, several examples of scientific evidence has demonstrated that an incorrect use of currently available antimicrobials in the clinical setting may greatly increase the risk of developing resistant genotypes [3–6].

Taking into account these considerations, the enormous scientific (and commercial) interest in discovering and developing novel classes of molecules exhibiting more or less pronounced antimicrobial properties has oriented the work of a growing part of the scientific community towards large-scale screening programs [2, 7].

The occurrence in some plants of secondary metabolites characterized by an *O*-heterocyclic structure and exhibiting antimicrobial properties is a well-known phenomenon [2, 8–10]. Among them, catechins and proanthocyanidins are two classes of compounds exhibiting antimicrobial properties towards both prokaryotic and eukaryotic microorganisms. Yet, despite the large number of studies published so far, the real potentialities and limitations given by the use of this class of molecules as antiviral or antimicrobial (antibacterial, antimycotic, antiprotozoal) agents have not been critically evaluated. The present chapter represents an overview of the recent literature on the antimicrobial properties exhibited by such class of compounds.

2
Chemistry of Catechins and Proanthocyanidins

Proanthocyanidins, otherwise labelled as condensed tannins, are a class of polyphenolic molecules constituted by oligomers of flavan-3-ol monomer units (catechins). They occur widely in the plant kingdom and are considered the second most abundant group of natural phenolics after lignin. In relation to their function in plants, it has been postulated that they accumulate in many different organs and tissues to provide protection against predation. The presence of these compounds in foods (e.g., cereals, fruits, vegetables, wines, etc.) affects their texture, color, and taste [11]. Proanthocyanidins, under the form of oligomers and polymers of prodelphinidins and procyanidins, are natural products frequently found in fruits (i.e., grapes, strawberry, blueberry, cocoa, coffee berry), vegetables, nuts, seeds (i.e., grape seeds, coffee seeds), beans, as well as medicinal plants [2, 12–16].

The chemistry of catechins and proanthocyanidins has been studied for many decades. The size of proanthocyanidin molecules vary as a function of the degree of polymerization. The three rings of flavan-3-ols are denoted as A, B, and C (Fig. 1). They differ structurally according to the number of hydroxyl groups on both aromatic rings and to the stereochemistry of the asymmetric carbons of the heterocycle, which may occur in different configurations. With only a few exceptions, the configuration of C2 is R. Flavan-3-ols with 2S configuration are distinguished by the prefix enantio (*ent*-). The stereochemistry of the C2–C3 linkage may be either *trans* (2R, 3S) or *cis* (2R, 3R), as observed in (+)-catechin and (–)-EC. The interflavan linkages can be α or β. The most common classes are the procyanidins, which are chains of catechin, EC, and their gallic acid esters (Fig. 1A), and the prodelphinidins (Fig. 1B), which consist of GC, EGC, and their galloyl derivatives as the monomeric units [17].

The B-type procyanidins include a mixture of oligomers and polymers composed of flavan-3-ol units linked mainly through C4 $\rightarrow$ C8 and/or C4 $\rightarrow$ C6 bonds, and represent the dominant class of natural proanthocyanidins. Among the dimers, procyanidins B1, B2, B3 and B4 (Fig. 2a) are the most frequently occurring in plant tissues. Procyanidin B5 (EC-(4$\beta \rightarrow$ 6)-EC), B6 (catechin-(4$\alpha \rightarrow$ 6)-catechin), B7 (EC-(4$\beta \rightarrow$ 6)-catechin) and B8 (catechin-(4$\alpha \rightarrow$ 6)-EC) are also widespread (Fig. 2b) [17–19].

Analogues of procyanidin B1 and B2 exhibiting EC chain extension units (2R, 3R-2,3-*cis* configuration) are very commonly represented in the plant kingdom, whereas many plants also produce analogues of procyanidin B3 to B8 [18, 19].

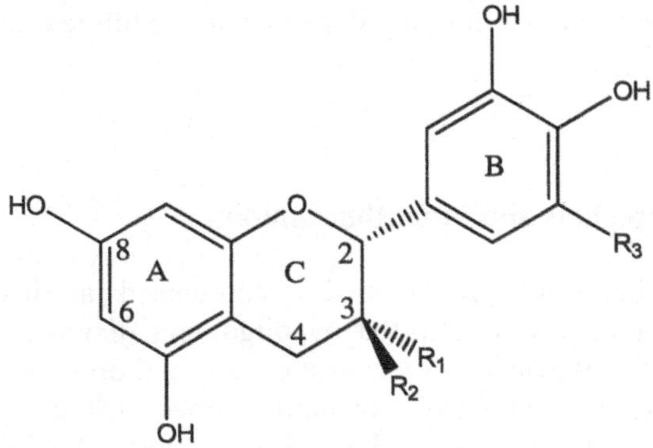

A) Procyanidins:
R_1=OH; R_2=H; R_3=H; EC
R_1=H; R_2=OH; R_3=H; Catechin
R_1=gallic acid ester; R_2=H; R_3=H; ECG
R_1=H; R_2= gallic acid ester; R_3=H; CG

B) Prodelphinidins:
R_1=OH; R_2=H; R_3=OH; EGC
R_1=H; R_2=OH; R_3=OH; GC
R_1=gallic acid ester; R_2=H; R_3=OH; EGCG
R_1=H; R_2= gallic acid ester; R_3=OH; GCG

Fig. 1 Structures of procyanidins (**A**) and prodelphinidins (**B**)

On the other hand, the flavan-3-ol units can also be doubly linked by an additional ether bond between C2 → O7 (A-type). Structural variations occurring in proanthocyanidin oligomers may also occur with the formation of a second interflavanoid bond by C–O oxidative coupling to form A-type oligomers (Fig. 3) [17, 20]. Due to the complexity of this conversion, A-type proanthocyanidins are not as frequently encountered in nature compared to the B-type oligomers.

The A-type proanthocyanidins are characterized by a second ether linkage between an A-ring hydroxyl group of the lower unit and C-2 of the upper unit. Since they are less frequently isolated from plants than the B-types, they have been considered unusual structures [18, 19]. The first identified A-type proanthocyanidin was procyanidin A2 isolated from the shells of fruit of *Aesculus hippocastanum*. Since then, many more A-type proanthocyanidins have been found in plants, including dimers, trimers, tetramers, pentamers and ethers [18, 21].

Furthermore, C-type procyanidins include trimers composed of flavan-3-ol units linked mainly through C4 → C8 (Fig. 4).

Interestingly, since all constitutive units and linkages can be distributed at random within a polymer, the number of possible isomers increases expo-

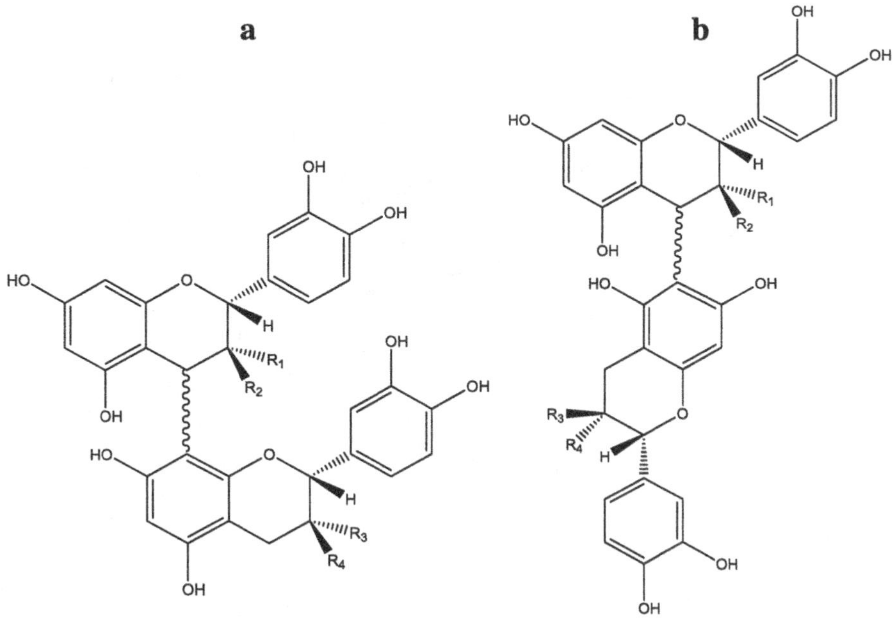

B1: R_1=OH; R_2=H; R_3=H; R_4=OH
B2: R_1=OH; R_2=H; R_3=OH; R_4=H
B3: R_1=H; R_2=OH; R_3=H, R_4=OH
B4: R_1=H; R_2=OH; R_3=OH; R_4=H

B5: R_1=OH; R_2=H; R_3=H; R_4=OH
B6: R_1=H; R_2=OH; R_3=OH; R_4=H
B7: R_1=OH; R_2=H; R_3=OH, R_4=H

Fig. 2 Structures of some proanthocyanidin dimers of the B-type

nentially with the chain length (e.g., 8 dimers, 32 trimers, and 128 tetramers in the B-type procyanidin series). On the whole, it has been calculated that a given n-mer with x types of constitutive units and y types of linkages is able to generate $n^x\ (n-1)^y$ possible isomers [22].

3
Biosynthesis of Catechins and Proanthocyanidins

In spite of the recent progress in understanding the biosynthesis of the major building blocks of proanthocyanidins, (+)-catechin and (–)-EC, some important questions still remain to be elucidated (e.g., the exact nature of the molecular species that undergo polymerization and the mechanisms of assembly). The biosynthetic pathways for proanthocyanidins have been extensively reviewed [23–28]. A general scheme summarizing proanthocyanidin biosynthesis adapted from [27] and [28] is reported in Fig. 5.

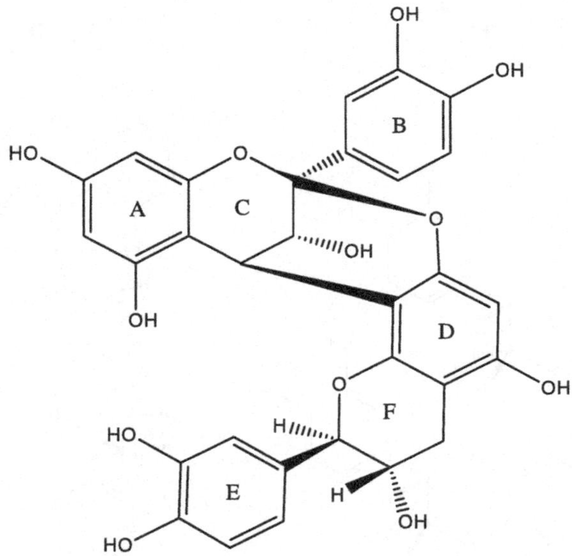

A2: R_1=OH; R_2=H; R_3=OH; R_4=H

Fig. 3 Structures of some proanthocyanidin dimers of the A-type

Flavan-3-ols are originated from a branch pathway of anthocyanin biosynthesis. Early studies covering enzymological aspects of proanthocyanidin biosynthesis were carried out on *Ginkjo biloba* and *Pseudotsuga menziesii* and reported that dihydroflavonols [(+)-dihydromyricetin or (+)-dihydroquercetin] are converted to the corresponding flavan-3,4-diols and catechin derivatives [(+)-GC or (+)-catechin, respectively] [29, 30]. The enzymatic basis for the formation of the 2,3-*trans*, catechin-derived series of flavan-3-ols has been postulated to be related to the consecutive action of a dihydroflavonol reductase (DFR) (which produces a leucoanthocyanidin) and a leucoanthocyanidin reductase (LAR). Genetic determinants encoding LAR have been investigated [31]. More recent literature regarding *Vitis vinifera* L. cv Shiraz cells [28] reported that two enzymes, leucoanthocyanidin reductase (LAR) and anthocyanidin reductase (ANR), can produce the flavan-3-ol monomers required for formation of proanthocyanidin polymers.

Xie and Dixon [27] underlined that although some models of proanthocyanidin biosynthesis show the extension units arising from condensation of an leucoanthocyanidin-derived electrophile with the nucleophilic 8 or 6 position of the starter unit, this scheme fails because of stereochemistry of leucoanthocyanidin is most likely 2,3-*trans*, whereas, in many cases, the extension units are 2,3-*cis*. One possible solution for this stereochemical para-

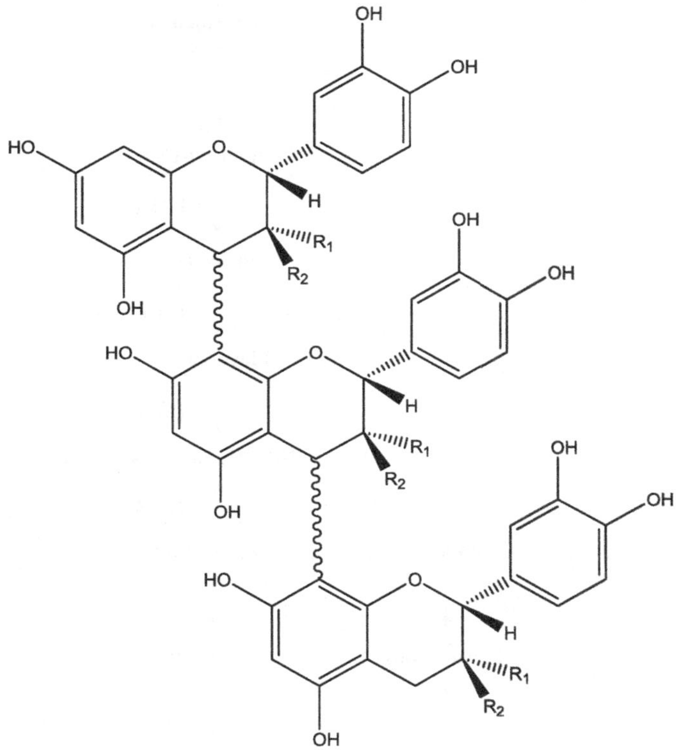

C1: R_1=OH; R_2=H
C2: R_1=H; R_2=OH

Fig. 4 Structures of some proanthocyanidin trimers of C-type

dox has been proposed by Xie et al. [32], who discovered that the BANYULS gene from *Arabidopsis thaliana* and the model legume *Medicago truncatula* encode anthocyanidin reductases that convert cyanidin to 2,3-*cis*-(–)-EC. Consistently with the proposed model, mutations in anthocyanidin synthase (ANS) resulted in a deficiency in proanthocyanidin accumulation in *Arabidopsis* sp. [32, 33].

4 Synthetic Strategies for Catechins and Proanthocyanidins

Although both catechins and proanthocyanidins are currently found in plant tissue composition, as reported above, their isolation requires sometime time-consuming procedures and purity and reproducibility are often suspect.

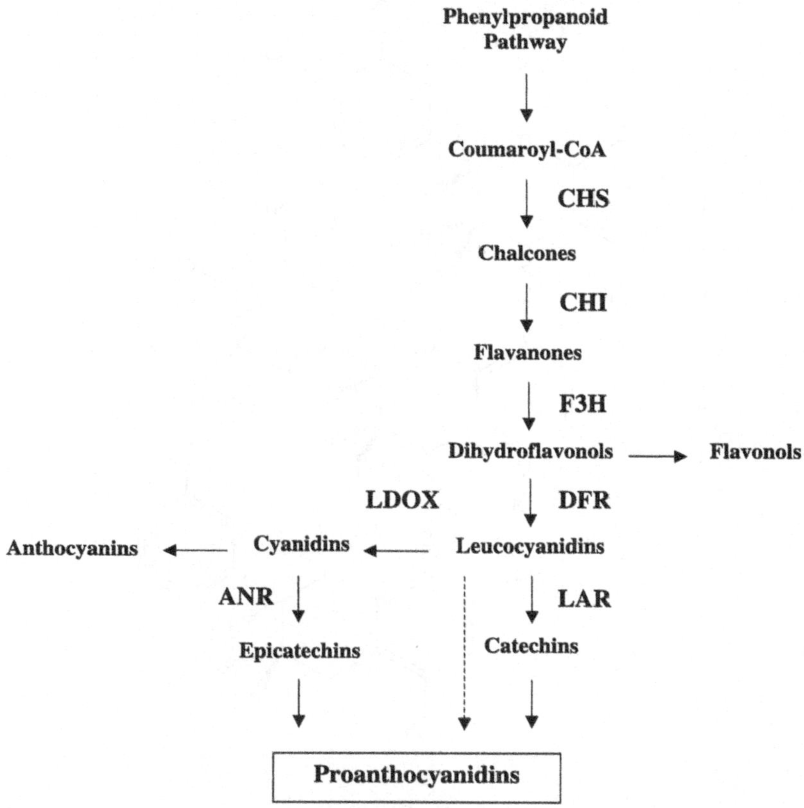

Fig. 5 Scheme of the flavonoid pathway leading to synthesis of proanthocyanidins. The enzymes involved in the pathway are shown as follows: CHS = chalcone synthase; CHI = chalcone isomerase; F3H = flavanone-3B-hydroxylase; DFR = dihydroflavonol-4-reductase; LDOX = leucoanthocynidin dioxygenase; LAR = leucoanthocyanidin reductase; ANR = anthocyanidin reductase adapted from [27] and [28]

Indeed, the well-documented presence of mixtures of closely related compounds and the frequent lack of commercially available pure compounds represent undoubtedly a problem in the case of studies aimed at individuating the nature of active compound(s).

The above-mentioned difficulties, coupled with the possibility of improving their biological activities, collude to discover an attractive challenge in organic synthetic strategies for supplying valuable pure compounds [34]. A number of papers dealing with the synthesis or chemical conversions of catechins or proanthocyanidins have been published since the 1980s [35–37].

Recently published literature described either the improvement of the strategies devoted to the synthesis of either catechins derivatives or oligomeric proanthocyanidins or the use of a synthetic program as a way to obtain nature-identical or even novel molecules often characterized by an improved biological activity. Unfortunately, to the authors knowledge, no study has been carried out so far on the antimicrobial activity of synthetic catechins or proanthocyanidins.

A recently described procedure, aimed at obtaining the oxidation of catechin derivatives through the introduction of a new hydroxyl group at C-6, was applied for the synthesis of elephantorrhizol, a natural flavan-3-ol exhibiting a fully substituted cycle A [38]. On the other hand, lipophilic catechins, characterized by an increased cell absorption, were synthesized by Jin and Yoshioka [39] as a way of obtaining new antioxidants effective in lipid bilayers. The hydroxyl groups of (+)-catechin was acylated randomly using lauroyl chloride, and the mixture was separated by preparative HPLC. 3-Lauroyl-, 3′,4′-dilauroyl- and 3,3′,4′-trilauroyl-catechins (3-LC, 3′,4′-LC, and 3,3′,4′-LC) were obtained. The radical scavenging activity of 3-lauroyl-catechin was similar to that of (+)-catechin, whereas those of 3′,4′-dilauroyl- and 3,3′,4′-trilauroyl-catechins were smaller. These authors concluded that the blocking of phenolic OH groups in the catechin B ring apparently decreased this biological activity.

In a few cases, the structural modification of the catechin skeleton has been the preliminary approach for the synthesis of modified polymeric structures. The regioselective introduction of substituents at C-8 of (+)-catechin has been reported by Beauhaire et al. [40]. These authors described the synthesis of several catechin derivatives characterized by the presence of various substitution patterns, as monomers to be used for the further synthesis of modified proanthocyanidins. Accordingly, a new 3-*O*-4 ether-linked procyanidin-like derivative was selectively achieved through condensation of 4-(2-hydroxyethoxy)tetra-*O*-benzyl catechin with the 8-trifluoroacetyl adduct of tetra-*O*-benzyl catechin. Analogously, Ohmori et al. [34] reported that the bromo-capping of the C(8) position of the flavan skeleton (enabling the equimolar coupling of electrophilic and nucleophilic catechin derivatives) allowed an efficient synthetic strategy to complex catechin oligomers.

In a few cases, the synthesis was directed towards well-defined oligomers (dimers, trimers, etc.). The synthesis of bis(5,7,3′,4′-tetra-*O*-benzyl)-EC 4β,8-dimer from 5,7,3′,4′-tetra-*O*-benzyl-EC and 5,7,3′,4′-tetra-*O*-benzyl-4-(2-hydroxyethoxy)-EC was described by Kozikowski et al. [41]. This compound exhibited the ability to inhibit the growth of several breast cancer cell lines through the induction of cell cycle arrest in the G_0/G_1 phase. Analogously, procyanidin-B3, a condensed catechin dimer, has been obtained through condensation of benzylated catechin with various 4-*O*-alkylated flavan-3,4-diol derivatives in the presence of a Lewis acid. This reaction led to protected procyanidin-B3 and its diastereomer. In particular, octa-*O*-benzylated procyanidin-B3 has been produced with high levels of stereoselectivity and in excellent isolation yields [42].

On the other hand, the synthesis of four natural procyanidin trimers {[4,8 : 4″,8″]-2,3-*cis*-3,4-*trans*: 2″,3″-*cis*-3″,4″-*trans*: 2⁗,3⁗-*trans*-(−)-EC-(−)-EC-(+)-catechin, [4,8 : 4″,8″]-2,3-*cis*-3,4-*trans*: 2″,3″-*cis*-3″,4″-*trans*: 2⁗,3⁗-*cis*-tri-(−)-EC: procyanidin C1, [4,8 : 4″,8″]-2,3-*cis*-3,4-*trans*: 2″,3″-*trans*-3″,4″-*trans*: 2⁗,3⁗-*trans*-(−)-EC-(+)-catechin-(+)-catechin: procyanidin C4, and [4,8 : 4″,8″]-2,3-*cis*-3,4-*trans*: 2″,3″-*trans*-3″,4″-*trans*: 2⁗,3⁗-*cis*-(−)-EC-(+)-catechin-(−)-EC} has been described by Saito et al. [43].

Oligomeric procyanidins containing 4*α*-linked EC units are rare in nature. Kozikowski et al. [44] reported the preparation of EC-4*α*,8-EC through the reaction of the protected 4-ketones with aryllithium reagents derived by halogen/metal exchange from the aryl bromides. On the other hand, four synthetic processes aimed at preparing protected EC oligomers exhibiting (4*β*,8)-interflavan linkages and characterized by anticancer activity have been recently patented by Kozikowski et al. [45]. In the first process, a tetra-*O*-protected EC monomer or oligomer has been coupled with a protected, C-4 activated EC monomer in the presence of an acidic clay, whereas in the second one, a 5,7,3′,4′-benzyl protected or a 3-acetyl-, 5,7,3′,4′-benzyl protected EC or catechin monomer or oligomer has been reacted with 3-*O*-acetyl-4-[(2-benzothiazolyl)thio]-5,7,3′,4′-tetra-*O*-benzyl-EC in the presence of silver tetrafluoroborate. The third process was characterized by the reaction between two 5,7,3′,4′-benzyl protected EC monomers activated with 2-(benzothiazolyl)thio groups at the C-4 positions in the presence of silver tetrofluoroborate, whereas the fourth one was designed for reacting an unprotected EC or catechin monomer with 4-(benzylthio)-EC or catechin [45].

The polymerization of (+)-catechin producing catechin-aldehyde polycondensates (characterized by an ethyl bridge through C-6 and C-8 positions of catechin A ring) has been achieved by Kim et al. [46]. Analogously, Chung et al. [47] assessed the antioxidant and enzyme inhibitory activity of synthetic poly(catechin)s condensed through acetaldehyde and characterized by different molecular weights. When compared to monomeric catechins, the poly(catechin)s exhibited great amplification of superoxide scavenging activity, xanthine oxidase inhibitory activity, and inhibition effects on human low-density lipoprotein oxidation. These activities were apparently proportional to their molecular weights.

5 Biological Activity of Catechins and Proanthocyanidins Other than Antimicrobial Activity

In light of increasing evidence associating these compounds with a wide range of potential health benefits, catechins and proanthocyanidins have attracted commercial attention [11]. Accordingly, the proanthocyanidins in grape seed extracts have been suspected to have free-radical scavenging abilities and to decrease the susceptibility of healthy cells to toxic and carcinogenic agents [48–50]. Analogously, in vitro and in vivo studies have demonstrated that these compounds exhibit the ability to scavenge reactive oxygen and nitrogen species [51], as well as inhibiting the progression of arteriosclerosis and preventing the increase of LDL cholesterol in high-cholesterol-fed rats [52].

Additional in vivo studies on the biological activity of proanthocyanidins investigating a series of behavioral activities (motility, body weight gain, body temperature, motoric coordination, anticonvulsant effects and central analgesic activities) showed no or only moderate pharmacological effects [53]. On the other hand, dietary supplementation with cocoa procyanidin supplements can dose-dependently prevent the development of hyperglycemia in diabetic obese mice [54].

Although the above profusion of in vivo studies evidence their health potentialities, the problem of the bioavailability of proanthocyanidins supplied by dietary supplementation has still not been completely resolved since unequivocal evidence for absorption is missing so far [11]. However, studies carried out using radio-labelled procyanidins revealed that dimers and trimers may be absorbed by intestinal cells, whereas a recent study demonstrated that procyanidin oligomers are readily adsorbed in rats [55], while it has been shown that colon microflora may be able to degrade proanthocyanidins to low-molecular-weight aromatic compounds [56].

The assessment of in vivo biological actions of these polymers is often influenced by unspecific mechanisms of protein denaturation [57]. Ku et al. [58] reported that purified proanthocyanidins from *Pinus radiata* bark exhibit a pH-, contact time- and temperature-dependent binding affinity to bovine Achilles tendon collagen (type I). The ability of proanthocyanidins to bind collagen and elastin is the rationale behind the introduction into commerce of proanthocyanidin-based products as skin protecting agents [59]. In particular, a series of commercial proanthocyanidin-based extracts have been recently described. Passwater [59] wrote about a product derived from pine bark of *Pinus pinaster* and characterized by a content of 65–75% procyanidins (consisting of catechin and EC subunits at various degree of polymerization); other minor constituents are polyphenolic monomers, phenolic or cinnamic acids and their glycosides. Furthermore, other products, commercialized as dietary supplements and including catechin, GC and proanthocyanidins (from 20 to 95% of total polyphenols) have been obtained from green tea, grape seeds, cocoa, coffee berry fruits and *Pelargonium sidoides* roots [50].

6 Antimicrobial Activity of Catechins and Proanthocyanidins

6.1 Antibacterial and Antiviral Activity of Catechins and Proanthocyanidins

Analytical determination of the polyphenolic fraction of green tea revealed that this product contains approximately 30% catechins, whereas completely fermented black tea typically contains approximately only 9% catechin, 4% theaflavin, 3% flavonols, and 15% undefined catechin condensate products [60]. Accordingly, catechins extracted from teas characterized by a different degree of fermentation were screened for their antimicrobial activity against Gram-positive and Gram-negative bacteria of the species *Bacillus subtilis*, *Escherichia coli*, *Proteus vulgaris*, *Pseudomonas fluorescens*, *Salmonella* sp. and *Staphylococcus aureus*. On the whole, the highest activity was observed on *P. fluorescens* cells. Antibacterial activity decreased as an apparent consequence of the fall of catechin concentration occurring during tea fermentation [61]. Analogously, water extracts obtained from *Cocos nucifera* husk fiber (containing catechin, EC and B-type proanthocyanidins) revealed a pronounced antimicrobial activity against strains of the species *S. aureus* [62], whereas Otake et al. [63] report that a commercially available preparation of catechins both prevented the attachment of a cariogenic *Streptococcus mutans* strain to hydroxyapatite and inhibited its glucosyltransferase activity.

Among catechins, the activity of selected molecules has been quantified. The antibacterial activity of EGCG against strains of *Staphylococcus* spp. and Gram-negative bacteria (including *E. coli*, *Klebsiella pneumoniae*, and *Salmonella* spp.) was determined by Yoda et al. [64]. In this study, 50–100 $\mu g\, mL^{-1}$ of this compound inhibited the growth of *Staphylococcus* spp. strains, whereas concentrations higher than 800 $\mu g\, mL^{-1}$ were required for the inhibition of Gram-negative bacteria. Similarly, it has also been reported that EGCG exhibited a pH-dependent bactericidal activity against various pathogenic species [65–71].

Concerning the mechanism of action of catechins, studies carried out on *S. aureus* and *E. coli* cells by Ikigai et al. [72] reported that their bactericidal effect is primarily involved in the damage of bacterial membranes: catechins induce a rapid leakage of small molecules entrapped in the intraliposomal space, determining the aggregation of the liposomes. These actions cause damage in the membrane lipid bilayer and cell death (Table 1).

In some cases, catechins can also act in synergistic mode when used in association with currently used antibiotic molecules (Table 2). EGCG exhibited synergy with β-lactams. Sudano Roccaro et al. [73] found that this compound is able to reverse tetracycline resistance in *Staphylococcus epidermidis* and *S. aureus* isolates. This synergistic interaction has been explained by inhibition of tetracycline efflux pump activity in microbial cells resulting in an increased intracellular retention of the antibiotic. On the other hand, the formation of a binding between EGCG and the cell wall peptidoglycan and the inhibition of penicillinase activity have also been used as rationale for explaining the observed synergy between this catechin and penicillin [74, 75].

Similar to catechins, several studies have reported that proanthocyanidins exhibit a more or less pronounced antibacterial activity. Chung et al. [76] reported that proanthocyanidins determine the growth inhibition of strains of *Aeromonas* spp., *Bacillus* spp., *Clostridium botulinum*, *Clostridium perfringens*, *Enterobacter* spp., *Klebsiella* spp., *Proteus* spp., *Pseudomonas* spp., *Shigella* spp., *S. aureus*, *Streptococcus* spp., and *Vibrio* spp.

The water extracts of green-, black- and red-colored azuki beans (*Vigna angularis* var. Dainagon) showed evident antibacterial effects against strains of the species *S. aureus*, *Aeromonas hydrophila* and *Vibrio parahaemolyticus*. The reduction of the number of viable cells of *S. aureus* after 24 h exposure to extracts of colored azuki beans has been related to their high concentration of proanthocyanidins [77]. Analogously, Amarowicz et al. [78] showed that proanthocyanidins of green tea extracts have antibacterial activity against *E. coli*.

Puupponen-Pimiä et al. [79] studied the antimicrobial activity of eight berry extracts against selected Gram-positive and Gram-negative bacteria, including probiotic bacteria and intestinal pathogens. The berry extracts (particularly cranberry and blueberry) mainly inhibited the growth of Gram-negative bacteria but had no effect on Gram-positive bacteria. In agreement with previous findings [80], Puupponen-Pimiä et al. [79] suggested that proanthocyanidins might be the class of compounds responsible for this behavior. Analogously, Taylor et al. [81] considered these molecules at least partially responsible for the antibacterial activity of methanolic extracts of the bark of *Terminalia alata* collected in Nepal.

The antibacterial activity of proanthocyanidins was also observed with regard to bacterial groups of veterinary interest. Purified proanthocyanidins from three different plants (*Schinopsis balansae*, *Desmodium ovalifolium*, and *Mirtus communis*) were investigated as antimicrobial compounds towards ruminal bacteria of the species *Streptococcus bovis*, *Ruminococcus albus*, *Fibrobacter succinogenes*, and *Prevotella ruminicola*. Proanthocyanidins from *D. ovalifolium*, and *M. communis* (characterized by a differential polymerization degree) exhibited the highest antimicrobial activity [82]. Analogously, Jones et al. [83] found that proanthocyanidins from leaves of *Onobrychis viciifolia* inhibited growth and protease activity of strains of the species *Butyrivibrio fibrisolvens* and *S. bovis*, but had only a scarce effect on those belonging to the species *P. ruminicola* or *Ruminobacter amylophilus*.

In a few cases, proanthocyanidins act by inhibiting adhesion of bacterial populations to human cells. Clinical, epidemiological and mechanistic studies supported the role of cranberry in maintaining the health of the urinary tract thorough the prevention of adhesion of P-fimbriated uropathogenic *E. coli*

Table 1 Mechanism of action hypothesized for catechins and proanthocyanidins

Compound	Microbial target	Cell target	Mechanism of action	Refs.
Catechins	Bacteria	Intraliposomal space of membrane lipid bilayer	Leakage of small molecules aggregation of liposomes	[63]
Catechins	Filamentous fungi	Cell membrane	Lysis of conidia and hyphae	[88]
Proanthocyanidins	Bacteria	Extracellular enzymes	Enzyme inhibition	[79]
Proanthocyanidins	Bacteria	Cell coat polymers	Inhibition of cell-associated proteolysis inhibition of cell wall synthesis	[76]
Proanthocyanidins	Bacteria		Iron depletion	[81]

Table 2 Mechanisms of synergistic action towards currently used antibiotics hypothesized for catechins and proanthocyanidins

Compound	Microbial target	Cell target	Antibiotic	Mechanism of synergy	Refs.
Catechins	Bacteria	Tetracycline efflux pump	β-lactams	Increased intracellular retention of tetracycline	[66]
Catechins	Bacteria	Cell wall peptidoglycan	β-lactams	Inhibition of penicillinase activity	[67]
Catechins	Yeasts	Cell membrane	Polyenes	Stimulation membrane permeability increased intracellular catechin concentration	[53]

to uroepithelial cells [80, 84]. Results obtained by these authors suggest that the presence of A-type proanthocyanidins in cranberry tissues may be responsible for their bacterial anti-adhesion activities. Moreover, Foo et al. [85] found that purified proanthocyanidins from American cranberry (*Vaccinium macrocarpon*) exhibited a potent activity by inhibiting adherence of *E. coli* uropathogenic isolates. The proanthocyanidins involved in these activities consisted predominantly of EC monomers with a degree of polymerization from 4 to 5 units and containing at least one A-type linkage. The procyanidin labelled as A2 (Fig. 3) was the most common terminating unit, occurring about four times as frequently as the EC monomer.

Although a few mechanisms have so far been proposed to explain the antimicrobial properties exhibited by proanthocyanidins (e.g., inhibition of extracellular enzymes) [86], Jones et al. [83] postulated that their ability to bind bacterial cell coat polymers and their ability to inhibit cell-associated proteolysis might be considered responsible for the observed activity (Table 1). Accordingly, despite the formation of complexes with cell coat polymers, proanthocyanidins penetrated to the cell wall in sufficient concentration to react with one or more ultra-structural components and to selectively inhibit cell wall synthesis. Decreased proteolysis in these strains may also reflect a reduction of the export of proteases from the cell in the presence of proanthocyanidins [83].

Additional hypotheses for their mechanism of action have more recently been proposed. It is well known that proanthocyanidins are able to complex metals through their *ortho*-diphenol groups. This property is often viewed as imparting negative traits (e.g., reduction of the bioavailability of essential mineral micronutrients, especially iron and zinc) [87]. Since iron depletion causes severe limitation to microbial growth, their ability to bind iron has been suggested as one of the possible mechanisms explaining the antimicrobial activity of proanthocyanidins [88] (Table 1).

Catechins and proanthocyanidins have a documented antiviral activity. Catechins from an extract of *Cocos nucifera* husk fibre exhibited a strong inhibitory activity against acyclovir-resistant herpes simplex virus type 1 (HSV-1-ACVr) [62]. The use of 10 to 20 $ng\,ml^{-1}$ of ECG and EGCG has been reported to cause 50% inhibition of human immunodeficiency virus reverse transcriptase [89], while Hara and Nakayama [90] reported that a patented chewing gum containing tea catechins is claimed to prevent viral infections against influenza and to inhibit dissemination of this virus.

6.2 Antimycotic Activity of Catechins and Proanthocyanidins

The antimycotic activity of catechins occurring in green tea composition against yeasts and filamentous fungi has been studied. Hirasawa and Takada [66] reported the susceptibility of *Candida albicans* to EGCG. Analogously, Park et al. [70] observed that the same molecule has antimycotic activity against 21 clinical *Candida* spp. isolates. Among them, the strains belonging to the species *Candida glabrata* were the most susceptible. Accordingly, the use of catechins as antimycotic agents in candidiasis has been recently proposed [91]. Likewise to the above reported activity towards bacteria, the antimycotic properties of catechins against *C. albicans* strains was apparently pH dependent: the minimum inhibitory concentration (MIC) of EGCG increased several folds as the pH was reduced from 7.0 to 6.0. Independently from this, catechins always displayed fungicidal action [92].

The mode of action of catechins on eukaryotic microorganisms has been studied little. In early investigations, Toyoshima et al. [93] suggested that catechins are able to attack the cell membrane of *Trichosporon mentagrophytes* causing lysis of the conidia and hyphae (Table 1).

Catechins sometimes exhibited synergy with molecules currently used as antimycotics (Table 2). Hirasawa and Takada [66] observed that the combined use of EGCG and sub-inhibitory concentration of amphotericin B inhibited the growth of *C. albicans*. Since sub-inhibitory concentrations of this polyene stimulate yeast membrane permeability, these authors suggested that the combined use of amphotericin B and EGCG might stimulate catechin uptake into the cell. As a consequence, the increased intracellular catechin concentration might act as a fungicidal agent. Accordingly, the use of EGCG as an adjuvant with antifungal agents in candidiasis has been proposed [91].

Differently from catechins, only a handful of studies have been so far devoted to the study of antimycotic properties of proanthocyanidins. *Pinus maritima* L. extracts exhibited a broad activity towards yeasts belonging to the genera *Candida*, *Cryptococcus*, *Filobasidiella*, *Issatchenkia*, and *Saccharomyces* [94]. Considering the results of HPLC/DAD and MS analyses, these authors concluded that proanthocyanidins occurring in *P. maritima* L. extracts can apparently be considered responsible for the observed antimycotic activity. Analogously, two proanthocyanidin-containing preparations were isolated and partially purified from the pulp of ripe *Coffea canephora* [95]. These proanthocyanidin-rich extracts inhibited the in vitro germination of *Hemileia vastatrix* uredospores: a greater proanthocyanidin concentration was associated with greater activity.

On the contrary, both crude and water extracts (containing both catechins and proanthocyanidins) obtained from *Cocos nucifera* L. husk fibre were inactive against both the yeasts *C. albicans*, *Cryptococcus neoformans* and the filamentous fungus *Fonsecaea pedrosoi* [62].

6.3
Antiprotozoal Activity of Catechins and Proanthocyanidins

In a similar way, several studies have been devoted to the study of antiprotozoal activity of catechins occurring in green tea extracts. The trypanocidal activity of green tea catechins against two different developmental stages of *Trypanosoma cruzi* has been reported by Paveto et al. [96]. Active compounds were identified by HPLC-DAD and GC-MS as a mixture of catechins (especially catechin, EC, EGC, ECG, GCG and EGCG). Concentrations from 0.12 to 85 pM of purified compounds killed more than 50% of the protozoa present in the blood of infected mice. The most active compounds were GCG and EGCG, with MICs of 0.12 and 0.53 pM, respectively. Interestingly, the activity of *T. cruzi* arginine kinase (a key enzyme in the energy metabolism of this parasitic species) was inhibited by about 50% by these two catechins, whereas the others were less effective [96].

Calzada et al. [97] studied the antiprotozoal activity of the dichloromethane-methanol extract obtained from the roots of *Geranium mexicanum* (as well as of the different fractions and pure compounds) against strains of the species *Entamoeba histolytica* and *Giardia lamblia*. The results indicated that the organic fraction was active against both protozoa with MICs of about 79 and 100 μg/ml, respectively. (–)-EC was the most active compound. Similarly, (–)-EC and (+)-catechin, extracted from the aerial parts of *Rubus coriifolius* (a medicinal plant used by the Maya communities in southern Mexico to treat bloody diarrhea) were tested for their antiprotozoal activity against *E. histolytica* and *G. lamblia*. (–)-EC was suggested to be the main responsible for the antiprotozoal activity exhibited by the extract from this plant against both protozoa [98].

Only a few studies have reported the antiprotozoal activity of proanthocyanidins. Bioassay-guided fractioning of the extract of *Geranium niveum* led to the isolation of two new A-type proanthocyanidins: *epi*-afzelechin-($4\beta \rightarrow 8,2\beta \rightarrow O \rightarrow 7$)-afzelechin and EC-($4\beta \rightarrow 8,2\beta \rightarrow O \rightarrow 7$)-afzelechin (labelled with the trivial names of geranins A and B, respectively). Both compounds exhibited antiprotozoal activity against *G. lamblia* and *E. histolytica* [97]. Analogously, a recently purified proanthocyanidin induced a dose-dependent growth arrest and cell lysis of *Trypanosoma brucei* [99]. On the contrary, Kolodziej and Kiderlen [100] found that the antiprotozoal activity of proanthocyanidins was generally less pronounced than that of hydrolyzable tannins. This study revealed that dimers and trimers showed similar moderate antiprotozoal activity, while an increased degree of polymerization was apparently less favorable.

7
SAR Studies on Catechins and Proanthocyanidins

Studies focused on the determination of SAR have revealed that the different enantiomeric form of catechins apparently affects their antibacterial activity. Bais et al. [101] found that (+)-catechin inhibited soil-borne bacteria of the species *Xanthomonas campestris*, *P. fluorescens*, and *Erwinia caro-*

tovora. In contrast, (–)-catechins showed no antibacterial activity against soil-borne pathogens. Analogously, Veluri et al. [102] compared the structure of several synthetic derivatives of (+)- and (±)-catechin with their antimicrobial activity against strains of the species *E. carotovora*, *Erwinia amylovora*, *X. campestris* pv. *vesicatoria*, and *P. fluorescens*. (+)-Catechin exhibited significant activity against all four microorganisms. In addition, the (+)-derivative was more active than the (±)-derivative. Fluorescence microscopy of bacterial cultures separately treated with the different synthetic catechin derivatives revealed that all of the tested structures exhibited bactericidal activity. These results were in contrast with those obtained by Palma et al. [103], who reported that (+)-catechin exhibited no antibacterial activity towards strains of *Bacillus cereus*, *S. aureus*, *Staphylococcus coagulans*, and *Citrobacter freundii* and only minimal activity against *E. coli*. Similarly, very weak activity was observed when these compounds were used against food-associated bacteria of the species *Lactobacillus rhamnosus*, and even none when used against other *Lactobacillus* spp., *E. coli*, *Salmonella enterica*, *Enterococcus faecalis*, and *Bifidobacterium lactis* [79]. The antibacterial activities of catechins has been related to the gallic acid moiety and the hydroxyl group member. Mabe et al. [65] screened six tea catechins (EGCG, ECG, EGC, EC, crude catechin, and crude theaflavin) against the ulcer-causing bacterium *Helicobacter pylori*. The EGCG showed the strongest activity. Analogously, Yee and Koo [104] found that EC and EGCG inhibited the growth of *H. pylori*. However, the latter compound exhibited a higher activity than the former, while Yanagawa et al. [105] observed that only EGCG and ECG exhibited antibacterial activity against this species.

A recent SAR study on different fractions from stem-bark extracts from *Stryphnodendron obovatum* Benth has been carried out by Conegero-Sanches et al. [106]. These authors observed that only the fractions containing monomeric and dimeric EGC exhibited a good activity against strains belonging to the species *Candida parapsilosis* and *C. albicans*. In addition, analogous to the antibacterial activity, the different enantiomeric forms of catechins apparently affected their antimycotic properties. (+)-Catechin was active against root-colonizing filamentous fungi of the species *Trichoderma reesi*, *Trichoderma viridans*, *Fusarium oxysporum*, *Aspergillus niger*, and *Penicillium* sp., whereas its tetramethoxy- and/or acetylated-derivatives considerably reduced their activity. On the contrary, (+)-catechin was reported to be inactive against the crop plant filamentous fungi of the species *Botrytis cinerea*, *Cladosporium echinulatum*, and *Penicillium griseoflulvum* [103]; neither (+)-catechin nor (±)-catechin inhibited the growth of the yeasts *C. albicans*, *Saccharomyces cerevisiae*, or that of the filamentous fungus *A. niger* [107].

A SAR study has been the base for assessing the therapeutics potency of the polyphenols of Central American indigenous plants against infectious diarrhoea in children caused by protozoa [108]. Accordingly, (–)-EGC and (–)-EGCG, isolated from *Helianthemum glomeratum* roots, were tested for their in vitro antiprotozoal effects. The former compound also acted by suppressing the growth of *G. lamblia*. These authors suggested that the antiprotozoal activity observed in these plant extracts might be due to the presence of (–)-EGC, but in combination with other polyphenolic molecules (e.g., flavonoids).

SAR studies were carried out by de Bruyne et al. [92] on a series of dimeric procyanidins, considered as model compounds for antiviral therapies. On the whole, proanthocyanidins containing EC dimers exhibited more pronounced activity against herpes simplex virus (HSV) and human immunodeficiency virus (HIV), while the presence of *ortho*-trihydroxyl groups in the B-ring appeared to be essential in all proanthocyanidins exhibiting anti-HSV effects. Galloylation and polymerization reinforced the antiviral activities markedly.

8 Concluding Remarks

The current literature reported herein underlines that the majority of studies on the antimicrobial activity of catechins and proanthocyanidins has been carried out by using pathogenic or environmental Gram-positive and Gram-negative bacteria, while the use of eukaryotic microorganisms (yeasts, filamentous fungi, protozoa) as target microorganisms has been less frequent. Similarly, the antiviral activity of both compounds has been only sporadically considered. In addition, SAR studies have revealed that the sometimes observed differential activity might depend either by the enantiomeric structure of catechins [102, 107] or on the polymerization degree of proanthocyanidins [92].

In spite of this redundancy of results, discrepancies among different data sets obtained from different laboratories on antimicrobial activity of these *O*-heterocycles against both prokaryotic and eukaryotic microorganisms have been sometimes observed. This fact is probably due to various causes.

First, the qualitative and quantitative variability of the amount of catechins and proanthocyanidins present in plant extracts used for different studies is probably the most significant. This might be due to the use of different procedures of extraction, quantification and structural elucidation. In most cases, even the lack of rigorous phytochemical characterization and quantification of active compound(s) constitutes a severe limitation on the reliability of the results. The lack of commercially available pure standards (particularly for some proanthocyanidins) represents an additional problem that has so far hampered the execution of rigorous SAR studies. This limitation means that although a number of in vitro or in vivo studies have been carried out by using more or less pure standards of catechins or with plant extracts containing both catechins and proanthocyanidins, only a handful of authors have so far used pure proanthocyanidins or catechin-free extracts characterized by a well-defined proanthocyanidins composition. The scarcity (or even lack) of rigorous analytical determination of these molecules in plant polyphenolic fractions is probably the consequence of the limited number of appropriate analytical methodologies currently in use in laboratories worldwide. Current analytical procedures aimed at evaluating the proanthocyanidin content of food or medicinal plant preparations are in large part based on the estimation of total polyphenols (e.g., the measurements of light absorbance at 280 nm, or the use of colorimetric reactions involving the use of more or less specific reagents). Although each of these methods can be successfully used to compare samples with similar compositions, they failed for assays of extracts characterized by different polyphenol profiles, because of different responses of the various polyphenol structures or for interference by a few non-phenolic compounds [22]. The state of the art has revealed that in recent years, limited progress in analytical techniques has considerably improved the quantification of proanthocyanidin-containing products. Anderson et al. [109] used chromatographic fractionation of proanthocyanidin oligomers from aqueous ethanolic extracts for the separation of the components of a mixture of EGC and GC based A- and B-type oligomers. On the other hand, Romani et al. [94] observed that reverse phase chromatographic profiles gave only complicated pictures of extracts characterized by a mixture of proanthocyanidin, because of the extreme difficulty to obtain a good separation of hydrolyzable tannins from oligomers of proanthocyanidins (trimers, tetramers, and higher polymerized oligomers). Furthermore, in more recent studies, Schotz and Noldner [53] optimized a mass spectroscopy technique for the detection of monosubstituted proanthocyanidin oligomers in extracts from *Pelargonium sidoides* roots. The MS spectrum revealed the presence of structures characterized by up to 16 repeating monomers.

Another cause which contributes to giving a fragmentary and sometimes confusing picture of the antimicrobial activity of these *O*-heterocycles are the well-documented strain-related differences in susceptibility of microbial genotypes towards different molecules. In this context, the current use of only a handful of strains (or even of a single strain representative for each target species) represents a severe limitation for most reviewed studies and should be avoided. Accordingly, larger sets of target strains should be used by including clinical or environmental isolates and microbial germplasm deposited in worldwide culture collections.

It is clear that all these limitations have thus far hampered large-scale pharmaceutical exploitation of these compounds, limiting their scaling-up from the laboratory to the industrial scale. Consequently, although an increasing number of laboratories worldwide are focusing efforts on discovering novel structures exhibiting in vitro inhibitory effects towards all types of microorganisms, the number of these compounds which are presently undergoing in vivo studies to determine their effectiveness and safety are quite low [2].

As stated above, the difficulty in obtaining pure standards might be circumvented by the use of organic synthetic strategies, which could also be a valuable tool for supplying homogeneous compounds for biological (e.g., antimicrobial) tests [34].

In this context, a multidisciplinary approach (phytochemical + synthetic + microbiological + pharmacological) aimed at evaluating the effectiveness of different *O*-heterocyclic structures should be emphasized and encouraged, also in view of the evidence reporting the existence of synergistic effects among these compounds with currently used antibiotic compounds.

References

1. Clark AM (1996) Pharm Res 13:1133
2. Cowan MM (1999) Clin Microbiol Rev 12:564
3. Rex JH, Walsh TJ, Sobel DJ (2000) Clin Infect Dis 30:662
4. Pfaller MA, Diekema DJ, Jones RN, Sader HS, Fluit AC, Hollis RJ, Messer SA (2001) J Clin Microbiol 39:3254
5. Masiá Canuto M, Gutiérrez Rodero F (2002) Lancet Infect Dis 2:550
6. Sanglard D, Odds FC (2002) Lancet Infect Dis 2:73
7. Cassidy A, Hanley B, Lamuela-Raventos RM (2000) J Sci Food Agr 80:1044
8. Etkin NL (1996) Int J Pharmacognosy 34:313
9. Johns T (1999) Plant constituents and the nutrition and health of indigenous peoples. In: Nazarea VD (ed) Ethnoecology. Situated Knowledge, Located Lives. The University of Arizona Press, Tucson, p 157
10. Pieroni A (2000) J Ethnopharmacol 70:235
11. Santos-Buelga C, Scalbert A (2000) J Food Sci Agric 80:1094
12. Cai Y, Evans FJ, Roberts MF, Phillipson JD, Zenk MH, Gleba YY (1991) Phytochemistry 30:2033
13. Tits M, Angenot L, Poukens P, Warin RD (1992) Phytochemistry 31:971
14. Lin LC, Kuo YC, Chou CJ (2002) J Nat Prod 65:505
15. Ming DS, Lopez A, Hillhouse BJ, French CJ, Hudson JB, Towers GH (2002) J Nat Prod 65:1412
16. Svedstrom U, Vuorela H, Kostiainen R, Tuominen J, Kokkonen J, Rauha JP, Laakso I, Hiltunen R (2002) Phytochemistry 60:821
17. Porter LJ (1989) Tannins. In: Harborne JB (ed) Methods in Plant Biochemistry. Plant Phenolic. Academic Press, San Diego, p 389
18. Porter LJ (1993) Flavans and proanthocyanidins. In: Harborne JB (ed) The Flavonoids: Advances in Research Since 1986. Chapman & Hall, London, p 23
19. Ferreira D, Nel RJJ, Bekker R (1999) Condensed tannins. In: Barton DHR, Nakanishi K, Meth-Cohn O, Pinto BM (eds) Comprehensive Natural Products Chemistry. Elsevier, Oxford, p 791
20. Porter LJ (1988) Flavans and Proanthocyanidins. In: Harborne JB (ed) The Flavanoids Advances in Research since 1980. Chapman & Hall, New York, p 21
21. Ferreira D, Slade D (2002) Nat Prod Rep 19:517
22. Cheynier V (2005) Am J Clin Nutr 81:223S
23. Shirley BW (1996) Trends Plant Sci 1:377
24. Winkel-Shirley B (2001) Plant Physiol 126:485

25. Winkel-Shirley B (2001) Plant Physiol 127:1399
26. Saito K, Yamazaki M (2002) New Phytol 155:9
27. Xie DY, Dixon RA (2005) Phytochemistry 66:2127
28. Bogs J, Jaffe FW, Takos AM, Walker AR, Robinson SP (2007) Plant Physiol 143:1347
29. Stafford HA, Lester HH (1984) Plant Physiol 76:184
30. Stafford HA, Lester HH (1985) Plant Physiol 78:791
31. Tanner GJ, Francki KT, Abrahams S, Watson JM, Larkin PJ, Ashton AR (2003) J Biol Chem 278:31647
32. Xie D, Sharma SR, Paiva NL, Ferreira D, Dixon RA (2003) Science 299:396
33. Abrahams S, Lee E, Walker AR, Tanner GJ, Larkin P, Ashton AR (2003) Plant J 35:624
34. Ohmori K, Ushimaru N, Suzuki K (2004) Proc Nat Acad Sci 101:12002
35. Botha JJ, Ferreira D, Roux DG (1981) J Chem Soc Perkin Trans 1:1235
36. Kawamoto H, Nakatsubo F, Murakami K (1991) Mokuzai Gakkaishi 37:488
37. Gross GG, Hemingway RW, Yoshida T (2000) Plant Polyphenols 2. Chemistry, Biology, Pharmacology, Ecology. Kluwer Academic Publishers, New York
38. Boyer FD, Es-Safi NE, Beauhaire J, Le Guernevé C, Ducrota PE (2005) Bioorg Med Chem Lett 15:563
39. Jin G, Yoshioka H (2005) Biosci Biotechnol Biochem 69:440
40. Beauhaire J, Es-Safi NE, Boyer FD, Kerhoas L, le Guernevé C, Ducrota PE (2005) Bioorg Med Chem Lett 15:559
41. Kozikowski AP, Tückmantel W, Böttcher G, Romanczyk LJ Jr (2003) J Org Chem 68:1641
42. Saito A, Nakajima N, Tanaka A, Ubukata M (2002) Tetrahedron 58:7829
43. Saito A, Doi Y, Tanaka A, Matsuura N, Ubukatae M, Nakajima N (2004) Bioorg Med Chem 12:4783
44. Kozikowski AP, Tückmantel W, Hu Y (2001) J Org Chem 66:287
45. Kozikowski AP, Tuckmantel W, Romanczyk LJ Jr (2006) US Patent 7 067 679
46. Kim YJ, Chung JE, Kurisawa M, Uyama H, Kobayashi S (2004) Biomacromolecules 5:11
47. Chung JE, Kurisawa M, Kim YJ, Uyama H, Kobayashi S (2004) Biomacromolecules 5:113
48. Bagchi D, Garg A, Krohn RL, Bagchi M, Tran MX, Stohs SJ (1997) Res Commun Mol Pathol Pharmacol 95:179
49. Waterhouse AL, Walzem RL (1997) Nutrition of grape phenolics. In: Rice-Evans CA, Packer L (eds) Flavonoids in Health and Disease. Marcel Dekker, New York, p 359
50. Joshi SS, Benner EJ, Balmoori J, Bagchi D (1998) FASEB J Abstr 12:4484
51. Hagerman AE, Riedl KM, Jones GA, Sovik KN, Ritchard NT, Hartzfeld PW, Riechel TL (1998) J Agric Food Chem 46:1887
52. Ursini F, Tubaro F, Rong J, Sevanian A (1999) Nutr Rev 57:241
53. Schotz K, Noldner M (2007) Phytomedicine 14:32
54. Sharma SD, Meeran SM, Katiyar SK (2007) Mol Cancer Ther 6:995
55. Shoji T, Masumoto S, Moriichi N, Akiyama H, Kanda T, Ohtake Y, Goda Y (2006) J Agric Food Chem 54:884
56. Koga T, Moro K, Nakamori K, Yamakoshi J, Hosoyama H, Kataoka M, Ariga T (1999) J Agric Food Chem 47:1892
57. Wink M (2005) Z Phytother 26:262
58. Ku CS, Sathishkumar M, Mun SP (2007) Chemosphere 67:1618
59. Passwater RA (2005) Pycnogenol nature's most versatile supplement. Jack Challem Series Editor, Basic Health Publications Inc, North Bergen, NJ, USA
60. Wiseman SA, Balentine DA, Frei B (1997) CRC Crit Rev Food Sci Nutr 37:705

61. Chou CC, Lina LL, Chung KT (1999) Int J Food Microbiol 48:125
62. Esquenazi D, Wigg MD, Miranda MMSF, Rodrigues HM, Tostes JBF, Rozental S, da Silva AJR, Alviano CS (2002) Res Microbiol 153:647
63. Otake S, Makimura M, Kuroki T, Nishihara Y, Hirasawa M (1991) Caries Res 25:438
64. Yoda Y, Hu ZQ, Zhao WH, Shimamura T (2004) J Infect Chemother 10:55
65. Mabe K, Yamada M, Oguni I, Takahashi T (1999) Antimicrob Agents Chemother 43:1788
66. Hirasawa M, Takada K (2004) Multiple J Antimicrob Chemother 53:225
67. Jodoin J, Demeule M, Beliveau R (2002) Biochim Biophys Acta 1542:149
68. Blanco AR, La Terra Mule S, Babini G, Garbisa S, Enea V, Rusciano D (2003) Biochim Biophys Acta 1620:273
69. Higdon JV, Frei B (2003) CRC Crit Rev Food Sci Nutr 43:89
70. Park BJ, Park JC, Taguchi H, Fukushima K, Hyon SH, Takatori K (2006) Biochem Biophys Res Comm 347:401
71. Taguri T, Tanaka T, Kouno I (2006) Biol Pharm Bull 29:2226
72. Ikigai H, Nakae T, Hara Y, Shimamura T (1993) Biochim Biophys Acta 1147:132
73. Sudano Roccaro A, Blanco AR, Giuliano F, Rusciano D, Enea V (2004) Antimicrob Agents Chemother 48:1968
74. Zhao WH, Hu ZQ, Hara Y, Shimamura T (2002) Antimicrob Agents Chemother 46:2266
75. Zhao WH, Hu ZQ, Okubo S, Hara Y, Shimamura T (2001) Antimicrob Agents Chemother 45:1737
76. Chung KT, Wei CI, Johnson MG (1998) Trends Food Sci Technol 9:168
77. Hori Y, Sato SS, Hatai A (2006) Phytother Res 20:162
78. Amarowicz R, Pegg BR, Bautista AD (2000) Nahrung 44:60
79. Puupponen-Pimiä R, Nohynek L, Meier C, Kähkönen C, Heinonen M, Hopia A, Oksman-Caldentey KM (2001) J Appl Microbiol 90:494
80. Howell AB, Vorsa N, Marderosian AD, Foo LY (1998) New Eng J Med 339:1085
81. Taylor RSL, Edel F, Manandhar NP, Towers GHN (1999) J Ethnopharmacol 50:97
82. Nelson KE, Pell N, Doane H, Giner-Chavez BI, Schofield P (1997) J Chem Ecol 23:1175
83. Jones GA, McAllister TA, Muir AD, Cheng KJ (1994) Appl Environ Microbiol 60: 1374
84. Howell AB, Reed JD, Krueger CG, Winterbottom R, Cunningham DG, Leahy M (2005) Phytochemistry 66:2281
85. Foo LY, Lu Y, Howell AB, Vorsa N (2000) Phytochemistry 54:173
86. Scalbert A (1991) Phytochemistry 30:3875
87. House WA (1999) Field Crops Res 60:115
88. Dixon RA, Xie DY, Sharma SB (2005) New Phytol 165:9
89. Nakane H, Ono K (1990) Biochemistry 29:2841
90. Hara Y, Nakayama M (2001) US Patent 6 248 346
91. Kobayashi T, Miyazaki Y, Yanagihara K, Kakeya H, Ohno H, Higashiyama Y, Hirakata Y, Mizuta Y, Tomono K, Tashiro T, Kohno S (2005) Intern Med 44:1191
92. de Bruyne T, Pieters L, Witvrouw M, de Clercq E, Vanden Berghe D, Vlietinck AJ (1999) J Nat Prod 62:954
93. Toyoshima Y, Okubo S, Toda M (1993) Kansenshogaku Zasshi 68:295
94. Romani A, Ieri F, Turchetti B, Mulinacci N, Vincieri FF, Buzzini P (2006) J Pharm Biomed Anal 41:415
95. González de Colmenares N, Ramírez-Martínez JR, Aldana JO, Ramos-Niño ME, Clifford MN, Pékerar S, Méndez B (1998) J Sci Food Agric 77:368

96. Paveto C, Güida MC, Esteva MI, Martino V, Coussio J, Flawiá MM, Torres HN (2004) Antimicrob Agents Chemother 48:69
97. Calzada F, Cervantes-Martínez JA, Yépez-Mulia L (2005) J Ethnopharmacol 98:191
98. Alanís AD, Calzada F, Cedillo-Rivera R, Meckes M (2003) Phytother Res 17:681
99. Kubata BK, Nagamune K, Murakami N, Merkel P, Kabututu Z, Martin SK, Kalulu TM, Haq Mustakuk H, Yoshida M, Ohnishi-Kameyama M, Kinoshita T, Duszenko M, Urade Y (2005) Int J Parasitol 35:91
100. Kolodziej H, Kiderlen AF (2005) Phytochemistry 66:2056
101. Bais HP, Walker TS, Stermitz FR (2002) Plant Physiol 128:1173
102. Veluri R, Weir TL, Bais HP, Stermitz FR, Vivanco JM (2004) J Agric Food Chem 52:1077
103. Palma M, Taylor LT, Varela RM, Cutler SJ, Cutler HG (1999) J Agric Food Chem 47:5044
104. Yee YK, Koo MWL (2000) Alim Pharmacol Ther 14:635
105. Yanagawa Y, Yamamoto Y, Hara Y, Shimamura T (2003) Curr Microbiol 47:244
106. Conegero-Sanches AC, Lopes GC, Nakamura CV, Dias Filho BP, Palazzo de Mello JC (2005) Braz J Pharm Sci 41:101
107. Rqauha JP, Remesa S, Heinonen M, Hopia A, Kähkönen M, Kujala T, Pihlaja K, Vuorela H, Vuorela P (2000) Int J Food Microbiol 56:3
108. Meckes M, Calzada F, Tapia-Contreras A, Cedillo-Rivera R (1999) Phytother Res 13:102
109. Anderson RA, Broadhurst CL, Polansky MM, Schmidt WF, Khan A, Flanagan VP, Schoene NW, Graves DJ (2004) J Agric Food Chem 52:65

Top Heterocycl Chem (2007) 10: 265–308
DOI 10.1007/7081_2007_064

Published online: 30 June 2007

Benzofuroxan and Furoxan. Chemistry and Biology

Hugo Cerecetto (✉) · Mercedes González

Laboratorio de Química Orgánica, Facultad de Ciencias/Facultad de Química,
Universidad de la República, Iguá 4225, Montevideo, 11400, Uruguay
hcerecet@fq.edu.uy

Abstract Since the middle of the 20th century, benzofuroxan and furoxan derivatives have been extensively studied as bioactive compounds. Maybe the most relevant reported biological property has been related to its capability to release nitric oxide; moreover, other significant activities have been identified, such as antibacterial, antifungal, antiparasite, cytotoxic, and herbicide properties. Recently, research and development in the medicinal chemistry field of these systems have produced hybrid compounds in which benzofuroxanyl or furoxanyl moieties together with classical drug moieties are present in a single molecule. Consequently, new anti-ulcer drugs, calcium channel modulators, vasodilator derivatives, antioxidants, among others, have been described and are currently under study. On the other hand, these kinds of compounds have also been reported as components in primary explosives, polymers, and propellants. This chapter includes older as well as more recent methods of synthesis of benzofuroxan and furoxan derivatives, chemical and biological reactivity, biological properties and mode of action, structure-activity studies and other relevant chemical and biological properties.

Keywords Benzofuroxan · Furoxan · Hybrid compounds · Nitric oxide-releasing agents · Ring-chain tautomerism

Abbreviations

ADP	Adenosine diphosphate
Bfx	Benzofuroxan
Bfz	Benzofurazan
CCM	Calcium channel modulator
cGMP	cyclic 3′,5′-guanosine monophosphate
CNS	Central nervous system
COL	Collagen
ESR	Electron spin resonance
Fx	Furoxan
Fz	Furazan
GTN	Glyceryl trinitrate
NO	Nitric oxide
NOS	NO synthase
PAF	Platelet activating factor
QSAR	Quantitative structure-activity relationship
SGC	Soluble guanylate cyclase
SNP	Sodium nitroprusside

1 Introduction

Although benzofuroxan (Bfx) and furoxan (Fx) were first synthesized over 100 years ago, it was not until the middle of the 20th century that its relevant bioactivities were identified. Bfx's and Fx's relevant biological behaviors convert these systems in one of the most current study heterocycles in medicinal chemistry. Formally, these compounds are 1,2,5-oxadiazole *N*-oxide derivatives, specifically Bfx should be named as benzo[1,2-*c*][1,2,5]oxadiazole 1-oxide and Fx as 1,2,5-oxadiazole 2-oxide, however, the common trivial name of the base heterocycle, benzofuroxan (Bfx) and furoxan (Fx), respectively, is still in common use. The position of the *N*-oxide moiety is indicated as 1-, *N*- or N^1-oxide in the case of Bfx or as 2-, *N*- or N^2-oxide for Fx. According to a convention [1], due to the rapid ring-chain tautomerism that Bfxs suffer at room temperature (see below), when the tautomeric pair of Bfxs is being considered it is referred to by assigning the lowest numerical value to the substituent (Scheme 1). The heterocycles are systematically numbered according to Scheme 1.

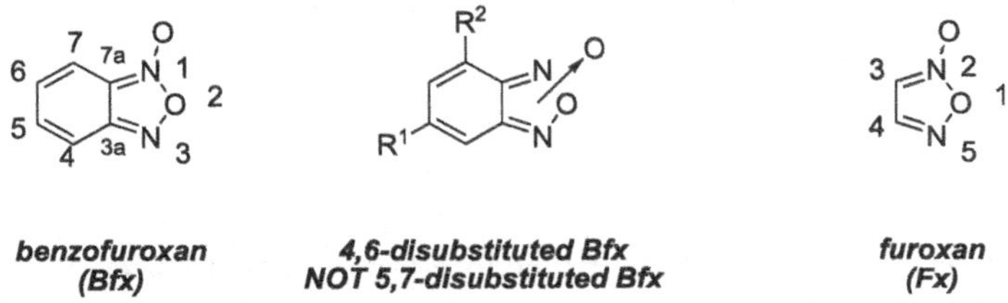

Scheme 1 Systematic numeration of Bfx and Fx

The *N*-oxide functional group is the result of the addition of an atomic oxygen to the lone pair electrons of the nitrogen atom. Formally this group is neutral, however nitrogen and oxygen possess positive and negative formal charges, respectively. Consequently, the correct representation should be $N^+ - O^-$, however in the formulas of this chapter, the *N*-oxide moiety will be represented as $N \rightarrow O$ [2].

2 Chemical Syntheses

Contrary to other *N*-oxide containing heterocycles, Bfx and Fx are not formed by direct oxidation of the parents systems, benzofurazan (Bfz) and furazan (Fz), respectively. Even though each kind of heterocycle possesses individual synthetic procedures, some common approaches should be mentioned, i.e., thermo or photochemical intramolecular cyclization of 1,2-azidonitro derivatives and oxidative cyclization of 1,2-dioximes. Some other synthetic procedures, such as oxidation of *o*-nitroanilines to Bfx or dimerization of nitrile oxides and dehydration of *α*-nitrooximes to Fx, have been depicted. The following sections provide the most recent descriptions.

2.1 Common Approaches

One of the best-known common approaches to generate the 1,2,5-oxadiazole *N*-oxide system involves cyclization of a nitro group unto a nitrene, arising by fragmentation of azides either thermo- or photochemically (Fig. 1) [3–8]. Singlet aryl nitrenes have been proposed as intermediates in these processes, but its too-short life have not allowed identifying either direct or indirectly. Nevertheless, it has been demonstrated theoretically [9, 10] and experimentally [11–13] that pyrolysis occurs through a concerted one-step mechanism and photolysis via a stepwise one.

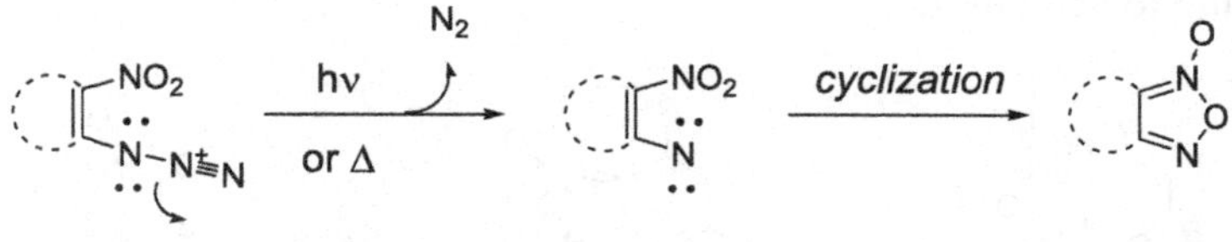

Fig. 1 Cyclization process of Bfx and Fx from 1,2-nitroazides

The other common synthetic procedure for Bfx and Fx preparation is the oxidative cyclization of 1,2-dioximes. 1,2-Dioximes are excellent starting materials for the syntheses of the 1,2,5-oxadiazole *N*-oxide system in presence of oxidizing conditions to promote the cyclization. Its utility is restricted for Bfxs syntheses because the restriction of *o*-quinone dioximes availability, contrarily α-glyoximes, which are useful to prepare Fx, are more easily to prepare. In Table 1, products, conditions, and comments for the most recent Fx synthesis using 1,2-dioximes are shown.

Table 1 General procedure to preparation of Bfx and Fx from α-glyoximes

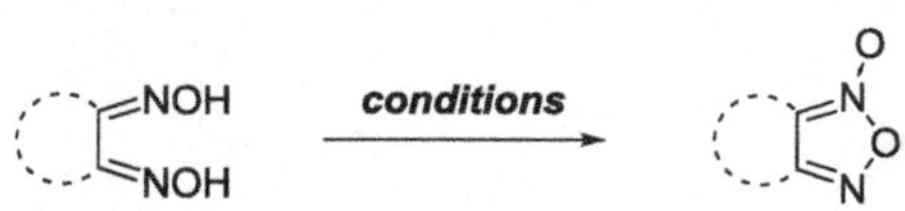

Product	Conditions	Comments	Refs.
Fx	Cl_2 or Br_2/AcONa/20–60 °C	Patent of invention	[14]
Fx	*t*-BuOCl/MeOH	–	[15]
Fx	KOH/EtOH – H_2O/5% NaClO	–	[16]

2.2
Other Synthetic Approaches for Benzofuroxans

Maybe the best well-established method for the preparation of Bfxs is the oxidative ring closure of *o*-nitroaniline derivatives. Alkaline hypochlorite, NaOCl/KOH, has been the most commonly used reagent but electrochemical synthesis [17] and bis(acetate-O-)phenyliodine as oxidizer have also been employed [18]. The procedure is particularly useful where the other methods do not carry out; however, it is restricted by the instability of some Bfx under the alkaline conditions. To avoid this problem, some experimental modifications have been studied, i.e., the use of solid-supported reaction [19] or soft alcoholic solvents [20]. The mechanism of the hypochlorite reaction has also been described as a singlet nitrene process via an initial *N*-chloride intermediate which after deprotonation losses chloride. In addition, the *o*-nitroaniline reactants could also be used to produce *o*-azidonitro derivatives starting material for the previously described synthetic procedure.

Klyuchnikov et al. have described an alternative substrate for the cyclization process, namely *o*-hydroxylaminonitro derivatives. These entities, previously synthesized or generated in situ, cycle in presence of a base, i.e., sodium bicarbonate and sodium acetate, producing the 1,2,5-oxadiazole *N*-oxide system (Fig. 2) [21–23].

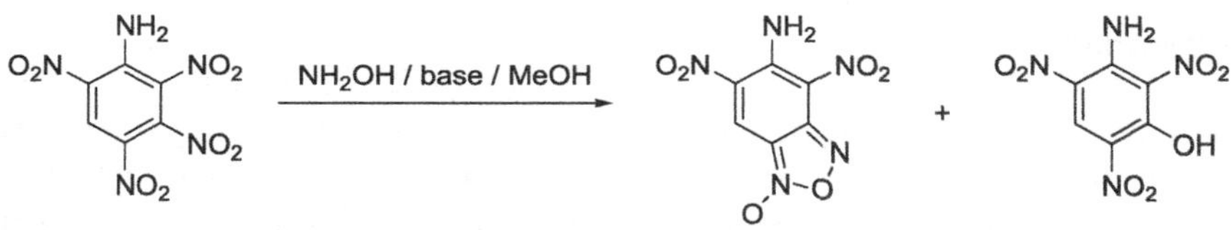

Fig. 2 Synthesis of Bfx using *o*-hydroxylaminonitro derivative

2.3 Other Synthetic Approaches for Furoxans

The intermolecular dimerization of nitrile oxides has been described as a procedure to prepare Fx with identical substituent both in the 3 and 4 position (Fig. 3). This procedure is a [3 + 2] cycloaddition where one molecule of nitrile oxide acts as 1,3-dipole and the other as dipolarophile [24–26]. Yu et al. has studied this procedure in terms of theoretical calculus [27,28]. Rearrangement of isocyanates competes with the bimolecular dimerization, with the former becoming dominant at elevated temperatures.

The earliest reported Fxs were the result of the reaction of nitrous acid with naturally occurring alkenes being the identified intermediate a α-nitrooxime that suffers dehydration with cyclization. Apart from these conditions, the most recent Fxs synthesis descriptions have involved reactions between alkenes and dinitrogen trioxide (Fig. 3), nitroalkanes and aluminum trichlo-

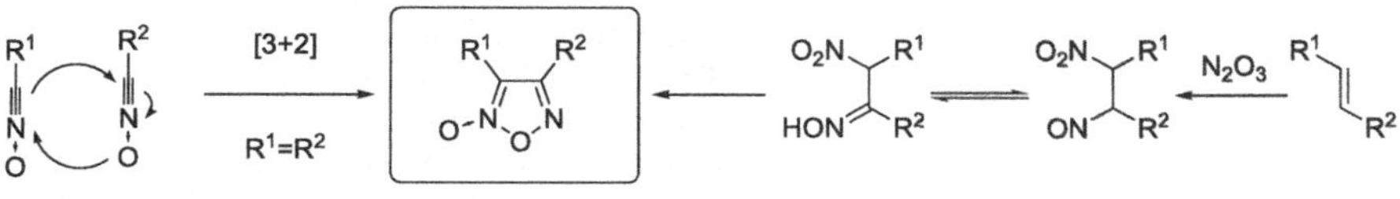

Fig. 3 Fx synthetic approaches. *Left*: nitrile oxides dimerization. *Right*: α-nitrooxime dehydration/cyclization

ride [29], alkenes and nitrogen monoxide in the presence of metallic complexes [30], alkenes with $AgNO_2$/TMSCl [31], or directly the α-nitrooxime in the presence of acidic alumina as dehydration/cyclization promoter [32].

3 Reactivity

The chemical aspects of Bfx and Fx systems have been reviewed in detail in the last years [33]. The chemistry of these heterocycles has been increasing in the last years due to their usefulness as precursors for the synthesis of other heterocycles. The present section outlines the well-known reactions described particularly in the last decade.

3.1 Deoxygenation

The reduction of the *N*-oxide moiety to produce the heterocycle parent compounds Bfz and Fz has been one of the most studied reactions for these systems. In general, the procedures describe mild conditions and regioselectivity. In Table 2, substrates, conditions, products, and comments for the most recent reported deoxygenations are shown.

Table 2 Reduction conditions used to obtain the parents compounds Bfz and Fz from Bfx and Fx and other reactions

Substrates	Conditions	Products	Comments	Refs.
Bfx	$P(OEt)_3$/toluene/55 °C	Bfz	Multikilogram scale process	[34]
Bfx	S/ethylene glycol/140–155 °C	Bfz	Patent of invention	[35]
Bfx	PPh_3	Bfz	Described for the synthesis of 1,2,5-oxadiazolo[3,4-*c*]pyridine system	[3]
Bfx	PPh_3/EtOH/85 °C	Bfz	Patent of invention	[36]
Fx	$P(OMe)_3$	Fz	–	[37]
Fx	Zn dust/NH_4Cl (30%)/THF/reflux	Fz	When 3-chloromethyl-Fx derivative was submitted to these reduction conditions a *bis*furazan was generated as a reductive dimerization Zn-promoted process	[38]
Fx	$Na_2S_2O_4$	Fx	A nitro moiety was selectively reduced in presence of the Fx heterocycle	[39, 40]

3.2 Heterocycle Transformations

The generation of other heterocycles from Bfx and Fx has been the subject of exhaustive investigation. The most important transformation of Bfx to other heterocycles has been described by Haddadin and Issidorides, and is known as the "Beirut reaction". This reaction involves a condensation between adequate substituted Bfx and alkene-type substructure synthons, particularly enamine and enolate nucleophiles. The Beirut reaction has been employed to prepare quinoxaline 1,4-dioxides [41], phenazine 5,10-dioxides (see Chap. "Quinoxaline 1,4-dioxide and Phenazine 5,10-dioxide. Chemistry and Biology"), 1-hydroxybenzimidazole 3-oxides or benzimidazole 1,3-dioxides, when nitroalkanes have been used as enolate-producer reagent [42], and benzo[*e*] [1, 2, 4]triazine 1,4-dioxides when Bfx reacts with sodium cyanamide [43–46] (Fig. 4).

In the case of Bfx bearing unsaturated groups at the 4-position, an important ring-opening ring-reclosure process leads to new heterocycles. This reaction, known as the Boulton-Katritzky rearrangement, has been studied for different 4-unsaturated moieties to produce new benzo-heterocycles, i.e.,

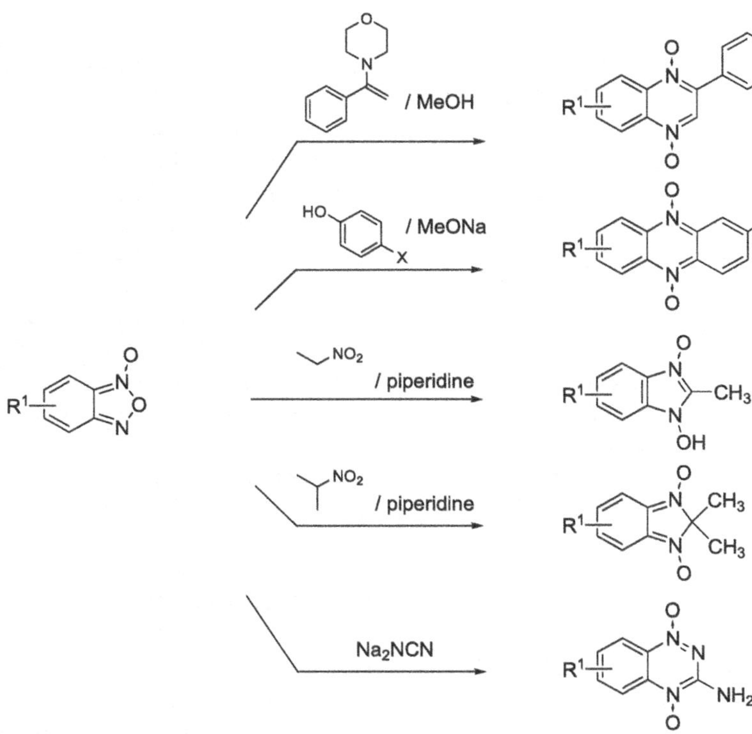

Fig. 4 Transformation of Bfx to different heterocycles using the Beirut reaction

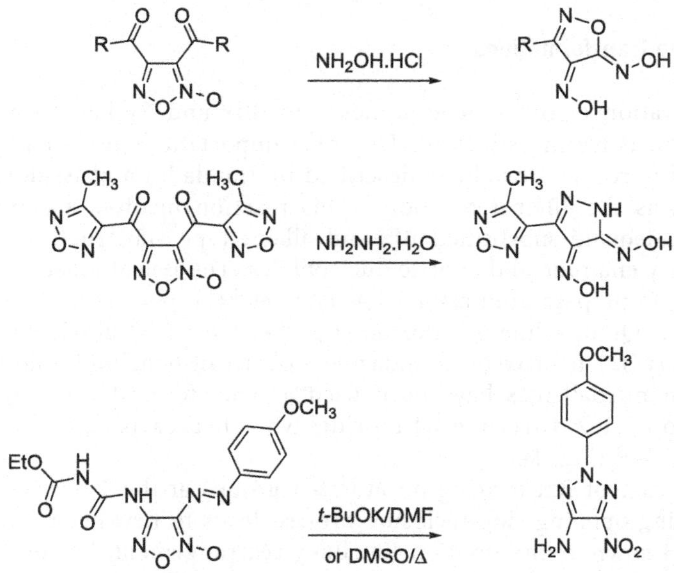

Fig. 5 Transformation of Fx into other heterocycles

C = O yields benzoisoxazoles, C = N yields indazoles, N = O yields benzofurazans, N = N yields benzotriazoles (see below, Fig. 6) [47].

Fx derivatives can be transformed into a variety of other heterocycles, including isoxazoles, pyrazoles, 1,2,3-triazoles, and furazans. An up-to-date assessment is provided in a recent review [48] and here follows only a summary of the most recent transformations. Two Russian teams have recently described relevant information in the Fx heterocyclic-transformations. On the one hand, Friedrichsen et al. have depicted the reaction of 3,4-diacylfuroxan derivatives with nitrogen nucleophiles producing isoxazol and pyrazol derivatives (Fig. 5) [49, 50]. On the other hand, Makhova et al. have studied successive-cascade-heterocyclic rearrangements where the final products are 4-amino-2-aryl-5-nitro-2*H*-1,2,3-triazol derivatives (Fig. 5) [51, 52]. The authors assume that this process includes two consecutive rearrangements forming, in a first step, a 1,2,4-oxadiazole ring, which is then transformed into a 1,2,3-triazole ring with the participation of the azo group.

3.3
Benzofuroxans and Furoxans as Nucleophilic and Electrophilic Entities

These heterocycles, mainly Fx, are especially deactivated toward electrophilic substitution (S_EAr) under standard conditions being the reaction with acid, H^+ as electrophile, also slow. However, some specific Bfxs have been depicted as nucleophile toward good electrophiles like $^+NO_2$ and $-N^+ \equiv N$ (Fig. 6) [47, 53].

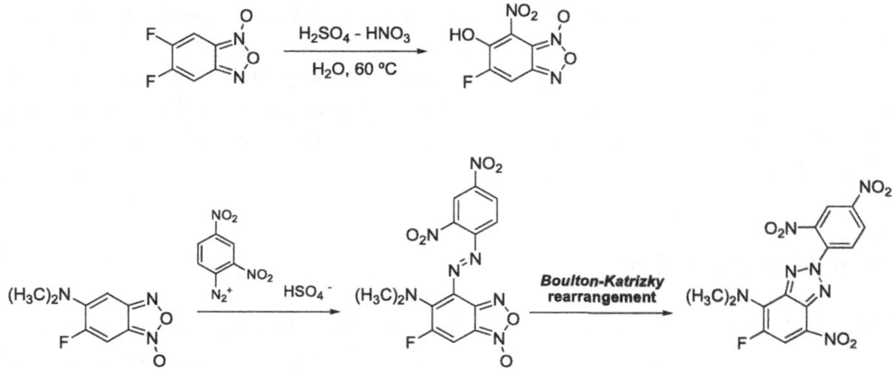

Fig. 6 Examples of Bfx as nucleophile

Nucleophilic reactions take place in the homocyclic ring, S_NAr or AE_a, when it is activated by electron-withdrawing substituents. It has been described that halides can be displaced by a great number of nucleophiles via a normal and "cine" substitution [54, 55]. Nitro containing Bfx has represented a class of neutral 10-π-electron-deficient system which exhibit an extremely high electrophilic character in many covalent nucleophilic addition and substitution processes. 4,6-Dinitrobenzofuroxan and others 4-nitro-6-substitutedbenzofuroxans (Scheme 2) have been defined as super-electrophiles and used as convenient probes to assess to the C-basicity of

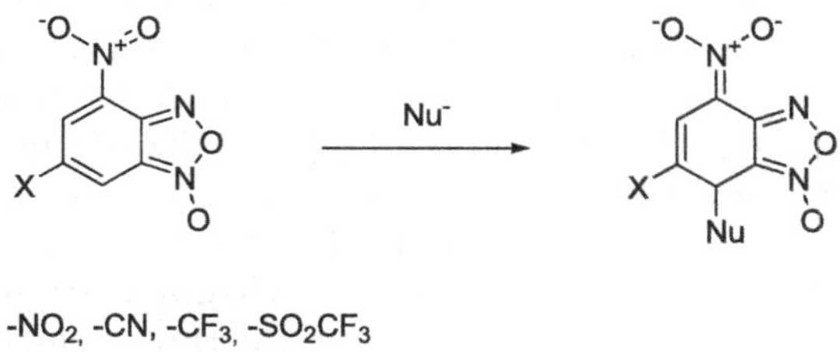

Scheme 2 Nucleophilic attack to 4-nitro-6-substituted Bfx

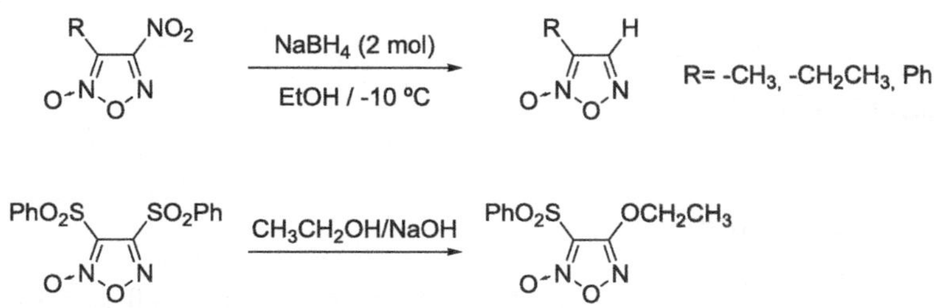

Fig. 7 Examples of nucleophilic displacements on Fx

carbon nucleophiles resulting in the formation of stable anionic σ-adducts, Meisenheimer complexes (Scheme 2) [56–59].

The Fx ring is more susceptible to nucleophilic attack than it is to react with electrophiles. Nitro, halides, and sulfonyl groups, in position 3- and 4-, have been reported as moieties that suffer nucleophilic displacement [60, 61]. Figure 7 depicts some recent examples.

3.4 Other Relevant Reactions

Other relevant reactions have been described for Bfx and Fx as reactants, among them cycloaddition processes, photochemical transformations, and complexation with metals.

Terrier et al. have widely described Bfxs 4-nitro-substituted participating in a series of Diels–Alder processes. Bfx 4-nitro-substituted system could participate as diene, $C=C-C=C$ system, or heterodiene, $C=C-NO_2$ system, in an inverse electron demand reaction [62, 63] or as dienophile, $C=C$ or $N=O$ systems, in a normal electron demand reaction [64–68]. Fxs have been reported as 1,3-dipole through the substructure $C=N^+-O^-$ reacting with a widely series of dipolarophiles yielding isoxazol derivatives [69, 70].

Hasegawa et al. have studied the photoreaction of benzofuroxan obtaining, like in other *N*-oxide photochemistry, lactames [71–73].

Preparation of metal complexes with Bfxs as ligands has been described by Fernández and Muro. These researchers have used in the coordination reactions a wide variety of metals, including Co(II), Fe(II), Al(III), V(III), V(IV), Zr(IV), Ti(IV), Mo(IV) and Cr(VI) [74, 75]. Interesting results have been obtained when 3,4-*bis*(2-pyridinyl)furoxan was coordinated with different transition metals [76]. In fact, coordination with Cu(II), Pd(II), and Ag(I) produced complexes with formula $[CuLCl_2]_2$, $[PdLCl_2]$, and $AgL(NO_3)$, being L the studied Fx, where the pyridine nitrogen acts as the coordinated atom. In contrast, reaction with the Ru(II) metallic precursors, used in this synthesis, leads to an Fx ring-opening ring-closing rearrangement producing a Ru(II) complex with 3-nitro-2-(2-pyridinyl)pyrazolo[1,5-*a*]pyridine as bidentate ligand.

4 Structural Studies

4.1 Ring-Chain Tautomerism

The substituted Bfx and Fx exhibit the isomerization or, well known, ring-chain tautomerism equilibrium depicted in Scheme 3. This phenomenon

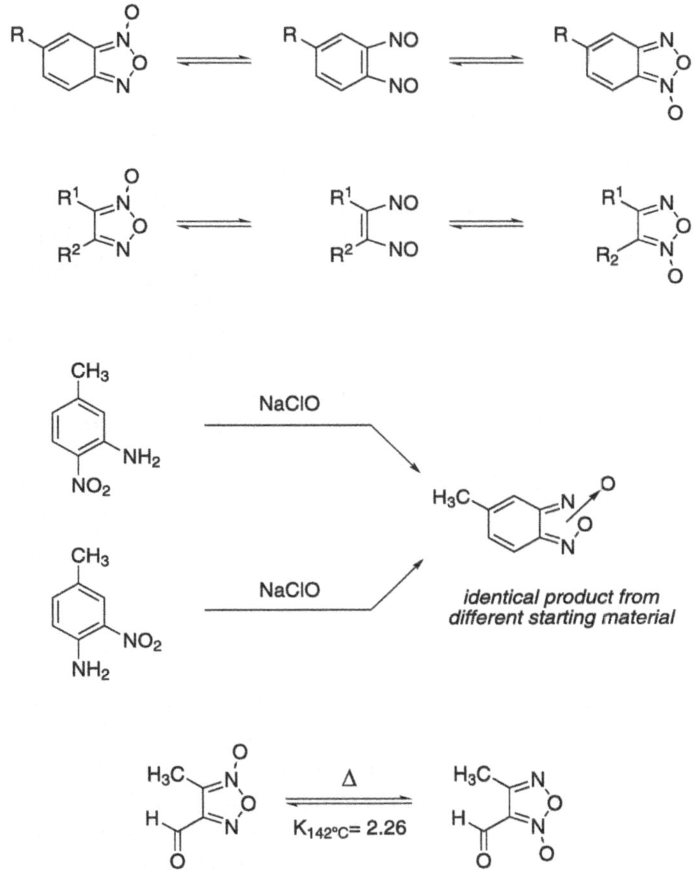

Scheme 3 Bfxs and Fxs ring-chain tautomerism and its implications

was evidenced when different aromatic precursors yielded identical Bfx (Scheme 3). This tautomerism, involving the corresponding 1,2-dinitroso as an intermediate, is energetically different for Bfx and Fx derivatives and depends on both the nature and position of substituents (in the case of Bfx), on the solvent and the temperature. While for Bfx a fast equilibrating system at room temperature has been described, for Fx, it has been demonstrated that the interconversion requires high energies (Scheme 3) [70]. This phenomenon has been intensively studied not only experimentally but also theoretically. The following sections will show some studies performed in order to structurally characterize the 1,2,5-oxadiazole *N*-oxide system.

4.1.1 Solid-State Studies

Structural studies in solid state have been performed for Bfx and Fx derivatives in terms of X-ray diffraction experiments [60, 77, 78]. Table 3 shows relevant atomic distance and angle values for some reported derivatives and parent compounds. According to these studies, the N5-O1 bond lengths of Bfxs and Fxs, ranged between 1.35 and 1.41 Å, similar for all the reported compounds. They are very close to the corresponding O1 – N2 bond lengths, which ranged between 1.44 and 1.47 Å. In the case of the parent Fz, both atomic distances are identical, showing absence of potential ring-chain tautomerism, while in the *N*-oxide derivatives both lengths are quite different. However, in any case, the C – N and N – O atomic distances for Bfx, Fx, and Fz correspond to nitroso moiety bond lengths. Compare the interatomic distances of Bfx and Fx derivatives ($d_{C\text{–}N} \approx 1.32$ Å, $d_{N5\text{–}O1} \approx 1.39$ Å) with the calculated bond lengths for nitroso moiety ($d_{C\text{–}N} \approx 1.45$ Å, $d_{N=O} \approx 1.20$ Å [84]) confirming the absence in the solid state of the *o*-dinitroso tautomer.

4.1.2 Spectroscopic Studies

Different spectroscopic studies have been performed in order to understand these tautomerisms. In this sense, ^{1}H, ^{13}C, ^{14}N, ^{15}N and ^{17}O NMR spectroscopies have been investigated. At room temperature, broad ^{1}H and ^{13}C NMR signals for Bfx derivatives have been reported, while at low temperature the ^{1}H and ^{13}C NMR signals are well resolved [81]. While only one signal of the benzofuroxan ^{14}N NMR spectra was observed at room temperature (– 19.7 ppm, in acetone), it becomes well resolved at 233 K (^{15}N NMR signals: – 6.7 and – 20.2 ppm, in acetone) [85]. This shows the dynamic average of both nitrogen atoms in the molecule that confirms fast tautomeric exchange in solution on NMR time-scale. For unsymmetrical derivatives, i.e., Bfx methyl-, methoxy-, and nitro-substituted, four different ^{15}N signals have been observed at low temperature [86]. In any case, NMR spectroscopy has evidenced the *o*-dinitroso entities. Other approaches [87–89] have identified *o*-dinitroso intermediate, via FTIR and UV, when benzofuroxan suffers photolysis at $\lambda = 366$ nm and 12 K. This intermediate regenerates benzofuroxan thermally or photochemically. Similar results have been obtained for 3,4-dimethylfuroxan, photolysis at $\lambda = 254$ nm and 12 K, however the 1,2-dinitroso intermediate decomposes upon prolonged photolysis times to give photostable acetonitrile *N*-oxide.

Table 3 Characteristic bond lengths and bond angles for Bfx and Fx derivatives

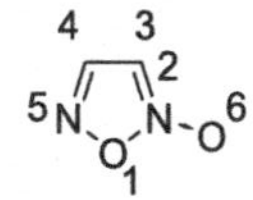

	Bond lengths (Å)						Bond angles (deg.)						
Derivative[a]	1-2	2-3	3-4	4-5	5-1	2-6	1-2-3	2-3-4	3-4-5	4-5-1	5-1-2	1-2-6	6-2-3
Bfx[b]	1.46	1.32	1.42	1.33	1.38	1.23	106.7	107.3	111.9	105.9	108.2	116.6	136.7
Bfx[c]	1.47	1.33	1.40	1.31	1.39	1.23	106.2	106.4	114.2	104.7	108.4	117.4	136.4
	1.44	1.31	1.42	1.32	1.41	1.22	105.7	109.3	110.7	105.4	108.9	118.3	136.0
Bfx[d]	1.44	1.33	1.41	1.33	1.38	1.23	106.6	107.1	111.7	105.6	109.0	118.0	135.4
Fx[e]	1.44	1.30	1.40	1.29	1.38	1.24	107.2	107.2	111.9	106.6	107.1	116.4	136.4
Fx[f]	1.46	1.32	1.42	1.31	1.35	1.18	105.1	108.3	110.6	107.3	108.7	115.2	139.7
Fz[g]	1.38	1.30	1.44	1.31	1.38	–	106.2	108.9	108.1	106.6	110.2	–	–

[a] See structures below; [b] [79]; [c] [80]; [d] [81]; [e] [82]; [f] [83]; [g] [38]

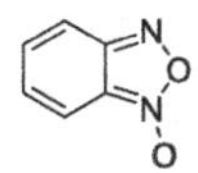

(b)

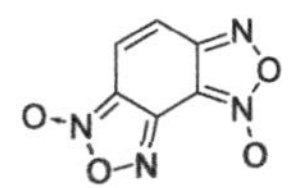

(c)

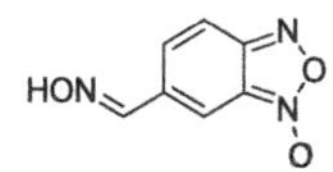

(d)

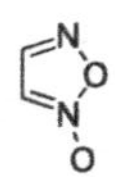

(e)

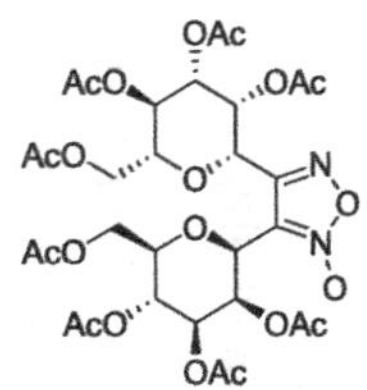

(f)

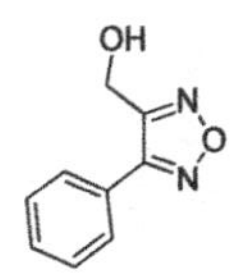

(g)

4.1.3 Theoretical Approaches

A great number of theoretical studies analyzing the ring-chain tautomerism have been described [90–102]. In these studies, the different 1,2-dinitroso conformers as intermediates in the equilibrium were analyzed, which found that the Bfx or Fx systems have lower energies and the reaction barrier is much higher for the rearrangement of Fx than for Bfx. The different energetic barrier has been explained taking into account the open-ring product of Bfx, the *o*-dinitrosobenzene, which is stabilized by strong conjugation effects that transform the quinoide six-membered ring into an aromatic benzene ring.

4.2 Other Structural Studies

Other Bfx and Fx structural studies involving different techniques have been reported, such as mass spectrometric [103–106], electrochemical [107, 108], and thermochemical studies [109].

5 Bioactivity

The Bfx and Fx derivatives have showed a wide spectrum of biological activities and its interest in medicinal chemistry has grown over the last two decades. Compounds incorporating the 1,2,5-oxadiazole *N*-oxide system in hybrid compounds have increasingly attracted the attention of medicinal chemists due to the fact that these compounds show a dual mechanism of action [110]. Maybe the first reported Bfx bioactivities have been described in the end of 1960s involving antileukemic and immunosuppressive properties [111–114], i.e., compounds **1** and **2** (Fig. 8), and in the 1980s the anti-infective properties of certain Fx derivatives [115, 116], i.e., compound **3** (Fig. 8). Currently, the interest in both kinds of bioactivities is still in progress and there are a great number of patent registrations and commercial formulations concerned with the Bfx or Fx uses. The next sections outline the most relevant bioactivities described in the recent years.

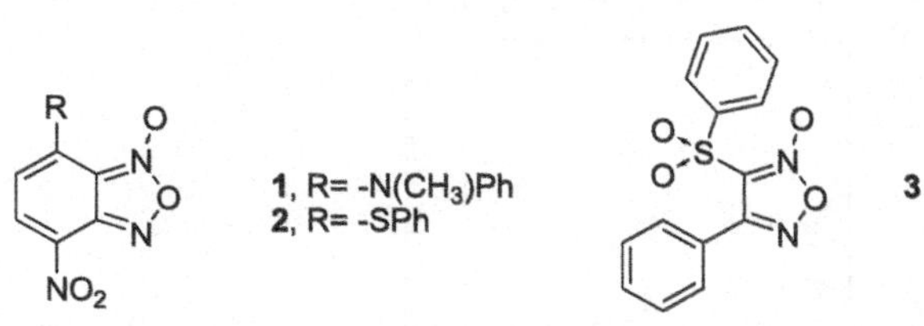

Fig. 8 First biactives Bfx and Fx reported in literature

5.1 Chemotherapeutic Activity

The findings that demonstrated the activity of some Fx as specific anti-microorganism agents [115, 116] opened the studies of Bfxs and other Fxs in this field. Currently, some patent registrations describe uses of Bfxs and Fxs as antibacterial and antifungal agents. In this sense, benzofuroxan has been employed as a microorganism growth inhibitor in nonsterile liquid meals [117] or some Bfxs have been claimed as effective agents against human and veterinary pathogens, i.e., gram-positive aerobic bacteria, anaerobic, and acid-fast organisms [118]. In 1998, Hwag et al. prepared a series of Fx derivatives as plant-fungicide agents using the nitrile oxide dimerization procedure [119] while in 2003 Nguyen et al. reported antimicrobial activity, against *Pseudomona aeruginosa* and *Bacilus subtillis*, of a series of 3-methyl-4-arylfuroxan [120].

As 1,2,5-thiadiazole analogues, potent HIV-1 reverse transcriptase inhibitors, some simple 1,2,5-oxadiazoles, compounds **4–6** (Fig. 9), have been synthesized using the traditional Wieland procedure as key for the heterocycle formation [121]. Such as thiadiazole parent compounds, derivative with chlorine atoms on the phenyl ring, i.e., **5**, showed the best anti-viral activity. Selectivity index (ratio of cytotoxic concentration to effective concentration) ranked in the order of **5** > **6** > **4**. The activity of Fz derivative **6** proved the *N*-oxide lack of relevance in the studied bioactivity. These products have been claimed in an invention patent [122]. On the other hand, compound **7** (Fig. 9) was evaluated for its nitric oxide (NO)-releasing property (see below) as modulator of the catalytic activity of HIV-1 reverse transcriptase. It was found that NO inhibited dose-dependently the enzyme activity, which is likely due to oxidation of Cys residues [123].

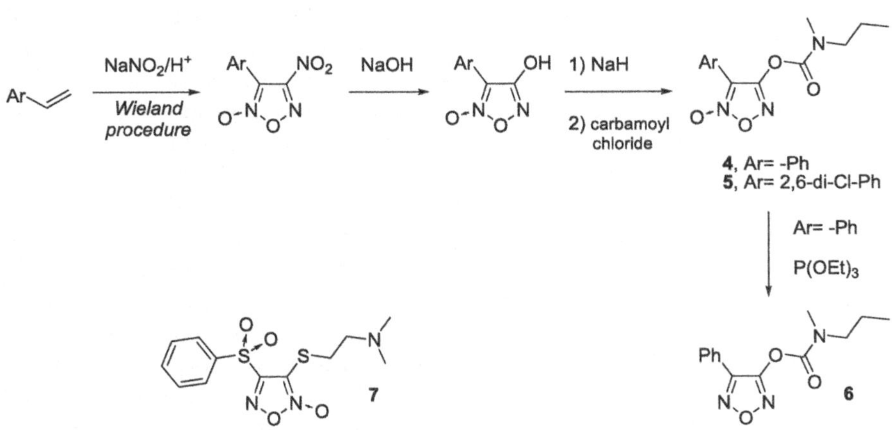

Fig. 9 Fx derivatives with antiviral activities

Significant advances in the cure of bacterial, fungal, and viral diseases have been made in the last century. However, compared to bacterial diseases, development of chemotherapy for the treatment of parasitic diseases has been hindered since parasitic protozoa are eukaryotic so they share many common features with their mammalian host making the development of effective and selective drugs a hard task. What is more, the spread of drug-resistant strains and the differences in drug susceptibility among different parasite strains lead to varied parasitological cure rates according to the geographical area [124]. As a result of the urgency to find new agents for specific parasitic diseases, much research with Bfxs and Fxs has been amassed in this field. The following section summarizes the most significant results in these approaches.

5.1.1 Antiparasite Activity

Parasitic diseases, such as trypanosomiasis, malaria, and leishmaniasis, affect hundreds of millions people around the world, mainly in underdeveloped countries. They are also the most common opportunistic infections that affect patients with acquired immunodeficiency syndrome (AIDS). Globally, malaria occupies the first place, but in Latin America, Chagas' disease (American Trypanosomiasis) is the most relevant parasitic disease that produces morbidity and mortality in low-income individuals.

Bfx and Fx derivatives have been evaluated against the parasites *Trypanosoma cruzi* (*T. cruzi*), which is responsible for American Trypanosomiasis and *Plasmodium falciparum* (*P. falciparum*) responsible for Malaria.

In a first approach, Cerecetto et al. evaluated a group of *N*-oxide containing heterocycle, i.e., **8–10** (Fig. 10), as in vitro anti-*T. cruzi* agents finding the best parasite-growth inhibition in some Bfx derivatives [125]. The trypanocidal activities of compounds depicted in Fig. 10 could be ranked in the order of **10** > **9** > **8** = **11**. The lack of activity of Bfz **11**, prepared as *N*-oxide-free counterpart, probed the relevance of the *N*-oxide moiety in the studied activity. Cytotoxicities, against mammalian fibroblasts, of the most active trypanocidal Bfxs were comparable to that of the reference drug, Nifurtimox. These results allowed the authors to select Bfx as the lead system for further structural modifications developing a second, a third, and a fourth generation of Bfx active against this protozoon, i.e., compounds **12–15** (Fig. 10) [108, 126–128]. The most active compounds have come from a "series design" methodology according to Hansch's substituent-selection clustering procedure [129]. With the aim of knowing the Bfx anti-*T. cruzi* mechanism of action, some studies have been performed by the authors attempting to probe if an oxidative stress operates as the main process. In order to demonstrate this hypothesis, the following studies were developed: electrochemical characterization [108, 130], intraparasite free radical production via electron spin

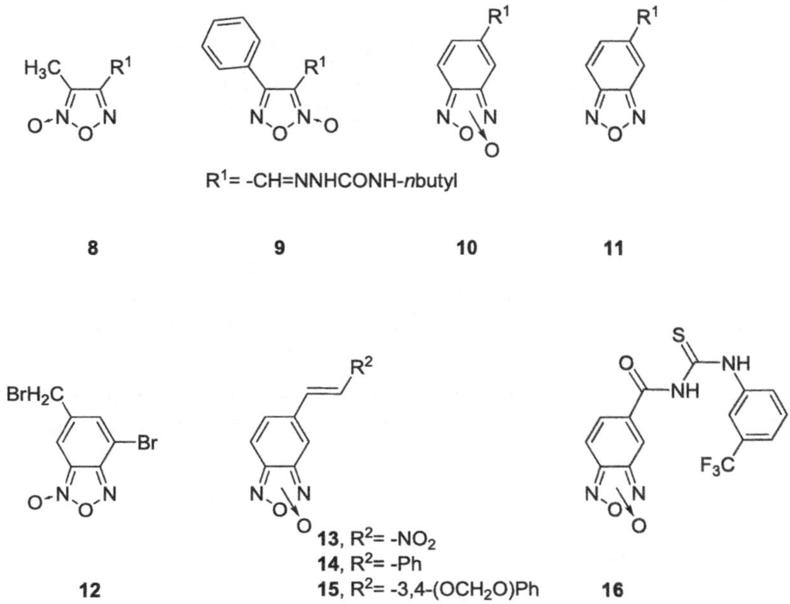

Fig. 10 Fx, Bfx and Bfz derivatives evaluated as anti-*T. cruzi* agents

resonance (ESR) experiments [131], and evaluation of the changes on the parasite oxygen uptake [127]. At the moment, it is clear that some active Bfxs perturb the mitochondrial electron chain, inhibiting parasite respiration after 5 days of incubation.

In 2000, Mc Kerrow et al. evaluated a compound-library as inhibitor of trypanosome cysteine proteinases, specifically cruzain (or cruzipain) and rhodesain, being the kind of enzymes one of the main biological targets for the development of effective anti-trypanosomal drugs [132]. The authors identified, among many other aryl ureas, the benzofuroxanyl thiourea derivative **16** (Fig. 10) as one interesting inhibitor [133]. Moreover, according to the authors, Bfx **16** presents the additional advantage that it adheres to Lipinski's "rule of 5" [134], so this compound is a reasonable starting point for drug discovery efforts. Further works of this research team found urea moiety as the pharmacophore for cruzain inhibitors. In another similar approach, Ascenzi et al. evaluated the Fx **7** (Fig. 9) as blocker of parasites cysteine proteinases, specifically *T. cruzi* cruzain, *P. falciparum* falcipain and *Leishmania infantum* cysteine proteinase [135]. In a different way of Mc Kerrow studies in this case the inactivation ability was explain by **7** NO-releasing capability (see below). The authors probed the *L. infantum* cysteine proteinase inactivation occurs via the NO-mediated *S*-nitrosylation of the Cys25 catalytic residue.

In another approach, Gasco et al. synthesized a series of Fx and its reduced analogues, i.e., **17–22** (Fig. 11), evaluating its in vitro as anti-*P. falci-*

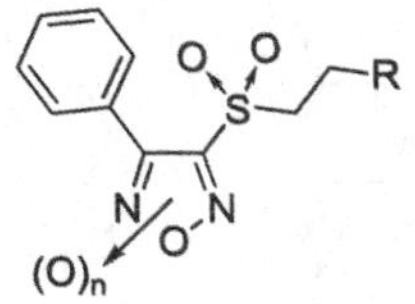
17, 3-Ph, R= -OH, n= 1
18, 4-Ph, R= -OH, n= 1
19, R= -OH, n= 0
20, 3-Ph, R= -SEt, n= 1
21, 4-Ph, R= -SEt, n= 1
22, R= -SEt, n= 0

Fig. 11 Sulfonyl Fxs and its deoxygenated analogues evaluated as anti-*P. falciparum* agents

parum agents. Two different parasite strains, including a drug resistant, were employed in this study [136]. The best biological results were observed in 3-SO_2R-substituted Fx, i.e., compounds **3** (Fig. 8), **18**, and **21**, and the worst in Fz series. Contrary to Chloroquine, the reference drug, the Fx activity was maintained between sensitive and resistant strains. The authors proposed that the activity is the result, at least in part, of Fx NO-releasing ability (see below).

5.2 NO-Releasing Activity

Perhaps the NO-releasing capacity of some Bfx and Fx represent the most interesting studied pharmacological property in the last years.

NO is a colorless gas at room temperature being a simple diatomic odd-electron species (•NO) moderately reactive. NO is an important messenger implicated in the regulation of numerous biological processes with physiological and pathological effects [137, 138]. Endogenous NO is almost exclusively generated via a five-electron oxidation of L-arginine, which is catalyzed by a family of isoforms of NO synthases (NOSs). The neuronal (nNOS or NOS1) and endothelial (eNOS or NOS3) forms are constitutive, whereas a third isoform (iNOS or NOS2) is induced by a variety of stimuli. In the heart, NOS enzymes have been identified in myocytes, vascular endothelium, endocardium, intracardiac nerves and macrophages. NO plays a crucial role either in vascular homeostasis dilating arterial blood vessels, inhibiting platelet adherence and aggregation, attenuating leukocyte adherence and activation, or in the neurotransmission facilitating the release of several neurotransmitters and hormones, stimulating the enzyme soluble guanylate cyclase (SGC), or in the immune response by its cytotoxic action to macrophages and leukocytes. NO is also potentially toxic by induction of genomic alterations. Other relevant biological functions of NO include their protection against cellular toxicity associated with oxidative stress [139]. This mechanism involves a radical-radical termination of NO with a propagating free radical. For example 2-acetamido-3-nitrosylmercapto-3-methylbutanoic acid, which releases NO in biological medium, is able to protect the lung epithelium from oxidants.

Consequently, NO-donor compounds that release NO in biological conditions could have special properties as drugs if they release NO in a controlled manner and in the adequate tissue. Classical examples of NO-releasing drugs used in cardiovascular diseases are the organic nitrates and nitrites, nitrosothiols, and nitrosyl complexes [139] being recently the 1,2,5-oxadiazole *N*-oxide system subject of study in this field. One of the first reports of 1,2,5-oxadiazol *N*-oxide as vasodilating agents, as result of its NO-releasing capability, was made by Ghosh et al. in 1974 [140]. In this case, compound **23** (Fig. 12), together with other 5-membered ring condensed heterocycle-benzofuroxans, was described as potent vasodilator. The vasodilatation effect was explained, at this time, in terms of topological structural similitude with the reference drug, glyceryl trinitrate (GTN) [141]. Nevertheless, recently, in a study including benzotrifuroxan **24** and Bfxs **25–29** (Fig. 12), it was demonstrated that the NO release is involved in the vasodilating action of compound **23** [142]. Compounds **24** and **29** showed complete vasodilatation at 30 μM, while the first is a NO-donor agent, Bfx **29** probably displays a moderate SGC stimulation. The increment of cyclic 3′,5′-guanosine monophosphate (cGMP) in the studied cells for benzofuroxan derivatives **25–29** could indicate the stimulation of SGC. The mechanism does not involve a thiol-induced NO production but probably involves an interaction of the Bfxs with the SGC heme site. Recently the preparation of furoxanopyrimidines derivatives as potential NO donors (**30–32**, Fig. 12) [143] have been described. Polarographic analysis showed that **32** is an NO donor.

In 1992, two independent groups reported Fxs capability to stimulated SGC [144–146]. In one case [144], working with 3-, 4-, and 3,4-dicarboxamide Fxs was confirmed the NO generation mediated by thiols identifying an excellent coronary vasodilator compound, namely ipramidil (**33**, Fig. 12). A complete lack of SGC stimulation, NO-releasing capability, and vasodilator activity was observed for the Fz analogues. The authors have proposed a speculative mechanism of NO release (Fig. 12) from the identification of main end products (**I** and **II**) in the reaction of **33** and L-cysteine and the measurement of the formation of *S*-nitrosothiols (**III**) at different pH. From the study on 3-furoxancarboxamides as NO-prodrugs was selected the compound **34** (Fig. 12) as a candidate for further pharmacological studies [15]. Fx **34**, which was synthesized using the reactions sequence depicted in Fig. 12 displayed in vivo an interesting haemodynamic profile without tolerance as other derivatives of the series [138, 147].

In the other case [145, 146], Gasco et al. described the capacity of Fx **35** (Table 4), prepared from the corresponding substituted glyoxime to activate the human platelet-SGC. Then in 1993, the Italian researchers have confirmed that analogues **3** (Fig. 8), **36**, and **37** (Table 4) generate NO by a thiol-mediated process [148] and they have evidenced that the maximum vasodilatory as well as antiaggregatory activity are displayed when the Fx ring is 3-sulfonyl-substituted. In the same year, they described the vasorelax-

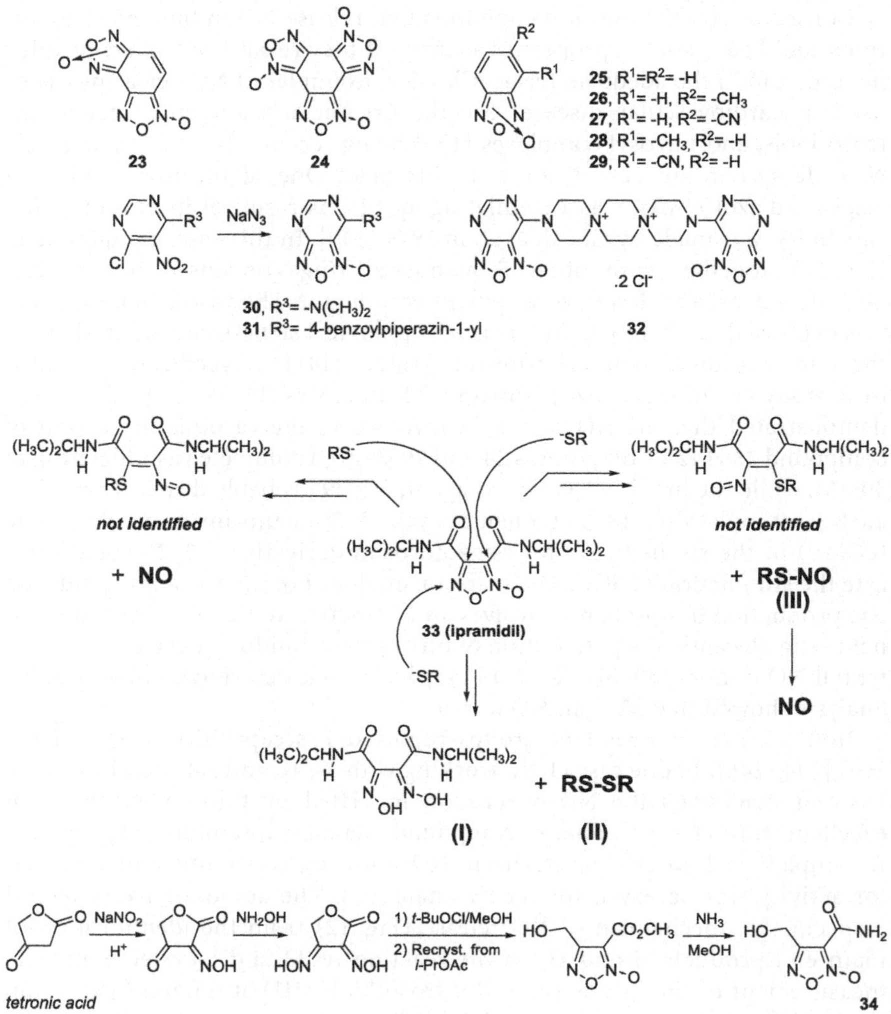

Fig. 12 *Top*: Bfx with NO-releasing capability and stimulation of SGC. *Middle*: Mechanism of NO release from Fx derivatives. *Bottom*: Preparation of 3-furoxancarboxamide with good haemodynamic profile in vivo

ant activity of the two furoxancarbonitrile isomers **38** and **39** (Table 4) [70]. With the NO-releasing capacity of derivative **39** and the identification of the main end products **IV–VI** (Fig. 13) in the reaction between **39** and thiophenol the authors have proposed a speculative mechanism (Fig. 13), which could account for the vasodilatory activity [149–151]. Other studies involving thiol-reactions were also performed [152, 153]. The end products in the reaction between Fx **40** (Fig. 13) and different thiols confirm that the attack

Table 4 Vasodilating activity of Fx derivatives and reference compound (GTN)

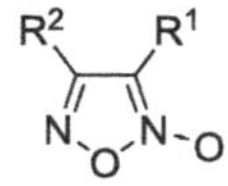

Derivative	R^1	R^2	$EC_{50} \pm s.e.$ (10^{-8} M) [a]
35	$-SO_2Ph$	$-CH_3$	190 ± 10 [b]
36	$-Ph$	$-SO_2Ph$	107 ± 1 [c]
3 [d]	$-SO_2Ph$	$-Ph$	43 ± 2 [c]
GTN	–	–	6.6 ± 0.5 [c]
37	$-SO_2Ph$	$-SO_2Ph$	5.5 ± 0.5 [c]
38	$-Ph$	$-CN$	0.95 ± 0.15 [e]
39	$-CN$	$-Ph$	0.64 ± 0.04 [e]

[a] EC_{50} represents the drug concentration required to cause 50% of the respective maximal relaxation of rabbit aorta, independently from its extent; [b] From reference [70]; [c] From reference [148]; [d] See Fig. 8; [e] From reference [149]

occurs at both the 3a and 7a positions, leading to subsequent ring opening and NO liberation. In 1998, in attempt to obtain more active compounds a new series of furoxancarbonitrile isomers has been synthesized and evaluated as NO-releasing agents, i.e., **41–44** (Fig. 13) [154]. Subsequently, Gasco et al. described the properties of some water soluble Fxs to release NO also in the absence of thiol cofactor, i.e., compounds **45–47** (Fig. 13) [155]. According to the authors, the first step on the route of NO production could be the result of the attack at the 3-carbon by a hydroxyl group (see example with compound **47** in Fig. 13). This report shows that the in vitro vasodilating potencies of Fxs are principally dependent on initial rates of NO release.

Over the last 15 years, the number of patent registrations and commercial products concerned with the Bfx and Fx use as NO donor agents has increased extremely [156–160].

5.2.1 Platelet Antiaggregatory Property

The study of Bfx and Fx antiaggregating activity is one of the biological evaluations that has been analyzed together with the NO-releasing properties of these compounds. For example, compound **3** (Fig. 8) reduces in a dose-dependent manner either the platelet aggregation induced by collagen (COL), adenosine diphosphate (ADP), or platelet activating factor (PAF), increasing the platelet cGMP levels [161]. Compound **48** (Fig. 14) [162] produces a concentration-dependent inhibition of the platelet aggregation induced

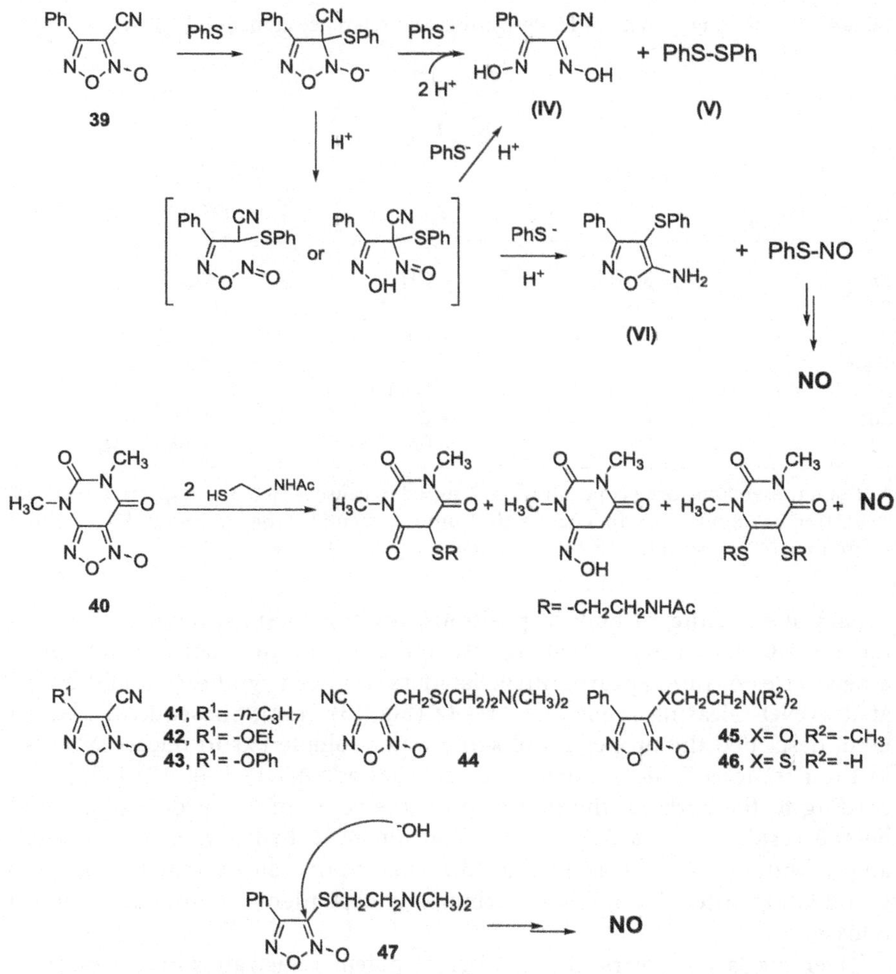

Fig. 13 *Top*: Mechanism of NO release proposed for 3-furoxancarbonitrile derivatives. *Middle*: Thiol induction of NO release from Fx **40**. *Bottom*: NO-releasing carbonitrile-Fxs and water-soluble-Fxs

either by COL, ADP, or PAF, elevating platelet cGMP [163], about five times more potent than sodium nitroprusside (SNP) in the aggregation assays being as potent as GTN in the vasorelaxant activity without in vitro cross tolerance. In 2000, it was demonstrated that Bfx **23** (Fig. 12) is a more potent inhibitor of ADP-induced human platelet aggregation than SNP [164], with about ten times more potent GC stimulation than SNP. Gasco et al. have described the synthesis [165] and the platelet antiaggregatory behavior [166] of ter-furoxan derivatives, i.e. **49–52** (Fig. 14). The biological effects, NO-releasing,

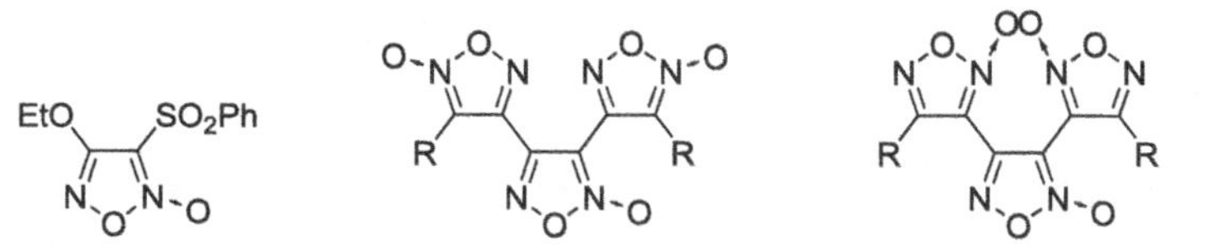

48

49, R= $-CH_3$
50, R= $-CH_2N(CH_3)_2 \cdot HCl$

51, R= $-CH_3$
52, R= $-CH_2N(CH_3)_2 \cdot HCl$

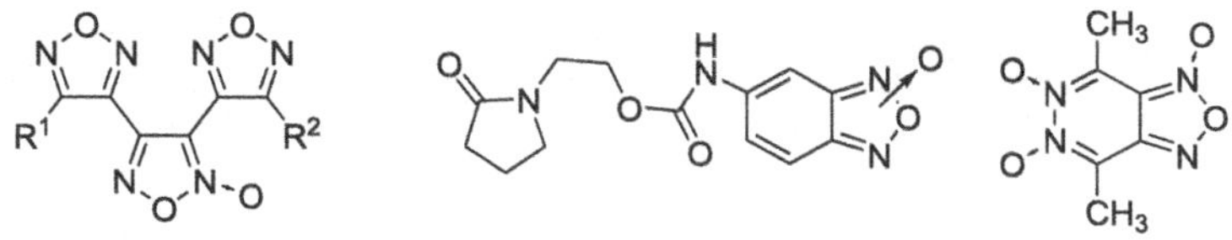

53, $R^1=R^2=$ -OH, $-NH_2$, $-N_3$, etc
54, $R^1=$ $-NH_2$, $R^2=$ -NH-alkanoyl, etc

55

56

Fig. 14 Fx and Bfx studied as inhibitors of platelet aggregation

vasodilating, and antiaggregating properties, are influenced both by the terfuroxan system structure and by the substituents. Compound **49** is eight times more potent than SNP on ADP-induced aggregation, while **50** and **51** are equipotent and Fx **52** is about five times less potent than SNP.

A recent patent of invention describes 3,4-bis(furazan-3-yl)furoxan, with general formula **53** and **54** (Fig. 14), able to generate NO and inhibit platelet aggregation [167]. Another invention has claimed Bfx with dual action as platelet aggregation inhibitors and antianginals, being compound **55** (Fig. 14) one of the presented example [168].

5.2.2 Vasodilating Property

The other pharmacological property evaluated together with the NO-release capacities of Bfxs and Fxs is the vasodilating activity. For example, some 1,2,5-oxadiazolo[3,4-*d*]pyridazine 1,5,6-trioxide derivatives, like compound **56** (Fig. 14), are indicated in cardiovascular and hypertension disorders [169] due to they activate SGC and produce a significant hypotensive effect without tolerance development. Specifically, compound **56** and its Fz analogue generate NO and react with thiols and haem [170]. The vasorelaxant activity of **56** is SGC-dependent and NO plays a predominant role at concentrations below 1 μM, while the platelet antiaggregatory property is partially involved in the SGC activation. Some recent patent of inventions related to compositions for treatment of cardiovascular diseases include Fx derivatives as vasodilator agents [171, 172].

5.3
Hybrid Compounds

The combination of a benzofuroxanyl or a furoxanyl moiety with another pharmacologically active substructure in a single molecule, known as hybrid drugs (Scheme 4), has recently received particular attention. Examples of these kinds of compounds are reported below.

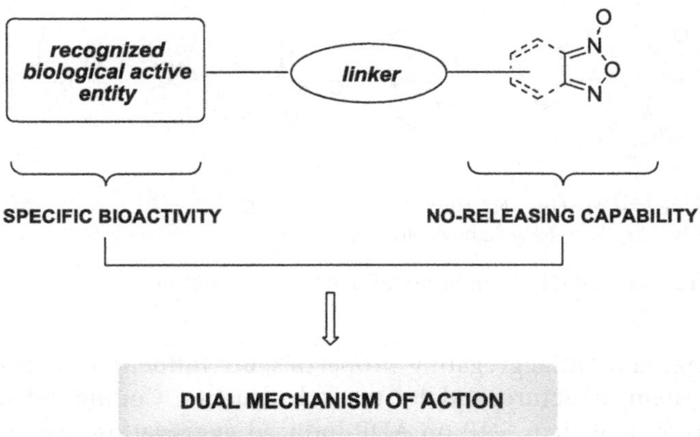

Scheme 4 Idea of Bfx and Fx hybrid drugs

5.3.1
Gastric Antisecretory and Protective Activities

In 1997, Sorba et al. described the synthesis and biological evaluation of the Fx derivative **57** (Fig. 15) as a potential antiulcer agent [173]. The authors choose some functional groups of Lamtidine, a well-known H_2-antagonist, and furoxanyl moiety as a NO-donor species. They combined the antisecretory activity of H_2-antagonist substructure with the NO-gastroprotective effects of this *N*-oxide. In another approach, Tiotidine-furoxan hybrid compounds derivatives, i.e., **58** (Fig. 15) were prepared [174] and evaluated for its NO-donor properties, H_2-antagonist properties and gastroprotective effects. Compound **58** is a more potent H_2-receptor antagonist than the corresponding Lamtidine hybrid (**57**), but it is only a partial gastroprotector. In another study, Lamtidine analogues were evaluated in different H_2 receptor assays and in the conscious rat against acid-induced gastric lesions [175]. Compound **57** is able to antagonize histamine-mediated responses at cardiac and gastric H_2 receptor, however, it is ten times less potent than the analogue lacking the NO-donor group, precursor **VIII** (Fig. 15). By contrast, compound **57** is a ten times more potent gastroprotective agent than **VIII**.

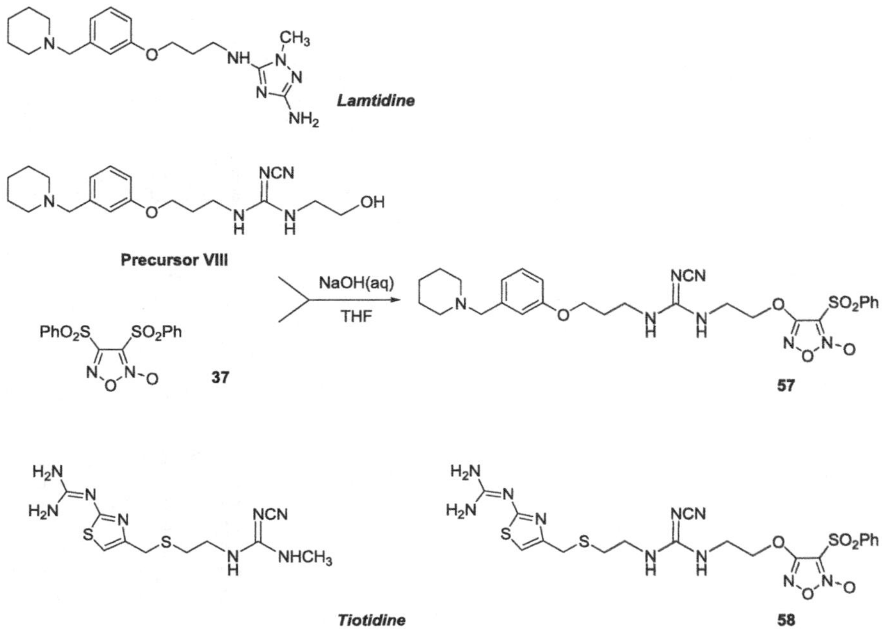

Fig. 15 Hybrid Fx developed as antiulcer agents

5.3.2
H_3-Antagonist Activity

Histamine produces its pharmacological actions by three subtypes of receptors: the postsynaptic H_1 and H_2 receptors and the presynaptic H_3 receptor. The H_3 receptor is mainly located in the central nervous system (CNS), where it acts as an inhibitory autoreceptor in the central histaminergic neuronal pathways [176]. A number of therapeutic applications have been proposed for selective H_3 receptor antagonists, including several CNS disorders such as Alzheimer's disease, Attention Deficit Hyperactivity Disorder, Schizophrenia, or for enhancing memory or obesity control.

Gasco et al. designed a series of hybrid compounds with both potential H_3-antagonist and NO-releasing effects. Firstly, they have reported hybrid furoxan-imidazole analogues of H_3-antagonists SKF 91486 and imoproxifan, i.e., **59–68** (Fig. 16), modifying the spacer between imidazole and Fx rings [177–180]. In the SKF 91486-analogues series, Fxs display good H_3-antagonist behavior and feeble partial H_2-agonist activity. Compound **61** possesses a dual NO-dependent muscle relaxation and H_3-antagonistic effect, with a H_3 receptor response higher than the corresponding H_2 agonist effect. In the imoproxifan-analogues series, examples **62–68**, the developed

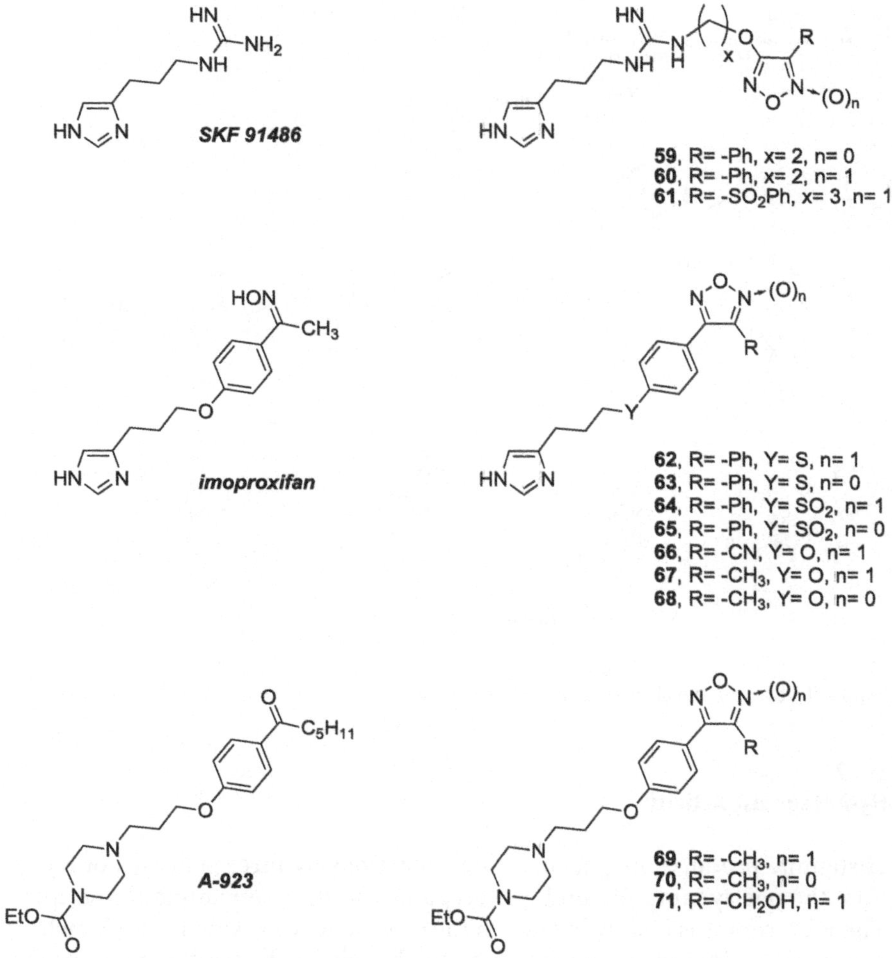

Fig. 16 Hybrid Fx developed as H_3-antagonist agents

compounds display lesser affinity for the H_3-receptor than the parent compound being the best derivative the Fz **68**. Recently, a series of non-imidazole NO-donor H_3-antagonists was analyzed by these researchers [181]. In this sense, the ligand A-923 was selected as the parent compound developing hybrid Fx with structures like compounds **69–71** (Fig. 16). The replacement of the imidazole ring results in a decreased H_3-antagonist activity with respect to the imidazolic series, compounds **59–68**, and in some cases, for example in compound **69**, induces relaxing effects on the electrically contracted guinea-pig ileum possibly as the result of the affinity for other receptor system.

5.3.3
α_1-, β-Adrenergic Antagonist Activity

In order to improve the properties of the well-known adrenergic antagonist, two chemical modulations on classical drugs including NO-donor moiety have been described. On the one hand, chemical modifications on Prazosin, an α_1-adrenergic antagonist, including furoxanyl moieties have been reported [70, 182, 183]. In this approach, the 2-furanylcarbonyl moiety of Pra-

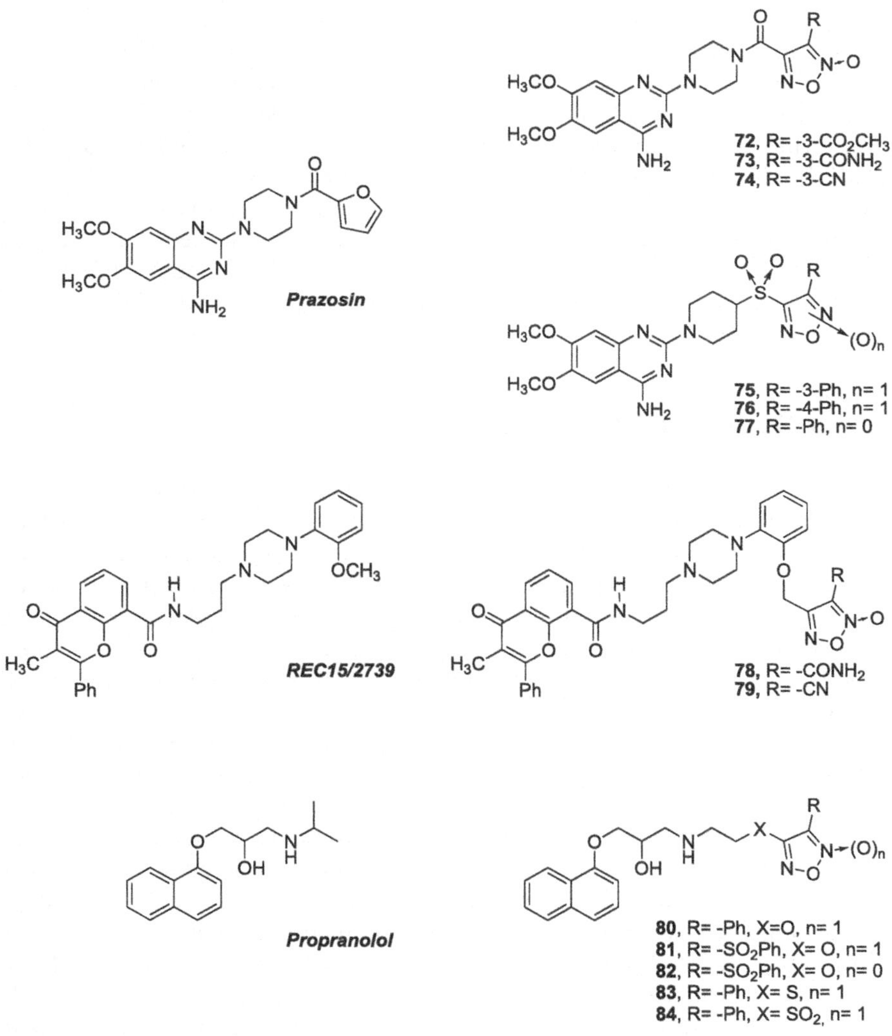

Fig. 17 Hybrid Fx and Fz developed as adrenergic antagonist agents

zosin was substituted by 1,2,5-oxadiazolecarbonyl or 1,2,5-oxadiazolesulfonyl groups, i.e., compounds **72–77** (Fig. 17). The Fxs are less potent, close to ten-fold less, than Prazosin in the α_X-antagonist activity being Fz analogue **77**, synthesized as NO-donor negative control, one of the most potent compound followed by Fx **75**. Regarding the NO-releasing properties, the vasodilating potency could be ranked on the order of **74** > **75** = **76** > **72** > **73**, displaying values close to SNP. Other hybrid Fxs have been reported as uroselective α_1-adrenoceptor antagonist using as drug template REC15/2739 developed by Recordati S.p.A. laboratories [184]. Compounds **78** and **79** (Fig. 17) are able to relax the prostatic portion of rat both for its α_{1A}-adrenergic antagonism and for its NO-releasing capability.

On the other hand, Gasco et al. have developed propranolol-like Fxs, i.e., compounds **80–84** (Fig. 17) as potential β-adrenergic antagonists [185]. These compounds show lower β_2 affinity than propranolol giving an increase in β_1/β_2 selectivity. The potency in the β_1-antagonism could be ranked in the order of propranolol = **81** > **82** > **80** = **84** > **83**. The hybridization leads to variation in NO-dependent vasodilating activity compared with the parent Fxs, without 3-napthyloxy-2-hydroxypropyl moiety. In all cases, except compound **81** and the corresponding furoxan (3-(phenylsulfonyl)-4-(2-aminoethyloxy)furoxan), this activity is greater than that of the corresponding simple furoxans.

5.3.4 Calcium Channel Modulator Properties

Calcium channel modulators (CCMs) are other recently studied Bfx and Fx hybrid compounds. Previously, some Bfzs with action as CCMs had been described, i.e., isradipine and compound **85** (Fig. 18) [186, 187]. Since 1992, Gasco et al. have described a series of studies involving the syntheses and biological evaluation of dihydropyridines containing benzofuroxanyl or furoxanyl moieties analogues of isradipine, nifedipine, or compound **85**, i.e., **86–111** (Fig. 18) [188–194]. In the nifedipine Bfx-analogues compounds **86** and **87** were the most active in the isolated rabbit basilar artery assay showing the same potency as the reference drug while 5-isomers **88** and **89** were ten-fold less potent. The presence of the *N*-oxide moiety does not modify substantially the range and the ratio of activity of the derivatives or the mechanism of their vasorelaxant action. The structure and distribution of C-4 rotamer of these compounds (in solution and in solid state) was investigated by NMR, X-ray diffraction, and theoretical studies [195]. Compound **85** Bfx-analogues, i.e., **90–93** (Fig. 18), were obtained as racemic mixtures and chiral HPLC was employed to resolve in its individual enantiomers. Each enantiomer displayed opposite effects on calcium cation currents through voltage-dependent L-type calcium channels, being the dextrorotatory potent calcium entry activators while the antipodes levorotatory are weak calcium

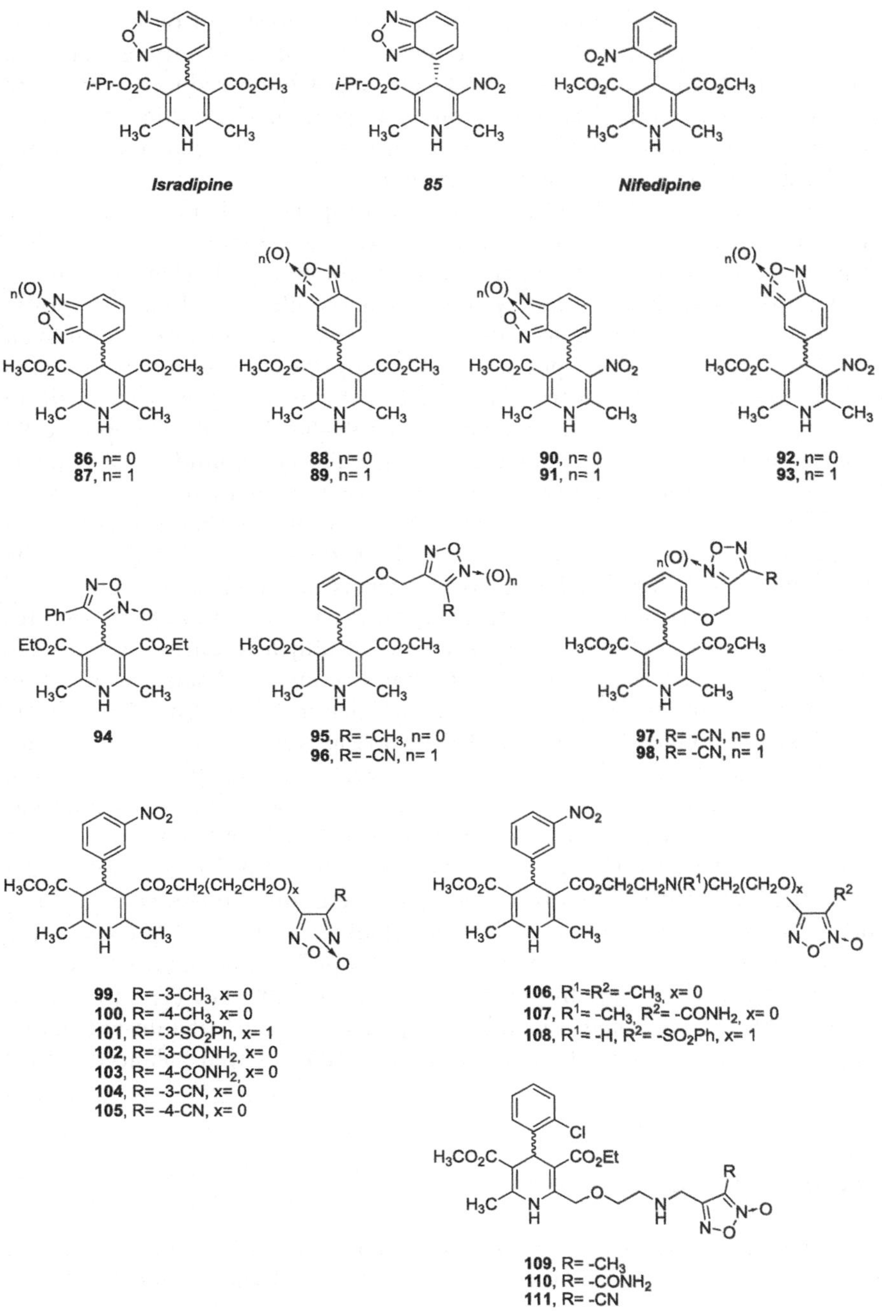

Fig. 18 Hybrid Bfx and Fx developed as calcium channel modulators

entry blockers. Compound (+)-**92** was the most interesting isomer, being capable of interfering with the voltage-dependent gating of L-type channels. However, the *N*-oxide analogue (+)-**93** exerted a strongly reduced capacity to alter the rate constant of calcium channel activation. The authors proposed that these differences could be the result of the different electronic distribution because **92** and **93** have similar lipophilicity.

In the first approach of furoxanyl-1,4-dihydropyridine hybrid compounds, the 4-phenyl substituent was replaced by a furoxanyl moiety finding modest activity in comparison with nifedipine where compound **94** (Fig. 18) was the most active. After this, attempting to improve the pharmacological response Gasco et al. have described four different series-linking furoxanyl moiety at different positions in the 4-phenyl-1,4-dihydropyridine system. Two options include the NO-donor substructure either at the *ortho* or the *meta* position of the 4-phenyl substituent, i.e., compounds **95–98** (Fig. 18). Compound **95** behaves as pure calcium channel blockers while compound **96** is equipotent with nifedipine with a vasodilator activity principally dependent on its NO-donor property. The *ortho*-derivatives **97** and **98** are the most active developed derivatives. The less potency of the *meta*-series with respect to the *ortho* one is explained by the authors in terms of the electronic and steric properties of substituents. In another study, the furoxanyl moieties were located on the lateral chain of the 3-ester of 1,4-dihydropyridine system, i.e., **99–108** (Fig. 18). Hybrid compounds **99, 100, 102** and **103** display vasodilating activity depending predominantly on their calcium-channel blocker properties while compounds **101, 104** and **105** behave as well-balanced hybrids with mixed calcium-channel blocking and NO-dependent vasodilating activities. Compounds **106–108** were designed as derivatives with improved pharmacokinetic properties by introducing a basic lateral chain at the 3-ester moiety. Compounds **106** and **107** are vasodilatators due to its calcium-antagonist properties while **108** is a well-balanced hybrid compound possessing calcium-channel blocking and NO-dependent vasodilating properties. The last study series involves the inclusion of basic and lipophilic furoxanyl moieties on the 2-position of 1,4-dihydropyridine system, i.e., **109–111** (Fig. 18). Compounds **109–110** are potent vasodilating agents due to its calcium-antagonist properties and compound **111**, like compound **108**, is a well-balanced hybrid derivative.

5.3.5
Other Hybrid Compounds

Other particular studies describing mixed compounds have been reported. Recently, special attention has been devoted to furoxan-nicorandil hybrid

Fig. 19 Hybrid Fxs and Bfx more recently developed and its parent compounds ▶

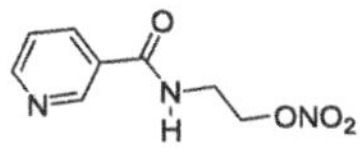

Nicorandil

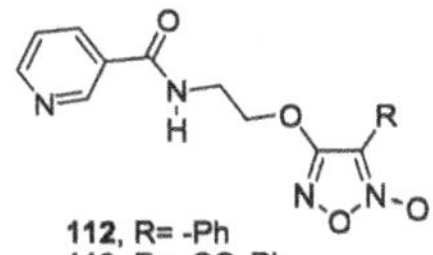

112, R= -Ph
113, R= -SO$_2$Ph

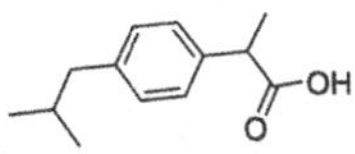

Ibuprofen

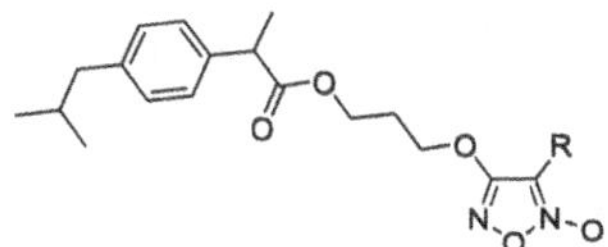

114, R= -SPh
115, R= -SO$_2$Ph

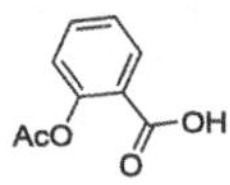

Aspirin

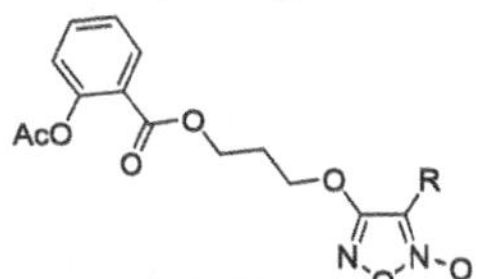

116, R= -Ph
117, R= -SO$_2$Ph

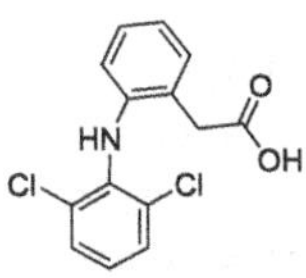

Diclofenac

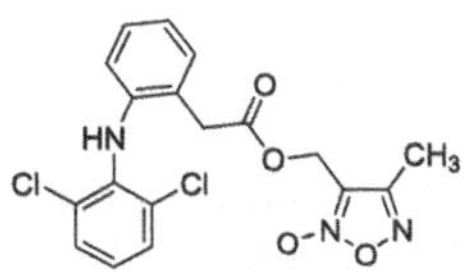

118

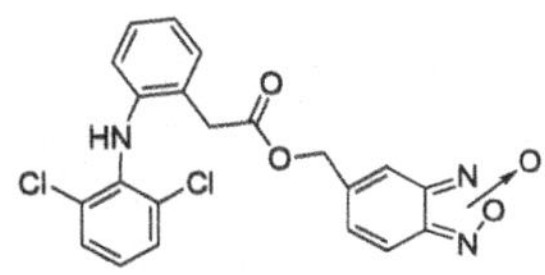

119

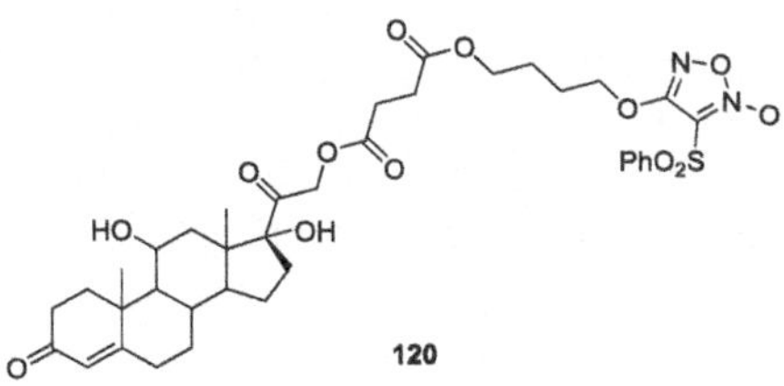

120

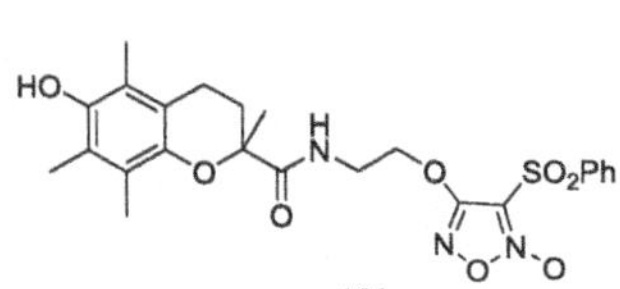

121

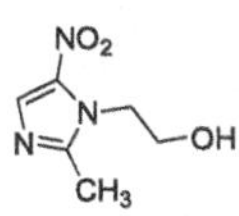

Metronidazol

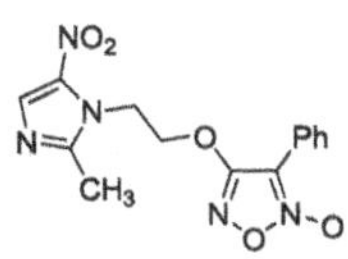

122

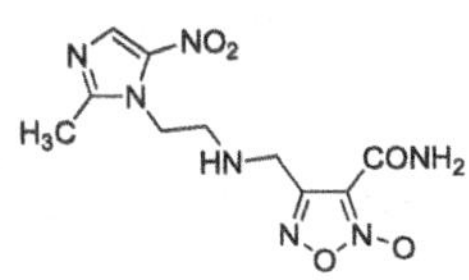

123

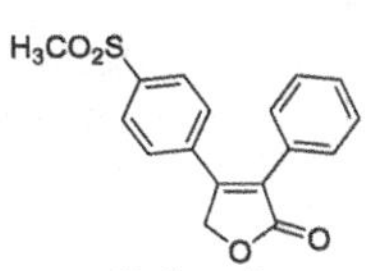

Rofecoxib

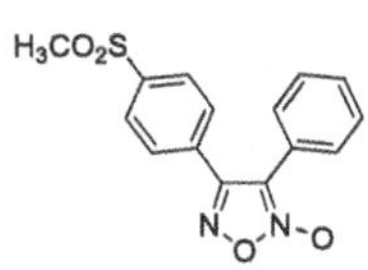

124

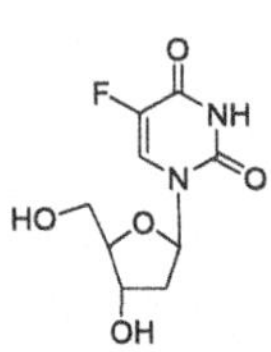

5-Fluoro-2'-deoxyuridine

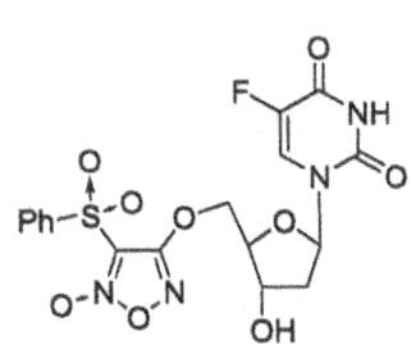

125

Table 5 Summary of hybrid Fxs recently developed

Parent drugs	Examples [a]	Extra bioactivities [b]	Refs.
Nicorandil	**112**, **113**	Potassium-channels activator	[196, 197]
Ibuprofen	**114**, **115**		[198–205]
Aspirin	**116**, **117**		
Diclofenac	**118**, **119**	Anti-inflammatory effect	
Glucocorticoid substructure	**120**		
Phenol substructure	**121**	Antioxidant activity	[206–209]
Metronidazol	**122**, **123**	Anti-*Helicobacter pylori*	[210, 211]
Rofecoxib	**124**	Selective COX-2 inhibition	[212–217]
5-fluoro- or 5-iodo-2′-deoxyuridine	**125**	Anticancer activity	[218]

[a] See structure in Fig. 19
[b] Apart of the Fx own bioactivity, in general NO-releasing capability

compounds as NO-donor and potassium-channels-activator agents with cardiovascular and cerebrovascular activities. Another effort describes mixing furoxan and nonsteroidal anti-inflammatory moieties, taking ibuprofen, aspirin, and diclofenac as the parent compound. In the same sense, hybrid steroidal anti-inflammatory Fxs have been recently described. Another approach has involved hybrid furoxan-antioxidant agents including different phenols as redox entities which potentially direct them to different tissues. Mixed Fx-nitroimidazole derivatives actives against *Helicobacter pylori* have been also described as gastroprotective agents. Another group of bioactivity extensively claimed in invention patents is the selective COX-2 inhibition. In Table 5 and Fig. 19 are listed examples, bioactivities, comments, and references for the most recent works.

5.4 Others Relevant Bioactivities

Besides the biological properties described in the previous section other significant bioactivities have been reported for Bfx and Fx derivatives. In Table 6 and Fig. 20 are listed examples, bioactivities, comments, and references for the most recent works.

6 Structure-Activity Relationships

A little number of works performing Bfx and Fx quantitative structure-activity relationship (QSAR) studies have been described. On the other hand, to gain insight into the biological behavior of Bfxs and Fxs some studies

Table 6 Summary of Bfx and Fx with diverse bioactivities

Family	Examples[a]	Bioactivity	Comments	Refs.
Bfx Fx	**126**	Hypoxic selective cytotoxicity	compound **126** DNA-interaction was studied showing poor affinity for this biomolecule	[219–222]
Fx	**39**[b]	Cytotoxic activity	compound **39** showed good in vivo antitumoral profile	[38, 223]
Bfx	–	Heme oxygenase inducer	patent of invention	[224]
Bfx	**23**[c]	Monoamine oxidases (A, B) inhibition	–	[225]
Bfx Fx	**127**	Herbicidal	–	[226]
Bfx	**128–130**	Preventive agents against *Piricularia oryzae*	patent of invention to use in agrochemistry (rice)	[227]
Bfx	**131, 132**	Agricultural arachnicide and bactericide	patent of invention	[228]
Bfx	**133**	Insect feeding deterrents	patent of invention	[229]
Fx	**134**	Co-ligand for ^{99m}Tc complexes synthesis	–	[230]

[a] See structure in Fig. 20
[b] See structure in Table 4
[c] See structure in Fig. 12

reporting dipole moment, lipophilicity and theoretical descriptors for these systems have been published [231–233]. These studies have shown that Fx and Fz rings posses similar lipophilicity, while electronic distribution resulted different. Fruttero et al. have determined by classical methodologies the Hammett's substituent constants (σ) for furoxan-3-yl, -4-yl, and furazanyl moieties [234]. These substituents are strong electron attracting groups by the inductive mechanism and they are very slightly conjugated as weak donors.

Kontogiorgis and Hadjipavlou-Litina [235] have reported QSAR studies for furoxancarbonitrile derivatives. In these studies they included compounds **38, 39, 41–44** (Table 4 and Fig. 13) among others. They found that its in vitro vasodilating activity is related to lipophilicity (expressed as clog P) and the position of *N*-oxide moiety (expressed as an indicator variable I_A). In the same study, these researchers found that the NO-releasing capability is related to the electronic properties of the Fx substituents (expressed as F, Swain-Lupton field parameter) and to the position of *N*-oxide moiety (in this case defined by the authors as I_B). On the other hand, Kontogiorgis and Hadjipavlou-Litina working with the water-soluble Fx developed by Gasco

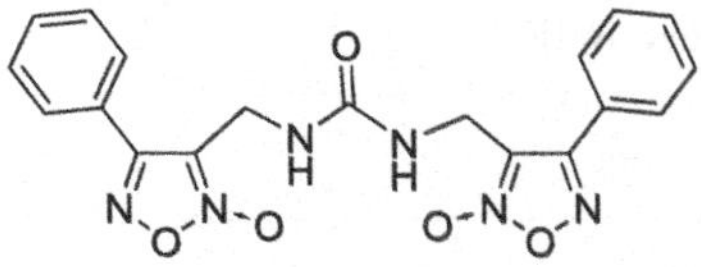

126

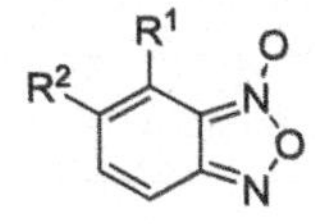

127, R^1= -H, R^2= -CONH-*n*-butyl
128, R^1=R^2= -H
129, R^1= -H, R^2= $-OCH_3$
130, R^1= $-NO_2$, R^2= -H

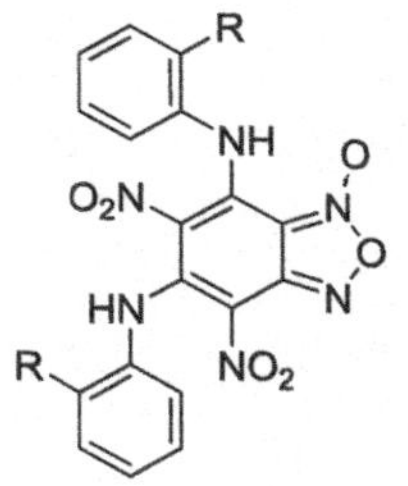

131, R= -Cl
132, R= -OH

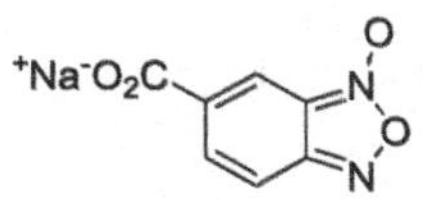

133

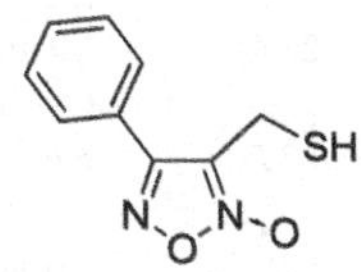

134

Fig. 20 Chemical structure of Bfx and Fx derivatives with bioactivities depicted in Table 6

et al. [155], i.e., compounds **45–47** (Fig. 13), have derivatized an equation that correlated the vasodilating potency (expressed as EC_{50}) to lipophilicity, volume and polarizability (expressed as MR descriptor) of alkylheteroyl substituent, and the presence of oxygen in the alkylheteroyl substituent (defined as indicator variable I_O). The expression of the equation is $\log(1/EC_{50}) = -1.296c\log P - 0.976I_O + 1.143MR + 6.150$ ($n = 14$, $r = 0.957$). The negative sign of lipophilicity is in agreement with the goal of the synthesis of water-soluble furoxan derivatives with potent NO-releasing properties.

Other studies performed on cytotoxic-Fx, like compound **126** (Fig. 20) [219–222], have revealed that the toxic effect is related to molecular LUMO energy, lipophilicity (theoretically calculated) and Mulliken charge on Fx 2-nitrogen [236]. A tentative mechanism of cytotoxicity was proposed using this correlation.

Anti-*T. cruzi* Bfx, including compounds **13–15** (Fig. 10), have been employed to develop 2D- and 3D-QSAR models [81, 126]. The 2D-QSAR model is generated using multiple regression analysis of tabulated substituents physicochemical descriptors and indicator variables. The derivatized equation correlates percentage of parasite growth inhibition to the preferential position of *N*-oxide moiety (expressed as indicator variable I_6), electronic properties (expressed as *F*, Swain-Lupton field parameter), volume and polarizability (expressed as MR descriptor), and donor hydrogen bond capability (defined as indicator variable I_{HBD}) of Bfx substituent. The indicator variable I_6 is established using low-temperature ^{1}H NMR experiments. In addition, a 3D-QSAR model is obtained using comparative molecular field analysis (CoMFA). Both QSAR models are in agreement with the structural requirements for optimal activity.

7 Pharmacological Behaviors

In this section, two aspects related to Bfx and Fx future use as drugs will be discussed, metabolic behavior and toxic effects.

7.1 Metabolic Studies

At the moment, only three in vitro studies have been performed on Bfx metabolic behavior. In one case, it has been shown that Bfxs are able to be reduced by oxyhemoglobin to the corresponding *o*-nitroaniline derivatives (Scheme 5) [237]. In the reaction between compound **135** and oxyhemoglobin compound **136** was generated as secondary product resulting from both nitrile hydrolysis and deoxygenation. This study indicates that blood is a possible site for metabolism of Bfxs with the consequent methemoglobinemia.

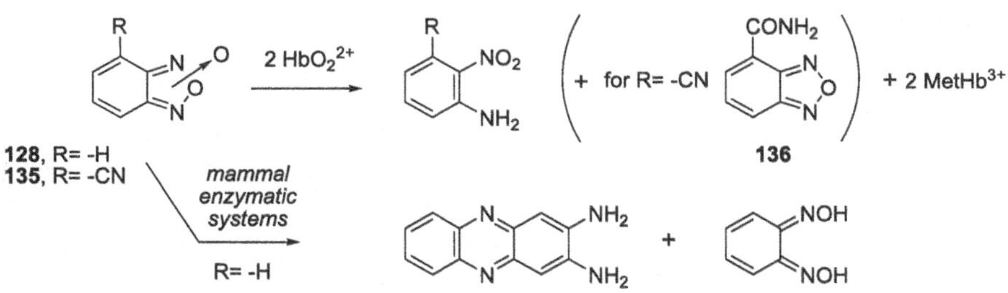

Scheme 5 Metabolic pathways for Bfx **128** and **135**

The second study was performed using either cytosolic or microsomal fractions from rat liver as the in vitro metabolic mammal models [238]. The studied compound, benzofuroxan (**128**, Fig. 20), is metabolized to *o*-quinone dioxime and 2,3-diaminophenazine (Scheme 4).

The third study has employed 4,6-dinitrobenzofuroxan and as metabolic systems the one-electron reductants NADPH:cytochrome P450 reductase and ferredoxin : NADP(+) reductase and the two-electron reductants DT-diaphorase and *Enterobacter cloacae* nitroreductase [239]. The compound is activated either by DT-diaphorase or nitroreductase.

7.2 Toxicity Studies

Some toxicity studies were performed analyzing Bfx and Fx mutagenic effects [240–242]. For example, Bfx and Bfz nitro-substituted were tested for mutagenicity in *Salmonella typhimurium* (*S. typhimurium*) TA 98 and TA 100 strains with and without metabolic activation (Ames test). All the Bfx (10/10) and some of the Bfz (9/15) are mutagenic without activation. Other study has demonstrated benzofuroxan (**128**, Fig. 20) is mutagenic in the Luria and Delbrueck's fluctuation test, with *Klebsiella pneumoniae*, and in the Ames test. In another study it has been found that compound **137** (Fig. 21) is not mutagenic to *S. typhimurium*.

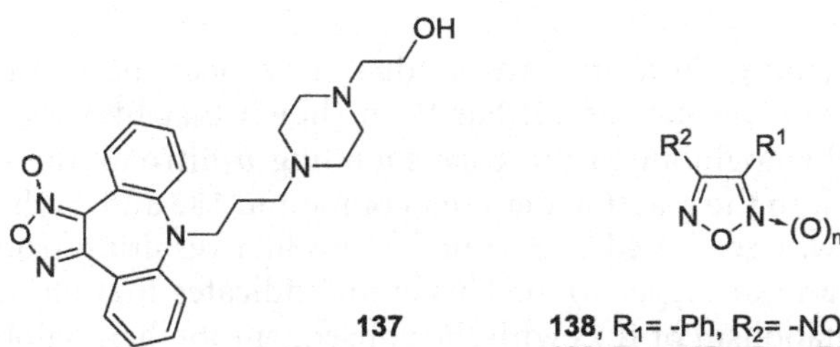

137

138, R_1= -Ph, R_2= $-NO_2$, n= 1
139, R_1= $-NO_2$, R_2= -Ph, n= 1
140, R_1= -Ph, R_2= $-NO_2$, n= 0

Fig. 21 Examples of Fx derivatives studied in the Ames test

Some Fx and Fz derivatives have been evaluated for its mutagenicity in the Ames test [115]. For example, Fx **138** and **139** and Fz **140** posses mutagenic activity while Fx **3** (Fig. 8) is not mutagenic in this test.

In vivo studies has been performed on selected Fx derivatives [223, 243, 244] finding toxic effects specially related to its capacity to release NO.

8 Others Applications

Bfx, Fx, and related compounds are the subject of a great number of invention patents especially related to its uses in material sciences. For example, they were included in the formulation as rubber additives [245, 246], as inhibitors in the polymerization of aromatic vinyl monomers [247], as components in the igniting composition for inflation of airbags [248, 249], as explosives [250–253], as solid propellants [254], as burn-rate modifiers [255], and as liquid-crystalline materials [256].

References

1. Boulton AJ, Halls PJ, Katritzky AR (1970) J Chem Soc B, p 636
2. Albini A, Pietra A (1991) Heterocyclic *N*-oxides. CRC Press, Boston
3. Stadlbauer W, Fiala W, Fischer M, Hojas G (2000) J Heterocycl Chem 37:1253
4. Takabatake T, Takei A, Miyazawa T, Hasegawa M, Miyairi S (2001) Heterocycles 55:2387
5. Lue L, Ou Y, Wang J (2004) Hecheng Huaxue 12:170
6. Wang J, Ou Y, Lue L, Chen B (2005) J Beijing Inst Technol 14:196
7. Al-Hiari YM, Khanfar MA, Qaisi AM, Shuheil MYA, El-Abadelah MM, Boese R (2006) Heterocycles 68:1163
8. Tyurin AY, Smirnov OY, Churakov AM, Strelenko YA, Tartakovsky VA (2006) Russ Chem Bull 55:351
9. Li J, Xiao H, Chen Z, Gong X, Dong H (1998) Fenzi Kexue Xuebao 14:193
10. Rauhut G, Eckert F (1999) J Phys Chem A 103:9086
11. Dyall LK (1975) Aust J Chem 28:2147
12. Takayama T, Kawano M, Uekusa H, Ohashi Y, Sugawara T (2003) Helv Chim Acta 86:1352
13. McCulla RD, Burdzinski G, Platz MS (2006) Org Lett 8:1637
14. Vladykin VI, Trakhtenberg SI (1980) RU Patent 721 430
15. Bohn H, Brendel J, Martorana PA, Schönafinger K (1995) Br J Pharmacol 114:1605
16. Armani V, Dell'Erba C, Novi M, Petrillo G, Tavani C (1997) Tetrahedron 53:1751
17. Rastogi R, Dixit G, Zutshi K (1984) Electrochim Acta 29:1345
18. Dyall LK, Harvey JJ, Jarman TB (1992) Aust J Chem 45:371
19. Avemaria F, Zimmermann V, Bräse S (2004) Synlett 7:1163
20. Fan L, Yang X (2005) Gongye Cuihua 13:43
21. Klyuchnikov OR, Khairutdinov FG, Golovin VV (2001) Chem Heterocycl Comp (NY) 36:1003
22. Klyuchnikov OR, Starovoitov VI, Khairutdinov FG, Golovin VV (2003) Chem Heterocycl Comp (NY) 39:135
23. Klyuchnikov OR, Starovoitov VI, Khairutdinov FG, Golovin VV (2004) Russ J Appl Chem 77:853
24. Itoh K, Horiuchi CA (2004) Tetrahedron 60:1671
25. Havasi B, Pasinszki T, Westwood NPC (2005) J Phys Chem A 109:3864
26. Krishnamurthy VN, Talawar MB, Vyas SM, Kusurkar RS, Asthana SN (2006) Def Sci J 56:551
27. Yu ZX, Caramella P, Houk KN (2003) J Am Chem Soc 125:15420
28. Yu ZX, Houk KN (2003) J Am Chem Soc 125:13825
29. Shilina MI, Smirnov VV (2000) Russ J Gen Chem 70:1687
30. Furusho Y, Sohgawa YH, Kihara N, Takata T (2002) Bull Chem Soc Jpn 75:2025
31. Demir AS, Findik H (2005) Lett Org Chem 2:602
32. Curini M, Epifano F, Marcotullio MC, Rosati O, Ballini R, Bosica G (2000) Tetrahedron Lett 41:8817
33. Sheremetev AB, Makhova NN, Friedrichsen W (2001) Adv Heterocycl Chem 78:66
34. Gut Ruggeri S, Bill DR, Bourassa DE, Castaldi MJ, Houck TL, Brown Ripin DH, Wei L, Weston N (2003) Org Process Res Dev 7:1043
35. Kondyukov IZ, Belyaev PG, Karpychev YV, Khisamutdinov GK, Valeshnii SI, Laishev VZ, Krotkova OV (2006) RU Patent 2 284 998
36. Das B, Rudra S, Salman M, Rattan A (2006) WO Patent 2 006 035 283
37. Mandal BK, Maiti S (1986) Eur Polym J 22:447

38. Boiani M, Cerecetto H, González M, Risso M, Olea-Azar C, Ezpeleta O, López de Ceráin A, Monge A (2001) Eur J Med Chem 36:771
39. Nguyen HD, Ngo TL, Le TTV (2004) J Heterocycl Chem 41:1015
40. Nguyen HD, Ngo TL, Pham VH (2006) J Heterocycl Chem 43:1657
41. Rong LC, Li XY, Yao CS, Wang HY, Shi DQ (2006) Acta Crystal E62:o1959
42. Boiani M, Boiani L, Denicola A, Torres de Ortiz S, Serna E, Vera de Bilbao N, Sanabria L, Yaluff G, Nakayama H, Rojas de Arias A, Vega C, Rolán M, Gómez-Barrio A, Cerecetto H, González M (2006) J Med Chem 49:3215
43. Suzuki H, Kawakami T (1997) Synthesis 8:855
44. Luo C, Huang R, Cui Y, Liang X, Li L, Gu Q, Shu R, Liang Q, Liu S, Chen Z (2003) CN Patent 1 412 187
45. Burgos AA (2003) WO Patent 2 003 042 192
46. Zhao S, Guo Q, Wang Y (2005) Zhongguo Yiyao Gongye Zazhi 36:457
47. Paton RM (2004) Science Synt 13:185
48. Makhova NN, Ovchinnikov IV, Kulikov AS, Molotov SI, Baryshnikova EL (2004) Pure Appl Chem 76:1691
49. Tselinsky IV, Melnikova SF, Pirogov SV, Shaposhnikov SD, Nather C, Traulsen T, Friedrichsen W (2000) Heterocycl Commun 6:35
50. Shaposhnikov SD, Pirogov SV, Mel'nikova SF, Tselinsky IV, Nather C, Graening T, Traulsen T, Friedrichsen W (2003) Tetrahedron 59:1059
51. Baryshnikova EL, Kulikov AS, Ovchinnikov IV, Solomentsev VV, Makhova NN (2001) Mendeleev Commun, p 230
52. Molotov SI, Kulikov AS, Strelenko YA, Makhova NN, Lyssenko KA (2003) Russ Chem Bull 52:1829
53. Chupakhin ON, Kotovskaya SK, Romanova SA, Charushin VN (2004) Russ J Org Chem 40:1167
54. Kurbatov S, Rodríguez-Dafonte P, Goumont R, Terrier F (2003) Chem Commun, p 2150
55. Kotovskaya SK, Romanova SA, Charushin VN, Kodess MI, Chupakhin ON (2004) J Fluorine Chem 125:421
56. Mokhtari M, Goumont R, Halle JC, Terrier F (2002) ARKIVOC 168
57. Goumont R, Jan E, Makosza M, Terrier F (2003) Org Biomol Chem 1:2192
58. Terrier F, Lakhdar S, Boubaker T, Goumont R (2005) J Org Chem 70:6242
59. Forlani L, Tocke AL, Del Vecchio E, Lakhdar S, Goumont R, Terrier F (2006) J Org Chem 71:5527
60. Fruttero R, Sorba G, Ermondi G, Lolli M, Gasco A (1997) Farmaco 52:405
61. Blinnikov AN, Makhova NN (1999) Mendeleev Commun, p 13
62. Hallé JC, Vichard D, Pouet MJ, Terrier F (1997) J Org Chem 62:7178
63. Sepulcri P, Goumont R, Hallé JC, Riou D, Terrier F (2000) J Chem Soc Perkin Trans 2:51
64. Vichard D, Hallé JC, Huguet B, Pouet MJ, Riou D, Terrier F (1998) Chem Commun, p 791
65. Sebban M, Goumont R, Hallé JC, Marrot J, Terrier F (1999) Chem Commun 11:1009
66. Goumont R, Sebban M, Sépulcri P, Marrot J, Terrier F (2002) Tetrahedron 58:3249
67. Kurbatov S, Goumont R, Lakhdar S, Marrot J, Terrier F (2005) Tetrahedron 61:8167
68. Goumont R, Sebban M, Marrot J, Terrier F (2004) ARKIVOC 2004 85
69. Shimizu T, Hayashi Y, Taniguchi T, Teramura K (1985) Tetrahedron Lett 41:727
70. Calvino R, Di Stilo A, Fruttero R, Gasco AM, Sorba G, Gasco A (1993) Farmaco 48:321

71. Hasegawa M, Takabatake T (1991) J Heterocycl Chem 28:1079
72. Takabatake T, Hasegawa M (1994) J Heterocycl Chem 31:215
73. Fukai Y, Miyazawa T, Kojoh M, Takabatake T, Hasegawa M (2001) J Heterocycl Chem 38:531
74. Muro C, Fernández V (1987) Inorg Chim Acta 134:215
75. Muro C, Fernández V (1987) Inorg Chim Acta 134:221
76. Richardson C, Steel PJ (2000) Aust J Chem 53:93
77. Wang J, Dong H, Huang Y, Li J (2006) Hecheng Huaxue 14:18
78. Pan YM, Zhang Y, Wang HS, Chen ZF, Wu Q, Ge CY, Zhang Y (2006) Jiegou Huaxue 25:1209
79. Britton D, Olson JM (1979) Acta Cryst B35:3076
80. Ammon HK, Bhattacharjee SK (1982) Acta Cryst B38:2498
81. Aguirre G, Boiani L, Boiani M, Cerecetto H, Di Maio R, González M, Porcal W, Denicola A, Piro OE, Castellano EE, Sant'Anna CMR, Barreiro EJ (2005) Bioorg Med Chem 13:6336
82. Godovikova TI, Golova SP, Strelenko YA, Antipin MY, Struchkov YT (1994) Mendeleev Commun 1:7
83. Baker KWJ, March AR, Parsons S, Paton RM, Stewart GW (2002) Tetrahedron 58:8505
84. Friedrichsen W (1994) J Phys Chem 98:12933
85. Cmoch P, Schilf W (1999) Magn Reson Chem 37:758
86. Cmoch P, Wiench JW, Stefaniak L, Webb GA (1999) Spectrochim Acta 55A:2207
87. Hacker NP (1991) J Org Chem 56:5216
88. Murata S, Tomioka H (1992) Chem Lett 57
89. Himmel HJ, Konrad S, Friedrichsen W, Guntram Rauhut G (2003) J Phys Chem A 107:6731
90. Seminario JM, Concha M, Politzer P (1992) J Comput Chem 13:177
91. Friedrichsen W (1994) J Phys Chem 98:12933
92. Ponder M, Fowler JE, Schaefer HF (1994) J Org Chem 59:6431
93. Rauhut G (1996) J Comput Chem 17:1848
94. Pasinszki T, Ferguson G, Westwood NPC (1996) J Chem Soc Perkin Trans 2:179
95. Klenke B, Friedrichsen W (1996) Tetrahedron 52:743
96. El-Azhary AA (1996) Spectrochim Acta 52A:33
97. Pasinszki T, Westwood NPC (1997) J Mol Struct 408/409:161
98. Rauhut G, Jarzecki AA, Pulay P (1997) J Comput Chem 18:489
99. Klenke B, Friedrichsen W (1998) Teochem 451:263
100. Li J, Xiao H, Dong H (2000) Huaxue Wuli Xuebao 13:55
101. Stevens J, Schweizer M, Rauhut G (2001) J Am Chem Soc 123:7326
102. Rauhut G, Werner HJ (2003) Phys Chem Chem Phys 5:2001
103. Dyall LK (1989) Org Mass Spectrom 24:465
104. Takakis IM, Tsantali GG, Haas GW, Giblin D, Gross ML (1999) J Mass Spectrom 34:1137
105. Hwang KJ, Jo I, Shin YA, Yoo S, Lee JH (1995) Tetrahedron Lett 36:3337
106. Cerecetto H, González M, Seoane G, Stanko C, Piro OE, Castellano EE (2004) J Braz Chem Soc 15:232
107. Hasiotis C, Gallos JK, Kokkinidis G (1993) Electrochim Acta 38:989
108. Olea-Azar C, Rigol C, Mendizábal F, Cerecetto H, Di Maio R, González M, Porcal W, Morello A, Repetto Y, Maya JD (2005) Lett Drug Des Discovery 2:294
109. Acree WE, Pilcher G, Ribeiro da Silva MDMC (2005) J Phys Chem Ref Data 34:553
110. Cerecetto H, Porcal W (2005) Mini-Rev Med Chem 5:57

111. Ghosh PB, Whitehouse MW (1968) J Med Chem 11:305
112. Whitehouse MW, Ghosh PB (1968) Biochem Pharmacol 17:158
113. Ghosh PB, Whitehouse MW (1969) J Med Chem 12:505
114. Ghosh PB, Ternai B, Whitehouse MW (1972) J Med Chem 15:255
115. Calvino R, Mortarini V, Gasco A, Sanfilippo A, Ricciardi ML (1980) Eur J Med Chem 15:485
116. Calvino R, Serafino A, Ferrarotti B, Gasco A, Sanfilippo A (1984) Arch Pharm 317:695
117. Ulrichsen BB, Castberg HB, Johnstone K (1996) WO Patent 9612646
118. Das B, Rudra S, Salman M, Rattan A (2006) WO Patent 2006035283
119. Hwang KJ, Park SM, Kim HS, Lee SH (1998) Biosci Biotechnol Biochem 62:1693
120. Nguyen HD, Ngo TL, Nguyen TY (2003) Tap Chi Hoa Hoc 41:115
121. Takayama H, Shirakawa S, Kitajima M, Aimi N, Yamaguchi K, Hanasaki Y, Ide T, Katsuura K, Fuijiwara M, Ijichi K, Konno K, Sigeta S, Yokota T, Baba M (1996) Bioorg Med Chem Lett 6:1993
122. Hanazaki Y, Ide T, Watanabe H, Katsura K, Ijichi K, Shigeta S, Baba M, Fujiwara M, Yokota T, Takayama H, Shirakawa S, Kitajima M, Aimi N, Yamaguchi K, Konno K (1996) JP Patent 08245384
123. Persichini T, Colasanti M, Fraziano M, Colizzi V, Medana C, Polticelli F, Venturini G, Ascenzi P (1999) Biochem Biophys Res Commun 258:624
124. WHO Thirteenth Program Report: UNDP/World Bank/World Health Organization program for research and training in tropical diseases (1997) Geneva
125. Cerecetto H, Di Maio R, González M, Risso M, Saenz P, Seoane G, Denicola A, Peluffo G, Quijano C, Olea-Azar C (1999) J Med Chem 42:1941
126. Aguirre G, Cerecetto H, Di Maio R, González M, Porcal W, Seoane G, Denicola A, Ortega MA, Aldana I, Monge A (2002) Arch Pharm 335:15
127. Aguirre G, Boiani L, Cerecetto H, Di Maio R, González M, Porcal W, Denicola A, Möller M, Thomson L, Tórtora V (2005) Bioorg Med Chem 13:6324
128. Porcal W, Hernández P, Aguirre G, Boiani L, Boiani M, Merlino A, Ferreira A, Di Maio R, Castro A, González M, Cerecetto H (2007) Bioorg Med Chem 15:2768
129. Hansch C, Unger SH, Forsythe AB (1973) J Med Chem 16:1217
130. Olea-Azar C, Rigol C, Mendizábal F, Briones R, Cerecetto H, Di Maio R, González M, Porcal W, Risso M (2003) Spectrochim Acta Part A 59:69
131. Olea-Azar C, Rigol C, Opazo L, Morello A, Maya JD, Repetto Y, Aguirre G, Cerecetto H, Di Maio R, González M, Porcal W (2003) J Chil Chem Soc 48:77
132. Cazzulo JJ (2002) Curr Top Med Chem 2:1261
133. Du X, Hansell E, Engel JC, Caffrey CR, Cohen FE, Mc Kerrow JH (2000) Chem Biol 7:733
134. Lipinski CA, Lombardo F, Dominy BW, Feeney PJ (1997) Adv Drug Delivery Rev 23:3
135. Ascenzi P, Bocedi A, Gentile M, Visca P, Gradoni L (2004) Biochim Biophys Acta 1703:69
136. Galli U, Lazzarato L, Bertinaria M, Sorba G, Gasco A, Parapini S, Taramelli D (2005) Eur J Med Chem 40:1335
137. Stamler JS, Singel DJ, Loscalzo J (1992) Science 258:1898
138. Schönafinger K (1999) Farmaco 54:316
139. Wang PG, Xian M, Tang X, Wu X, Wen Z, Cai T, Janczuk AJ (2002) Chem Rev 102:1091
140. Ghosh PB, Everitt BJ (1974) J Med Chem 17:203
141. Ghosh P, Ternai B, Whitehouse M (1981) Med Res Rev 1:159

142. Medana C, Di Stilo A, Visentin S, Fruttero R, Gasco A, Ghigo D, Bosia A (1999) Pharm Res 16:956
143. Granik VG, Kaminka MÉ, Grigor'ev NB, Severina IS, Kalinkina MA, Makarov VA, Levina VI (2002) Khimiko-Farmatsev Z 36:523
144. Feelisch M, Schönafinger K, Noack E (1992) Biochem Pharmacol 44:1149
145. Ghigo D, Heller R, Calvino R, Alessio P, Fruttero R, Gasco A, Bosia A, Pescarmona G (1992) Biochem Pharmacol 43:1281
146. Calvino R, Fruttero R, Ghigo D, Bosia A, Pescarmona G, Gasco A (1992) J Med Chem 35:3296
147. Hecker M, Vorhoff W, Bara AT, Mordvintcev PI, Busse R (1995) Naunyn-Schmiedeberg's Arch Pharmacol 351:426
148. Ferioli R, Fazzini A, Folco GC, Fruttero R, Calvino R, Gasco A, Bongrani S, Civelli M (1993) Pharmacol Res 28:203
149. Medana C, Ermondi G, Fruttero R, Di Stilo A, Ferretti C, Gasco A (1994) J Med Chem 37:4412
150. Ferioli R, Folco GC, Ferretti C, Gasco AM, Medana C, Fruttero R, Civelli M, Gasco A (1995) Br J Pharmacol 114:816
151. Di Stilo A, Cena C, Gasco AM, Gasco A, Ghigo D, Bosia A (1996) Bioorg Med Chem Lett 6:2607
152. Sako M, Oda S, Ohara S, Hirota K, Maki Y (1998) J Org Chem 63:6947
153. Nirode WF, Luis JM, Wicker JF, Wachter NM (2006) Bioorg Med Chem Lett 16: 2299
154. Gasco AM, Boschi D, Di Stilo A, Medana C, Gasco A, Martorana PA, Schönafinger K (1998) Arzneim Forsch 48:212
155. Sorba G, Medana C, Fruttero R, Cena C, Di Stilo A, Galli U, Gasco A (1997) J Med Chem 40:463
156. Herrmann F, List N (1994) DE Patent 4305881
157. Narayanan AS (2002) IN Patent 188020
158. Ellis JL (2005) WO Patent 2005070006
159. Garvey DS, Ranatunge RR (2006) WO Patent 2006055542
160. Ellis JL, Garvey DS (2007) WO Patent 2007016677
161. Civelli M, Caruso P, Bergamaschi M, Razzetti R, Bongrani S, Gasco A (1994) Eur J Pharmacol 255:17
162. Sorba G, Ermondi G, Fruttero R, Galli U, Gasco A (1996) J Heterocycl Chem 33:327
163. Civelli M, Giossi M, Caruso P, Razzetti R, Bergamaschi M, Bongrani S, Gasco A (1996) Br J Pharmacol 118:923
164. Bussygina OG, Pyatakova NV, Khropov YV, Ovchinnikov IV, Makhova NN, Severina IS (2000) Biochemistry 65:457
165. Gasco AM, Cena C, Di Stilo A, Ermondi G, Medana C, Gasco A (1996) Helv Chim Acta 79:1803
166. Cena C, Visentin S, Gasco A, Martorana PA, Fruttero R, Gasco A (2002) Farmaco 57:417
167. Pirogov SV, Mel'nikova SF, Tselinskii IV, Romanova TV, Spiridonova NP, Betin VL, Postnikov AB, Kots AY, Khropov YV, Gavrilova SA, Grafov MA, Medvedeva NA, Pyatakova NV, Severina IS, Bulargina TV (2004) RU Patent 2240321
168. Sankaranarayanan A (2003) WO Patent 2003053439
169. Prous JR (1999) Annu Drug Data Rep 21:787
170. Kots AY, Grafov MA, Khropov YV, Betin VL, Belushkina NN, Busygina OG, Yazykova MY, Ovchinnikov IV, Kulikov AS, Makhova NN, Medvedeva NA, Bulargina TV, Severina IS (2000) Br J Pharmacol 129:1163

171. Garvey DS, Fang X, Subhash KP, Ranatunge RR, Wey SJ (2005) WO Patent 2 005 060 603
172. Garvey DS (2006) WO Patent 2 006 078 995
173. Sorba G, Gasco A, Coruzzi G, Adami M, Pozzoli C, Morini G, Bertaccini G (1997) Arzneim Forsch 47:849
174. Coruzzi G, Adami M, Morini G, Pozzoli C, Cena C, Bertinaria M, Gasco A (2000) J Physiol 94:5
175. Bertinaria M, Sorba G, Medana C, Cena C, Adami M, Morini G, Pozzoli C, Coruzzi G, Gasco A (2000) Helv Chim Acta 83:287
176. Hill SJ, Ganellin CR, Timmerman H, Schwartz JC, Shankley NM, Young JM, Schunack W, Levi R, Haas HL (1997) Pharmacol Rev 49:253
177. Bertinaria M, Stilo AD, Tosco P, Sorba G, Poli E, Pozzoli C, Coruzzi G, Fruttero R, Gasco A (2003) Bioorg Med Chem 11:1197
178. Bertinaria M (2003) Farmaco 58:279
179. Tosco P, Bertinaria M, Di Stilo A, Marini E, Rolando B, Sorba G, Fruttero R, Gasco A (2004) Farmaco 59:359
180. Tosco P, Bertinaria M, Di Stilo A, Cena C, Sorba G, Fruttero R, Gasco A (2005) Bioorg Med Chem 13:4750
181. Tosco P, Bertinaria M, Di Stilo A, Cena C, Fruttero R, Gasco A (2005) Farmaco 60:507
182. Di Stilo A, Fruttero R, Boschi D, Gasco AM, Sorba G, Gasco A, Orsetti M (1993) Med Chem Res 3:554
183. Fruttero R, Boschi D, Di Stilo A, Gasco A (1995) J Med Chem 38:4944
184. Boschi D, Tron GC, Di Stilo A, Fruttero R, Gasco A, Poggesi E, Motta G, Leonardi A (2003) J Med Chem 46:3762
185. Boschi D, Di Stilo A, Cena C, Lolli M, Fruttero R, Gasco A (1997) Pharmacol Res 14:1750
186. Neumann P (1980) DE Patent 2 949 464
187. Hof RP, Vogel A (1988) Drugs Future 10:746
188. Gasco AM, Fruttero R, Sorba G, Gasco A, Budriesi R, Chiarini A (1992) Arzneim Forsch 42:921
189. Gasco AM, Ermondi G, Fruttero R, Gasco A (1996) Eur J Med Chem 31:3
190. Visentin S, Amiel P, Fruttero R, Boschi D, Roussel C, Giusta L, Carbone E, Gasco A (1999) J Med Chem 42:1422
191. Di Stilo A, Visentin S, Cena C, Gasco AM, Ermondi G, Gasco A (1998) J Med Chem 41:5393
192. Cena C, Visentin S, Di Stilo A, Boschi D, Fruttero R, Gasco A (2001) Pharm Res 18:157
193. Boschi D, Caron G, Visentin S, Di Stilo A, Rolando B, Fruttero R, Gasco A (2001) Pharm Res 18:987
194. Rolando B, Cena C, Caron G, Marini E, Grosa G, Fruttero R, Gasco A (2002) Med Chem Res 11:322
195. Ermondi G, Visentin S, Boschi D, Fruttero R, Gasco A (2000) J Mol Struct 523:149
196. Boschi D, Cena C, Di Stilo A, Fruttero R, Gasco A (2000) Bioorg Med Chem 8:1727
197. Mu L, Feng SS, Go ML (2000) Chem Pharm Bull 48:808
198. Lolli ML, Cena C, Medana C, Lazzarato L, Morini G, Coruzzi G, Manarini S, Fruttero R, Gasco A (2001) J Med Chem 44:3463
199. Cena C, Lolli ML, Lazzarato L, Guaita E, Morini G, Coruzzi G, McElroy SP, Megson IL, Fruttero R, Gasco A (2003) J Med Chem 46:747
200. Turnbull CM, Cena C, Fruttero R, Gasco A, Rossi AG, Megson IL (2006) Br J Pharmacol 148:517

201. Garvey DS (2006) WO Patent 2 006 127 591
202. Wang W, Zhang Y, Ji H, Deng G, Feng X, Xu X, Peng S (2004) Zhongguo Yaowu Huaxue Zazhi 14:1
203. Xiaolin Y, Hui J, Yihua Z, Ruiwen L, Sixun P (2001) Zhongguo Yaoke Daxue Xuebao 32:301
204. Pedrazzoli JJ, De Carvalho PS, Gambero A (2006) WO Patent 2 006 042 387
205. Fang L, Zhang Y, Lehmann J, Wang Y, Ji H, Ding D (2007) Bioorg Med Chem Lett 17:1062
206. López GV, Batthyany C, Blanco F, Botti H, Trostchansky A, Migliaro E, Radi R, González M, Cerecetto H, Rubbo H (2006) Bioorg Med Chem 13:5787
207. Boschi D, Tron GC, Lazzarato L, Chegaev K, Cena C, Di Stilo A, Giorgis M, Bertinaria M, Fruttero R, Gasco A (2006) J Med Chem 49:2886
208. Cena C, Bertinaria M, Boschi D, Giorgis M, Gasco A (2006) ARKIVOC 301
209. Chegaev K, Lazzarato L, Rolando B, Marini E, López GV, Bertinaria M, Di Stilo A, Fruttero R, Gasco A (2007) ChemMedChem 2:234
210. Bertinaria M, Galli U, Sorba G, Fruttero R, Gasco A, Brenciaglia MI, Scaltrito MM, Dubini F (2003) Drug Dev Res 60:225
211. Sorba G, Galli U, Cena C, Fruttero R, Gasco A, Morini G, Adami M, Coruzzi G, Brenciaglia MI, Dubini F (2003) ChemBioChem 4:899
212. Guizhen A, Yihua Z, Hui J, Gang D (2004) Zhongguo Yaoke Daxue Xuebao 35:200
213. Velazquez C, Rao PNP, McDonald R, Knaus EE (2005) Bioorg Med Chem 13:2749
214. Del Grosso E, Boschi D, Lazzarato L, Cena C, Di Stilo A, Fruttero R, Moro S, Gasco A (2005) Chem Biodivers 2:886
215. Wey SJ, Garvey DS, Fang X, Richardson SK (2006) WO Patent 2 006 099 416
216. Garvey DS (2007) WO Patent 2 007 016 095
217. Garvey DS (2007) WO Patent 2 007 016 136
218. Moharram S, Zhou A, Wiebe LI, Knaus EE (2004) J Med Chem 47:1840
219. Monge A, López de Ceráin A, Ezpeleta O, Cerecetto H, Dias E, Di Maio R, González M, Onetto S, Risso M, Seoane G, Zinola F, Olea-Azar C (1998) Pharmazie 53:698
220. Monge A, López de Ceráin A, Ezpeleta O, Cerecetto H, Dias E, Di Maio R, González M, Onetto S, Seoane G, Suescun L, Mariezcurrena R (1998) Pharmazie 53:758
221. Cerecetto H, González M, Risso M, Seoane G, López de Ceráin A, Ezpeleta O, Monge A, Suescun L, Mombrú A, Bruno AM (2000) Arch Pharm 333:387
222. Zhang W, Wu Y, Lu W, Li W, Sun X, Shen X, Li X, Tang L (1995) Zhongguo Yaowu Huaxue Zazhi 5:242
223. Aguirre G, Boiani M, Cerecetto H, Fernández M, González M, León E, Pintos C, Raymondo S, Arredondo C, Pacheco JP, Basombrío MA (2006) Pharmazie 61:54
224. Tetsuya H, Masato H, Ryuichi K (2005) JP Patent 2 005 075 806
225. Severina IS, Axenova LN, Veselovsky AV, Pyatakova NV, Buneeva OA, Ivanov AS (2003) Biochemistry 68:1048
226. Cerecetto H, Dias E, Di Maio R, González M, Pacce S, Saenz P, Seoane G, Suescun L, Mombrú A, Fernández G, Lema M, Villalba J (2000) J Agric Food Chem 48:2995
227. Iwamoto R, Sakata H, Okumura K, Hongo A, Sekiguchi S (1977) JP Patent 52 007 055
228. Yusupova LM, Falyakhov IF, Spatlova LV, Garipov TV, Shindala MHTA, Ishkaeva DR (2005) RU Patent 2 255 935
229. Gibbs DE (1998) US Patent 5 773 454
230. Cerecetto H, González M, Onetto S, Risso M, Rey A, Giglio J, León E, León A, Pilatti P, Fernández M (2006) Archiv Pharm 339:59
231. Calvino R, Gasco A, Leo A (1992) J Chem Soc Perkin Trans 2 1643
232. Všeteèka V, Fruttero R, Gasco A, Exner O (1994) J Mol Struct 324:277

233. Caron G, Carrupt PA, Testa B, Ermondi G, Gasco A (1996) Pharm Res 13:1186
234. Fruttero R, Boschi D, Fornatto E, Serafino A, Gasco A, Exner O (1998) J Chem Res 495:2545
235. Kontogiorgis CA, Hadjipavlou-Litina D (2002) Med Res Rev 22:385
236. Boiani M, Cerecetto H, González M (2004) Farmaco 59:405
237. Medana C, Visentin S, Grosa G, Fruttero R, Gasco A (2001) Farmaco 56:799
238. Grosa G, Galli U, Rolando B, Fruttero R, Gervasio G, Gasco A (2004) Xenobiotica 34:345
239. Nemeikaite-Ceniene A, Sarlauskas J, Miseviciene L, Anusevicius Z, Maroziene A, Cenas N (2004) Acta Biochim Pol 51:1081
240. Macphee DG, Robert GP, Ternai B, Ghosh P, Stephens R (1977) Chem Biol Interact 19:77
241. Voogd CE, Van der Stel JJ, Jacobs JJJAA (1980) Mutat Res 78:233
242. Faigle JW, Blattner H, Glatt H, Kriemler HP, Mory H, Storni A, Winkler T, Oesch F (1987) Helv Chim Acta 70:1296
243. Fundaro A (1974) Boll Soc Ital Biol Sper 50:1654
244. Fundaro A, Cassone MC (1980) Boll Soc Ital Biol Sper 56:2364
245. Graves DF (1993) Rubber Chem Technol 66:61
246. Klyuchnikov OR, Deberdeev RY, Berlin AA (2005) Dokl Phys Chem 400:7
247. Friedman HS, Abruscato GJ, DeMassa JM, Gentile AV, Grossi AV (1997) US Patent 5 659 095
248. Zeuner S, Schropp R, Hofmann A, Roedig KH (2000) DE Patent 19 840 993
249. Sonti C (2002) US Patent 2 002 143 189
250. Pepekin VI, Smirnov AS (2001) Khimicheskaya Fizika 20:78
251. Sikder AK, Sikder N (2001) Indian J Heterocycl Chem 11:149
252. Sikder AK, Salunke RB, Sikder N (2002) J Energ Mater 20:39
253. Zhang C (2006) THEOCHEM 765:77
254. Hong W, Tian D, Liu J, Wang F (2001) Guti Huojian Jishu 24:41
255. Shinde PD, Salunke RB, Agrawal JP (2003) Propellants Explos Pyrotech 28:77
256. Bezborodov VS, Kauhanka MM, Lapanik VI (2004) Vestsi Nats Akad Navuk Belarusi, Ser Khim Navuk 3:61

Author Index Volumes 1–10

The volume numbers are printed in italics

Subject Index

The manufacturer's authorised representative in the EU is Springer Nature Customer Service Centre GmbH, Europaplatz 3, 69115 Heidelberg, Germany. If you have any concerns regarding our products, please contact ProductSafety@springernature.com

Printed and bound by CPI Group (UK) Ltd, Croydon, CR0 4YY
30/09/2026
02225561-0008